CONTROL AND DYNAMIC SYSTEMS

Advances in Theory and Applications

Volume 49

CONTRIBUTORS TO THIS VOLUME

TAE H. CHO

I. J. CONNELL

NDY N. EKERE

P. M. FINNIGAN

PAUL M. FRANK

JOSEPH C. GIARRATANO

ROGER G. HANNAM

A. F. HATHAWAY

W. E. LORENSEN

V. N. PARTHASARATHY

J. B. ROSS

JERZY W. ROZENBLIT

RALF SELIGER

FREDERICK E. SISTLER

DAVID M. SKAPURA

STELIOS C. A. THOMOPOULOS

IAN WHITE

BERNARD P. ZIEGLER

CONTROL AND DYNAMIC SYSTEMS

ADVANCES IN THEORY AND APPLICATIONS

Edited by
C. T. LEONDES

School of Engineering and Applied Science
University of California, Los Angeles
Los Angeles, California
and
College of Engineering
University of Washington
Seattle, Washington

VOLUME 49: MANUFACTURING AND
AUTOMATION SYSTEMS:
TECHNIQUES AND TECHNOLOGIES
Part 5 of 5

ACADEMIC PRESS, INC.
Harcourt Brace Jovanovich, Publishers
San Diego   New York   Boston
London   Sydney   Tokyo   Toronto

ACADEMIC PRESS RAPID MANUSCRIPT REPRODUCTION

Copyright © 1991 by ACADEMIC PRESS, INC.
All Rights Reserved.
No part of this publication may be reproduced or transmitted in any form or by any
means, electronic or mechanical, including photocopy, recording, or any information
storage and retrieval system, without permission in writing from the publisher.

Academic Press, Inc.
San Diego, California 92101

United Kingdom Edition published by
Academic Press Limited
24–28 Oval Road, London NW1 7DX

Library of Congress Catalog Number: 64-8027

International Standard Book Number: 0-12-012749-0

PRINTED IN THE UNITED STATES OF AMERICA
91 92 93 94   9 8 7 6 5 4 3 2 1

CONTENTS

CONTRIBUTORS

Numbers in parentheses indicate the pages on which the authors' contributions begin.

Tae H. Cho (191), *AI Simulation Group, Department of Electrical and Computer Engineering, The University of Arizona, Tucson, Arizona 85721*

I. J. Connell (289), *GE Corporate Research and Development Center, Schenectady, New York 12301*

Ndy N. Ekere (129), *University of Salford, Salford, United Kingdom*

P. M. Finnigan (289), *GE Corporate Research and Development Center, Schenectady, New York 12301*

Paul M. Frank (241), *Department of Measurement and Control, University of Duisburg, W-4100 Duisburg 1, Germany*

Joseph C. Giarratano (37), *University of Houston—Clear Lake, Houston, Texas 77508*

Roger G. Hannam (129), *University of Manchester, Institute of Science and Technology (UMIST), Manchester M60 1QD, United Kingdom*

A. F. Hathaway (289), *GE Corporate Research and Development Center, Schenectady, New York 12301*

W. E. Lorensen (289), *GE Corporate Research and Development Center, Schenectady, New York 12301*

V. N. Parthasarathy (289), *GE Corporate Research and Development Center, Schenectady, New York 12301*

J. B. Ross (289), *GE Aircraft Engines, Cincinnati, Ohio 45215*

Jerzy W. Rozenblit (191), *AI Simulation Group, Department of Electrical and Computer Engineering, The University of Arizona, Tucson, Arizona 85721*

Ralf Seliger (241), *Department of Measurement and Control, University of Duisburg, W-4100 Duisburg 1, Germany*

Frederick E. Sistler (99), *Department of Agricultural Engineering, Louisiana State University Agricultural Center, Baton Rouge, Louisiana 70803*

David M. Skapura (37), *Loral Space Information Systems, Houston, Texas 77058*

Stelios C. A. Thomopoulos (339), *Decision and Control Systems Laboratory, Department of Electrical and Computer Engineering, The Pennsylvania State University, University Park, Pennsylvania 16802*

Ian White (1), *Defence Research Agency, Portsdown, Cosham PO6 4AA, England*

Bernard P. Ziegler (191), *AI Simulation Group, Department of Electrical and Computer Engineering, The University of Arizona, Tucson, Arizona 85721*

PREFACE

At the start of this century, national economies on the international scene were, to a large extent, agriculturally based. This was, perhaps, the dominant reason for the protraction, on the international scene, of the Great Depression, which began with the Wall Street stock market crash of October 1929. In any event, after World War II the trend away from agriculturally based economies and toward industrially based economies continued and strengthened. Indeed, today, in the United States, approximately only 1% of the population is involved in the agriculture industry. Yet, this small segment largely provides for the agriculture requirements of the United States and, in fact, provides significant agriculture exports. This, of course, is made possible by the greatly improved techniques and technologies utilized in the agriculture industry.

The trend toward industrially based economies after World War II was, in turn, followed by a trend toward service-based economies; and, in fact, in the United States today roughly 70% of the employment is involved with service industries, and this percentage continues to increase. Nevertheless, of course, manufacturing retains its historic importance in the economy of the United States and in other economies, and in the United States the manufacturing industries account for the lion's share of exports and imports. Just as in the case of the agriculture industries, more is continually expected from a constantly shrinking percentage of the population. Also, just as in the case of the agriculture industries, this can only be possible through the utilization of constantly improving techniques and technologies in the manufacturing industries. As a result, this is a particularly appropriate time to treat the issue of manufacturing and automation systems in this international series. Thus, this is Part 5 of a five-part set of volumes devoted to the most timely theme of "Manufacturing and Automation Systems: Techniques and Technologies."

The first contribution to this volume is "Fundamental Limits in the Theory of Machines," by Ian White. This contribution reviews some of the fundamental limits of machines that constrain the range of tasks that these machines can be made to undertake. These include limitations on the computational process, limitations in physics, and limitations in the ability of their builders to define the processes to be undertaken. This contribution seeks to relate several developments in the theory of computer science, physics, and philosophy to the question of what machines can and cannot do. The question is appallingly hard, so the reader should seek in this contribution some insights rather than total enlightenment. Nonetheless, the question is more than a philosophical one. The relationship of machines to intelligent functioning is a central question in machine theory and is one that brings philosophy into the mainstream of what is far more empirical science than is generally acknowledged. As more and more is expected of increasingly capable manufacturing systems, the many fundamental issues raised in this contribution need to be recognized and taken into account.

The next contribution is "Neural Network Techniques in Manufacturing and Automation Systems," by Joseph C. Giarratano and David M. Skapura. A neural net is typically composed of many simple processing elements arranged in a massively interconnected parallel network. Depending on the neural net design, the artificial neurons may be sparsely, moderately, or fully interconnected with other neurons. Two common characteristics of many popular neural net designs are that (1) nets are trained to produce a specified output when a specified input is presented rather than being explicitly programmed and (2) their massive parallelism makes nets very fault tolerant if part of the net becomes destroyed or damaged. This contribution shows that neural networks have a growing place in industry by providing solutions to difficult and intractable problems in automation and robotics. This growth will increase now that commercial neural net chips have been introduced by vendors such as Intel Corporation. Neural net chips will find many applications in embedded systems so that the technology will spread outside the factory. Already, neural networks have been employed to solve problems related to assembly-line resource scheduling, automotive diagnostics, paint quality assessment, and analysis of seismic imaging data. These applications represent only the beginning. As neural network technology flourishes, many more successful applications will be developed. While not all of them will utilize a neural network to solve a previously intractable problem, many of them will provide solutions to problems for which a conventional algorithmic approach is not cost-effective. Based on the success of these applications, one looks forward to the development of future applications.

The next contribution is "Techniques for Automation Systems in the Agriculture Industry," by Frederick E. Sistler. The agriculture industry encompasses the growth, distribution, and processing of food and fiber, along with related suppliers of goods and services. This contribution presents techniques and control systems used in on-farm agriculture. It is applications-oriented rather than mathematically oriented because the primary contribution is seen to be in the unique applications of existing sensors, systems, and techniques to biological systems. The properties and behavior of plants and animals vary greatly both among and within species. The response of a biological system is greatly dependent upon its environment (moisture, temperature, relative humidity, soil, solar radiation, etc.), which itself can be highly variable and difficult to model. All of this makes biological systems more difficult to model than inorganic systems and materials. Automation is used in agriculture for machine control, environmental (building) control, water management, sorting and grading, and food processing. Farming has traditionally been associated with a very low level of automation. However, as noted at the beginning of this preface, more and more is expected of a diminishing percentage of the population, which can only be achieved through constantly improving automation techniques and technologies such as are presented in this contribution.

The next contribution is "Modeling and Simulation of Manufacturing Systems," by Ndy N. Ekere and Roger G. Hannam. A manufacturing system generally includes many linked processes, the machines to carry out those processes, handling equipment, control equipment, and various types of personnel. A manufacturing system for an automobile could include all the presslines to produce the body panels; the foundries to produce the engine blocks and transmission housing; forge shops to produce highly stressed parts such as suspension components and crankshafts; the machine shops that convert the forgings, castings, and other raw material to accurately sized components; and the subassembly and final assembly lines that result in the final product being produced. Many writers call each of these subsections a manufacturing system, although each is also a constituent of a larger manufacturing system. The machines and processes involved in manufacturing systems for mass production are dedicated to repetitive manufacture. The majority of products are, however, produced by batch manufacturing in which many different parts and products are produced on the same machines and the machines and processes are reset at intervals to start producing a different part. The techniques presented in this contribution apply to manufacturing systems that extend from a few machines (that are related—generally because they are involved in processing the same components) up to systems that might comprise the machines in a complete machine shop or complete processing line. The characteristics of batch manufacturing are often analyzed by simulation; mass production systems are analyzed more by mathematical analysis. This contribution is an in-depth treatment of these issues of modeling and simulation that are of major importance to manufacturing systems.

The next contribution is "Knowledge-Based Simulation Environment Techniques: A Manufacturing System Example," by Tae H. Cho, Jerzy W. Rozenblit, and Bernard P. Zeigler. The need for interdisciplinary research in artificial intelligence (AI) and simulation has been recognized recently by a number of researchers. In the last several years there has been an increasing volume of research that attempts to apply AI principles to simulation. This contribution describes a methodology for building rule-based expert systems to aid in discrete event simulation (DEVS). It also shows how expert systems can be used in the design and simulation of manufacturing systems. This contribution also presents an approach to embedding expert systems within an object-oriented simulation environment, under the basic idea of creating classes of expert system models that can be interfaced with other model classes. An expert system shell for the simulation environment (ESSSE) is developed and implemented in DEVS-scheme knowledge-based design and simulation environment (KBDSE), which combines artificial intelligence, system theory, and modeling formalism concepts. The application of ES models to flexible manufacturing systems (FMS) modeling is presented.

The next contribution is "Fault Detection and Isolation in Automatic Processes," by Paul M. Frank and Ralf Seliger. The tremendous and continuing progress in computer technology makes the control of increasingly complex manufacturing and automation systems readily possible. Of course, the issues of reliability, operating safety, and environmental protection are of major importance, especially if potentially dangerous equipment like chemical reactors, nuclear power plants, or aircraft are concerned. In order to improve the safety of automatic processes, they must be supervised such that occurring failures or faults can be accommodated as quickly as possible. Failures or faults are malfunctions hampering or disturbing the normal operation of an automatic process, thus causing an unacceptable deterioration of the performance of the system or even leading to dangerous situations. They can be classified as component faults (CF), instrument faults (IF), and actuator faults (AF). The first two steps toward a failure accommodation are the detection and the isolation of the fault in the system under supervision. The term *detection* denotes in this context the knowledge of the time at which a fault has occurred, while *isolation* means the determination of the fault location in the supervised system (i.e., the answer to the question "which instrument, actuator, or component failed?") This contribution is an in-depth treatment of this issue of fault detection and isolation and the role it can play in achieving reliable manufacturing and automation systems.

The next contribution is "CATFEM—Computer Assisted Tomography and Finite Element Modeling," by P. M. Finnigan, A. F. Hathaway, W. E. Lorensen, I. J. Connell, V. N. Parthasarathy, and J. B. Ross. Historically, x-ray computed tomography (CT) has been used for visual inspection of cross-sectional data of an object. It has been successfully applied in the medical field as a noninvasive diagnostic tool and in industrial applications for quality evaluation. This contribution presents a conventional look at CT and, in addition, details revolutionary approaches to the use of computed tomography data for engineering applications, with emphasis on visualization, geometric modeling, finite element modeling, reverse engineering, and adaptive analysis. The concept of a discrete solid model, known as a digital replica TM, is introduced. The digital replica possesses many of the same attributes intrinsic to a conventional CAD solid model, and thus it has the potential for broad applicability to many geometry-based applications, including those that are characteristic of steps that are involved in many manufacturing processes. This contribution discusses three-dimensional imaging techniques for the CT slice ensemble using surface reconstruction. Such capability provides the user with a way to view and interact with the model. Other applications include the automatic and direct conversion of x-ray computed tomography data into finite element models. The notion of reverse engineering a part is also presented; it is the ability to transform a digital replica into a conventional solid model. Other technologies that support analysis along with a system architecture are also described. This contribution provides sufficient background on CT to ease the understanding of the applications that build on this technology; however, the principal focus is on the applications themselves.

The final contribution to this volume is "Decision and Evidence Fusion in Sensor Integration," by Stelios C. A. Thomopoulos. Manufacturing and automation systems will, in general, involve a number of sensors whose sensed information can, with advantage, be integrated in a process referred to as sensor fusion. Sensor integration (or sensor fusion) may be defined as the process of integrating raw and processed data into some form of meaningful inference that can be used intelligently to improve the performance of a system, measured in any convenient and quantifiable way, beyond the level that any one of the components of the system separately or any subset of the system components partially combined could achieve. This contribution presents a taxonomy for sensor fusion that involves three distinct levels at which information from different sensors can be integrated; it also provides effective algorithms for processing this integrated information.

This volume concludes this rather comprehensive five-volume treatment of techniques and technologies in manufacturing and automation systems. The authors of this volume and the preceding four volumes are all to be commended for their splended contributions, which will provide a uniquely significant reference source for workers on the international scene for years to come.

FUNDAMENTAL LIMITS
IN THE THEORY OF MACHINES

IAN WHITE
Defence Research Agency
Portsdown, Cosham, PO6 4AA, ENGLAND.

1. INTRODUCTION

This Chapter reviews some of the fundamental limits of machines which constrain the range of tasks which we can make them undertake. These include limitations on the computational process, limitations in physics and limitations in the ability of their builders to define the processes to be undertaken. The chapter seeks to relate several developments in the theory of computer science, physics and philosophy to the question of what machines can, and cannot do. The question is appallingly hard, so the reader should seek in these few pages some insights rather than total enlightenment. Nonetheless the question is more than a philosophical indulgence. The relationship of machines to intelligent functioning is a central question in machine theory, and one which brings philosophy into the mainstream of what I believe is a far more empirical science that is generally acknowledged.

In many branches of physics and engineering the role of theory is not just to predict situations which have yet to be realized, but to check if these situations violate known physical limits. It is both expedient and commonplace in the domains of computer and automation applications to presume that there are no absolute limits; that all that restricts our ambitions are time, money and human resource. If these are supplied in sufficient abundance a solution will appear in due course. The presumption of some solution is a common trait in military thinking, and in science and engineering the parallel is of a *tough nut* to crack - a problem to be *defeated*. This is often a productive approach of course, but it is not always the way forward. It applies in many physical situations, but as the problem becomes more abstract, and 'softer' (to use the term applied by some to pursuits such as psychology, cognition, and intelligence), the solutions become harder. The question left after years of research in many areas of psychology, and AI, is whether in the normal parlance of the harder sciences, in which I include the theory of computing machines, and much of computer science, is *is there a solution at all?* Is a comprehensive limitative theory for machine capability fundamentally unfathomable?

In examining these questions we first review briefly some of the classical limitations of machines, and then the scope of mathematical formalisms, and question the extent to which the latter can be used to determine the behaviour of a real world machine. The structure of this chapter is to:

i) briefly review some of the limitations of computer science, physics, and then to

ii) examine definitional questions of how machines relate to a real world environment. The thesis is developed that conventional mathematics requires many assumptions about the real world which our experience tells us we cannot easily make part of any universal theory.

iii) penultimately the nature of information is examined. A primary differential between us and inanimate objects is that we generate and control and use information in a very rich sense; but what is information? The definitions of information which we possess have limited applicability, and in many contexts do not have an accepted definition at all.

iv) finally some options not based on symbolic representations are cited.

II. LIMITS

The possible existence of limits is a feature often ignored in the development of computer based systems. What limits the achievement of computational systems? What are the limits of a system which interacts with its environment, via sensor and effector reactions? Are there such limits and can we determine what they are ? A range of limitative factors which apply to the process of computation was reviewed by the author in an earlier paper [1]. In this chapter the theme is taken somewhat further. First some of the limitative features are briefly reviewed. Reference [1] should be consulted for more details, and for a bibliography of further reading.

A. ABSOLUTE LIMITS

Absolute limits define bounds which no machine can ever transgress. These limits define sets of problems which are formally unsolvable, by any computer. The centerpieces here are:

TURING'S THEOREM: There can be no algorithm which can determine if an arbitrary computer program, running on a basic form of computer (the Turing Machine) will halt [2]. Because any computer can be emulated by a Turing Machine, and any programme translated to a Turing machine form, this limit applies to all computer programmes.

GODEL's THEOREM (Which is related to Turing's theorem) [3], [4]. A system of Logic L is said to be 'simply consistent' if there are no propositions U, such that U and $\neg U$ are provable[1]. A Theory T is said to be decidable if there exists an algorithm for answering the question *'does some sentence S belong to T ?'*

Theorem 1: For suitable L there are undecidable propositions in L, that is propositions such that neither U nor $\neg U$ is provable. As U and $\neg U$ express contradictory sentences, one of them must express a true sentence, so there will be a proposition U that expresses a true sentence, but nevertheless is not provable.

---

1. $\neg$ denotes logical NOT.

Theorem 2: For suitable L the simple consistency of L cannot be proved in L.

These results show us immediately that computers have real limitations. We cannot know in general if a computer programme will terminate; we cannot know if a system is totally consistent, without resort to some form of external system viewpoint; there will be truths which we cannot prove. Deriving from these limitative results are a wide range of similar limits, which can be proved to be equivalent to these statements. Examples and an elegant introduction to Turing machines are given in Minsky [5].

B. COMPLEXITY LIMITS

Complexity theory seeks to determine the resources needed to compute functions on a computer [6], [7]. The resources are time steps and and storage space. The major complexity classifications are for algorithms which run in:

   i) polynomial time, i.e. the run time is some polynomial function of the length of the input string. An algorithm of this form is said to be of time complexity *P*.

   ii) exponential time, i.e. the run time is some exponential function of the length of the input string. This class of algorithm is said to be of time complexity *E*.

Similar definitions apply to the memory requirements of algorithms, where we refer to *Space Complexity*. In computer science it is usual to regard exponential algorithms as intractable, and polynomial ones as tractable. It needs to be remembered however that although exponential algorithms are generally intractable, polynomial ones may also be, as the table 1 below shows. In practice high power polynomial algorithms are quite rare.

Time Complexity Function		Input Length (bits)		
		10	**30**	**60**
polynomial	n	0.01 millisec.	0.03 millisec.	0.06 millisec.
	n^6	1.0 second	12.0 minutes	13 hours
	n^{12}	277 hours	168 centuries	$6.9.10^7$ centuries
exponential	2^n	1 millisec.	17.9 minutes	366 centuries
	3^n	0.06 seconds	6.5 years	$1.3.10^{13}$ centuries

Table 1. 'Time-to-compute' limits, assuming 1 step = 1 microsecond.

There is an extension of these complexity measures, namely nondeterministic polynomial/exponential complexity denoted by *NP*, or *NE*. For these measures it is assumed that many parallel computations on the input occur, each of the same (deterministic) complexity. Put another way if one guesses the solution for an *NP* problem, the answer can be checked in polynomial time *P*. It transpires that a very wide range of common problems are characterized by being *NP*. As is clear from the table there will be problems whose complexity means that they can never be solved, even though we may know that the algorithm will in the Turing sense 'Stop'.

When a class of problems is defined as *NP* (or *P, E* etc.) we mean that the worst case in that set is of this complexity. In practice many solutions may be contained much more rapidly. There is very little theory yet which can provide a statistical summary of likely results for a given class of problem.

Related but interestingly different ideas have been suggested by Chaitin [8], Kolmogorof [9] and Martin Loff [10], in which the complexity of a computation is defined as the number of bits of the computer programme needed to define it on a Turing machine. These definitions are of particular interest in understanding the concept of information, which is discussed later in his Chapter.

C. INFINITY AND FINITY

The execution of a computer programme with a specific input can be regarded as an attempt to prove an assertion about the input. A computer program which stops (correctly!) can be regarded as a form of theorem proof, whilst one which does not stop, is' evidence' of an unprovable statement. This is merely an informal restatement of Turing's thesis. In any practical situation of course we do not have endless time to wait for computations to stop, nor do we have endless resources of tape for our Turing machine. Note that the statements:

stops {eventually},
never stops

cannot be given any real meaning in empirical terms, unless we can prove for a given class or example of a problem that it will *never stop* (e.g. a loop). It is by definition not empirically observable. Similarly *stops {eventually}* is also not sufficiently tangible. In any real world setting we will have to apply a finite time bound on any attempt to witness *stops [eventually]*. Where we are resource limited in time and / or space, any computation which transgresses these limits is *unprovable*. In this sense it seems appropriate to define a given input string as:

Resource(T,S) provable, or as
Resource(T,S) unprovable,

where *T* and *S* denote *Time* and *Space* resource limits respectively. Concepts along these lines are used in cryptography. Thus when talking of a computer never halting, this must always be in terms of some constraint. Similarly space resources (e.g. tape) can be supplied at a fixed upper rate, and any algorithm with this type of demand, will in effect translate space demand into time demand.

In many of the discussions of computation, and the underlying mathematics of artificial intelligence and logic, the concept of infinite sets, or infinite resources are often invoked. In any theory of machines in the real physical world it needs to be remembered that this is wrong. Mathematicians frequently invoke the 'axiom of infinity', for it solves many problems in mathematics. In the physical world it is an axiom. A generating algorithm for π means that the number is always finitely extendable. The real number π does not exist, and the distinction between finitely extendable' and infinity requires the *axiom of infinity*. In the physical world *Infinity* is a huge joke invented by mathematicians.

D. MACHINE LIMITATIONS BASED ON PHYSICS

1. IRREVERSIBLE AND REVERSIBLE MACHINES

This section follows the treatment of physical limits presented in [1]. All machines have to be used in the physical world, which poses the question 'what are the limits imposed by physics on machines?' The need for a physical theory of computing was made long ago by Landauer [11], [12], and Keyes [13] and has subsequently developed into an important part of fundamental computer theory. Each step of a machine is a decision, which must be perceived, by either one or more elements of the machine, or an external observer. This requires the expenditure of energy, communication in space, and the passage of time. The development of our understanding of physical limits suggests that machines be characterized as:

irreversible-classic,
reversible-classic, and
reversible quantum.

Any machine that can be played in reverse from its output state, back to its input state is reversible. 'Classic' and 'quantum' refer to the physical processes within the machine. Any computer can in principle include composites of these forms. Each of these types has important implications for the limits on what can be computed.

2. POWER DISSIPATION

The reversibility or otherwise of a machine may determine the fundamental limits on its economy of energy consumption. Any irreversible binary decision requires the expenditure of a minimum energy of kTlog2 joules [14], [15], [16]. Reversible processes by contrast do not appear in principle to require that any energy expenditure is required. The principles underlying this remarkable conclusion, that we can have 'computing for free', is that if a mechanical or electrical process can be made arbitrarily lossless, a network of these processes can be assembled which is reversible, and allows the output to be adroitly sampled and no other energy to be used. The argument is nicely exemplified by a ballistic computer in which totally elastic balls are bounced around a series of lossless reflecting boundaries [17]. It is shown that this type of computer can be made to enact the basic logic function needed in computing (AND, OR, NOT) A more realistic implementation of this type of computer can be approximated with a cellular automata version [18].

A similar lossless machine following a constrained Brownian random walk is Bennett's 'clockwork' Turing machine [19], which enacts the function of a reversible Turing machine, again with notionally lossless moving components. Bennet shows that this computer with a minute amount of biassing energy will migrate towards a solution. If the computation proceeds Brownian style, the length of the computation will naturally be a function of the complexity of the algorithm, and consequently, the longer, the more improbable is any outcome. Thus a problem of low order polynomial complexity will compute in a few steps, whereas one with exponential, or even transexponential complexity will take astronomic time! The point is not just that it will take longer, but that if there is an energy bound based on the diffusion rate of Brownian process for example, the complexity dictates a quite specific limit on the computation of the process.

Models of this sort cannot in reality be totally energy free, for the usual reasons that invalidate perpetual motion, and because of the need to prevent the propagation of small random errors. In fact Zurek has shown that these non-quantum models do dissipate energy of order kT/operation.

3. QUANTUM COMPUTERS

For quantum computers, in which the processes are in theory locally reversible, there appears to be no reason why the energy expended cannot be significantly lower than kT/decision [20]. Such lossless models have not yet been developed to a stage where definite limits can be stated, but the need to define the input state, and read the output state do require that energy of kT/bit be expended.

The thesis of reversible computing is generally accepted by physicists, who see no reason why reversible quantum effects cannot be exploited to achieve this type of computation. Problems which are posed by this type of computation which impinge on its energy needs are;

> entering the input,
> the decision to start the computation,
> the need to read the output,
> the need to know when to read the output.

An irreversible decision needs not less than kT joules of energy. It follows from this that a reversible computation must use not less than:

$$kT[(\text{no of changes in } i) + (\text{no of added/deleted bits})] \tag{1}$$

to perform a computation, where i is the input bit set. This is the number of bits in the input string which need to be changed, or introduced to define the output. Zurek develops this idea further, in the form of a theorem stating that the least increase in entropy in computing where the output, o replaces the input, i is [21]:

$$dS(i \rightarrow o) > |i^*| - |o^*| \tag{2}$$

where i^*, o^* are the minimum programs for generating the input and the output strings respectively.

When to read the output is problematic with energyless computers, because to check the output requires the expenditure of energy. If we sample the output but it is not yet available, the output must be read again. If the check is made too late the computation would need to be reversed, or repeated. The normal process of irreversible computation is enacted by a precisely controlled time incrementing system under the direction of s system clock. An <answer ready> register setting a 'flag' on completion of the computation can be employed which is repeatedly sampled by another process until the flag indicates that the answer must be read. With a reversible computer this problem would require that the computation took a known amount of time. This indeterminacy makes any real quantum computer difficult to design. It also suggests an energy bound related to the uncertainty of the run time.

Other difficulties with the reversible model are the tasks of specifying the input, and the program, (features explicitly stated in the Turing definition of computing). These essential elements appear to make the real physical formulation of a reversible machine very difficult. If the program itself is written by a reversible process, some energy store is needed to keep the energy until another programme is written! Because reversibility requires that the input and output states have the same number of bits, this class of programmes is significantly constrained, a concern discussed by Rothstein [22]. It should be noted that all known useful computers, man made, and biological are irreversible, and use energy considerably higher than kT/decision. The zero energy thesis has been challenged by Porod et al, on the grounds that it confuses logical irreversibility with physical reversibility [23], but the challenge is vigorously answered by Bennett et al [24], [25], and more recently was supported by Feynman, [26].

Quite aside from the energy considerations, quantum physics includes many phenomena which have no classical counterpart, and which cannot be emulated on a classical computer. Only a quantum computer could model these phenomena. Deutsch [27] has defined a universal quantum computer (UQC) on the very sound basis that classical physics is wrong! These computers are defined to take into account the fact that any computer has to be made in the real (quantum) world. Although as formulated by Deutsch, these computers can implement any reversible Turing machine function, there are other functions they can compute, that cannot be implemented on any Turing machine. Specifically, no Turing machine can generate a true random number. The UQC can, and by extension can also generate outputs according to any input defined density function, including some which are not Turing computable. It is perhaps obvious that a Turing machine cannot simulate the full range of quantum phenomena. A particular example of this UQC can simulate the Einstein-Podolski-Rosen (EPR) effect in quantum mechanics. This is the famous demonstration in quantum physics of two physically well separated measurements, which nonetheless are according to quantum theory (and verified experimentally) correlated by the action of seemingly independent acts of measurement [28]. No locally causal process can represent this effect, and attempts to do so lead to the need for negative probabilities [29].

Reversibility is not observed, and therefore reversible computing (permitted by quantum theory) is only a theoretical hypothesis. In examining reversible computation the Brussels school of Prigogene has argued that as reversibility is not observed in practice, the quantum theory is incomplete. Rather than looking for reversible computation, the search should be for a more accurate formulation of QT which does not permit any significant degree of reversibility [30].

Penrose ([4], Ch. 8) argues for a clear distinction between symmetry of the wave function and the function describing the 'collapse' of the wave function upon measurement. Whilst the former is symmetrical, the latter is not. Penrose cites the following simple example, shown in figure 1, which clearly demonstrates that a light-source to detector transfer of photons via a half silvered mirror is not time reversible in that the results do not correlate for forward and backward time.

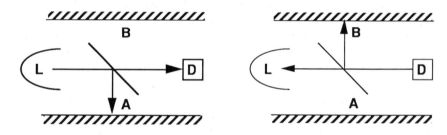

Fig. 1. Asymmetry of half silvered mirror (after Penrose, [4])

In this figure a photon from the light source L travels via a half silvered mirror to A or D. The probability that D will detect a photon is 0.5. The reverse probability that a photon left L, given a photon is detected at D is 1, whereas if we fired a photon from D under a reversed experiment, the probability of a photon at L would be 0.5. A similar asymmetry is evident in the probabilities of photons at points A and B. This seems to be no different logically from the arguments used about direction reversal in computers as an argument for energy conservation. We appear to say L to D collapses to probability P1, whilst D to L collapses to probability P2, which is clearly asymmetrical. There appears to be no essential difference between L to D as time reversal and L to D as a reversible computer operation, in which case this simple model is not reversible. The detector here plays the role of the element reading the output for a reversible computer. Although the schema of figure 1 is not reversible, (because of the walls) it indicates a difficulty which could be encountered in making a detection and seeking to regenerate a reversed energy flow after detection.

4. EXAMPLE 1: LIMITS DUE TO PROPAGATION TIME

We consider in more detail the space/time limitations imposed by the computation for two different models, the first a von Neumann machine, and the second a massively parallel machine.

Here the computational process is a set of sequential operations by a processor applying a stored program to a data base of stored elements (The von Neumann model) . Thus one cycle of the machine comprises:
- Read data element from Input Store,
- Execute operation on data element,
- Return result to Internal Store.

Within this computer:
- the *Internal Store* is regular cubic lattice, with the storage elements at an interstice distance l_e.
- the processor to memory link has a length l_{ps}.
- although depicted as linear tapes, the input and output may (as intermediate forms) be presumed to be also cubic stores, similarly structured to the *Internal Store*.

To read a data element requires a two way communication to address memory and fetch its contents to the processor. If the memory size is M, the mean addressing distance L is given by,

$$L = l_e.M^{1/3}, \tag{3}$$

so that the mean addressing distance, L_m is given by:

$$L_m = 2[L + l_{ps}] \tag{4}$$

The corresponding addressing time, T_m is:

$$T_m = (2/c)[L + l_{ps}] \tag{5}$$

The single instruction time of the machine can be defined as T_i, so the total cycle time is given by:

$$T_c = (3/c)[L + l_{ps}] + T_i \tag{6}$$

Here the time to return the result to store is also included in the operation. If we distinguish between the input and internal stores, the first term of equ. (4) is merely divided into the appropriate recall times for each memory unit.

Taking $M = 10^6$, $l_e = 10^{-6}$m, $l_{ps} = 0$, $T_i = 0$, gives $T_c = 10^{-12}$ secs, i.e a rate of computation which cannot exceed 10^{12} operations per second (a teraflop).

5. EXAMPLE 2: MASSIVELY PARALLEL IRREVERSIBLE COMPUTER:

Under the assumption of a kT/decision machine an estimate of machine performance can be obtained by evaluation of the capability of the lattice of elements shown in Figure 2. We first cite some crude limits, and then proceed to more realistic ones. Thus a 10^{60} bit memory will always be impossible, because allowing 1 molecule/bit would use more mass than exists in the solar system. Thermodynamics places a limit of 10^{70} on the total number of operations of all computers over all time, because to exceed this limit would require a power in excess of that of the sun (assuming irreversibility), [31]. Less astronomic bounds can be obtained for the simple lattice structure of digital elements, shown in figure 2, by examining the limits on the power, speed, closeness, and interconnectivity of the lattice elements.

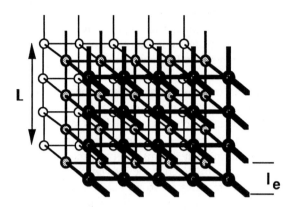

Fig. 2. Lattice of Digital Elements

Power: Considerations of state stability of extensive concatenations of digital elements in fact requires an energy of not less than $\approx 20kT$, and in devices today digital stability requires operation into the non-linear region of device transfer characteristics for which the inequality $V \gg kT/q$ must be satisfied. If we denote the energy per logical operation by $K_d kT$, K_d a constant $\gg 1$, the power needed for an element continuously operating at speed S will be, $P_d = K_d kTS$.

The power needed to energize and continuously communicate from one element to another is of the order of [13]:

$$P_c \gg \frac{(kT/q)^2 \, \lambda}{Z_0 \quad l_t} = \frac{P_0 \lambda}{l_t} \tag{7}$$

where k is Boltzman's constant, T the absolute temperature, and q the electron charge, Z_0 the impedance of free space, λ the distance between intercommunicating elements, and l_t the distance equivalent to the rise time, t (i.e. $l_t = c_i t$, and c_i is the speed of energy propagation in the computer). Equ. (7) is valid for short lines with high impedance levels, where the terminating impedance $Z \gg Z_i$, the line impedance. The total power needed therefore is $P_d + P_c$. In the following it is assumed $\lambda/l_t \cong 1$.

Inter-element Communication: If we assume that inter-element communication needs are random in each dimension within a cubic lattice of side L, then the mean path length of inter-element communication, $\lambda = L$. If the fastest possible speed of communication is needed, the distance equivalent rise time must be less than the inter-element transit time (i.e. $l_t < L$). With a richly connected lattice, λ is of the same order as the lattice side length L, however if the lattice is rectilinearly connected, (6 links per element) a communication must transit $m = L/l_e$ elements, taking m cycles. In this latter case each communication is m times slower, and $m-2$ of the elements serve a communication rather than a computing function.

Heat Dissipation: An upper limit on heat dissipation in electronic systems is 2×10^5 W/m [13] although a practical system of any size would be extremely difficult to build with this level of heat dissipation.

Inter-element Spacing: Inter-element spacing of digital elements is influenced by a variety of factors, including doping inhomogeneity, molecular migration and quantum mechanical tunnelling. If semiconductor devices are made too small the doping proportion becomes ill-defined and the device behavior is unpredictable. The limits of this process are where a doped semiconductor is reduced in size until it contains no doping element, and is pure! If very high potential gradients are created in very small junctions, the device can become structurally unstable due to migratory forces on the elements causing the junction to diffuse in relatively short time scales. Finally normal semiconductor action is subverted by quantum mechanical tunnelling. All these effects become significant at inter-element spacings of $<10^{-8}$m .

A Fundamental Limit Machine: With the information given above, some bounds can be obtained for a 'fundamental limit machine'. The total power consumption of this machine, P_m is bounded due to heat limitations by:

$$P_m = N(P_d + P_c) < QA \quad ; \quad N < QA/(P_d + P_c). \tag{8}$$

where N is the number of active elements, A the surface area of the computer, and Q the heat dissipation (W/m^2). For a cubic computer the available area is $6L^2$. Although this area size can be improved by multi-plane stacking and similar topological tricks, the cooling area remains proportional to L^2 (i.e. $A = K_a L^2$). Assuming that computer speed is set by the inter-element communication, then $l_t < L$ and the speed of operation of a single element, $S_c < c_i/L$. Applied to equ. (8), this gives for the speed limit of the total machine, S_m:

$$S_m = NS_c < QK_a Lc_i/(P_d + P_c) \tag{9}$$

If communication over distance L is achieved via $m = L/l_e$ intermediate elements, then to keep the speed limit defined by $S_c < c_i/L$, requires m times the power P_d to propagate the state, and m times the power P_c to propagate the state over m 'hops'. Furthermore only N/m of the total elements are directly computing; the remainder are supporting communication, so that in this case,

$$S_m < QK_a Lc_i/[m^2(P_d + P_c)]. \tag{10}$$

This equation shows the power of our computer to be directly proportional to the linear side length, speed of energy propagation and mean power dissipation/element, and inversely proportional to the power for changing state within an element, and the power needed to communicate that change of state. The factor m is a measure of the inadequacy of the inter-element communication within the computer. The above assumes a fully active totally parallel system with all elements in the computer active at any instance, and also takes a fairly cavalier view of how computing is defined!

Assuming that for a large computer with good interconnectivity   the communication limit will dominate, and taking the values $Q = 2.10^5$ W/m^2, $K_a = 1$, L = 1 metre, $P_o = 2$ x 10^{-6} W (minimal figure at 300^0 K [13]), and $c = 3$ x 10^8 m/s, gives a computer power $S_m = 3$ x 10^{19} operations/second at 300^0 K. At 30^0 K, the computer power would be increased 100 fold to $= 3$ x 10^{21} operations/second, but to make this 1m^3 computer 10^3 faster its side size would have to increase correspondingly by 10^3, to give a machine 1 km x 1 km x 1 km! It is clear therefore that around these limiting values, major advances are not possible by merely making the machine larger! Note that at these extremes inter-element spacing is not a problem. The speed limit defined by cellular density is:

$$S_m = (L/l_e)3c_i/l_t \quad \text{With } l_t = L \text{ this gives,} \tag{11}$$

$$S_m = L2c_i/l_e^3 \tag{12}$$

This computer's speed is much greater by this criterion than the value due to heat dissipation, with inter-element spacing l_e as large as 10^{-6} metres, so that heat dissipation will be the limiting factor in 300^0K systems.

Although these limits may seem bizarre, it does well to remember that the remorseless evolution of digital electronics does in fact have limits. Furthermore these figures are optimistic, assuming that all elements are switching at maximum rate. In any orthodox digital machine only a small proportion of elements are active at any time, although in highly parallel machines of the future, this proportion is bound to increase.

6. DISCUSSION

The above analyses are somewhat simplistic, and developments in connection topology, and molecular (quantum?), biological or optical and perhaps reversible computing elements and more precise definition of 'what is computation?' will lead to more refined limits. Combining physical limits and computational complexity limits will enable the difficulty of a problem to be stated in terms of space, time, and energy.

III. MODELS AND MEANING

Thus far we have shown that machines are restrained in algorithmic scope, and are restrained by physics. In addition machines are restrained by our inability to determine how to define and build them. The examination of the number of computational steps needed to perform an algorithm assumes that we know what we want to do, what we what to apply it to, and what the answer means.

This brings us into deeper water! The starting point is to examine the paradigm of *interpreted* mathematics. The motivation for a paradigm is to in some sense create a model of the process that concerns us. The process can be the understanding of mathematics, the meaning of set theory, the definition of some human process, such as speech understanding. For a wide variety of problems we commence by postulating a model. The question is *what limits apply to the process of using paradigms?*

A. PARADIGMS

Assume we have some entity (process etc.) that we wish to understand better. We observe it in some sense, and postulate an entity model. Note that this process *'postulate'* is undefined! The process is shown in figure 3.

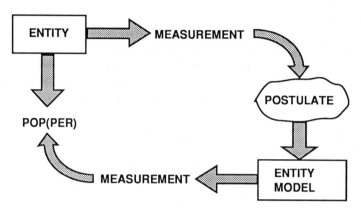

Fig. 3. The Life Cycle of a Paradigm

The model can be exercised to see if its behaviour agrees with that of the entity in question. If it does, according to some criteria of acceptance it is, to a degree, validated. Any real World model of this sort is in a strong sense only semi decidable. Although confirming evidence sustains it, any falsifying evidence negates the model, as Popper long ago observed. In our real world there are many entities which are not totally understood, and it has to be accepted that the entity model is virtually never complete. That its mechanism imitates the source-entity, *formally* infers little or nothing of the source-mechanism, particularly if the measurement is incomplete. It is an inductive inference that the longer the model survives, the more strongly it *is* the entity process.

The sting is that this measure (of completeness) itself follows a cycle similar to that of figure 3. If the measurements do not match our entity, then our paradigm is wrong. It is in this sense that Popper declares that a theory is never proved right, but can be unequivocally proved wrong [32].

The exactitude of mathematics has lead to its widespread use as the language of paradigms. Models of entities are specified mathematically. The advent of computer hardware and software now allows large arbitrarily structured models, which because they are enacted logically, are a form of mathematics, albeit one without any widespread acceptance.

A mathematical theory of reasoning is normally formulated in terms of *sets* of entities which are reasoned about in terms of *relationships* between sets, and a formalized logic is used for dealing with these. The sets and the logic may be uninterpreted or interpreted. In the symbolic machine theory whose ambit seeks to include robotics and artificial intelligence, it is axiomatic that there are clear definitions of sets, logic and their interpretation, all of which add up to a rational fruitful theory in the real world. This thesis is challenged in this Chapter.

B. MATHEMATICS

1. NUMBERS

In the physical world we can only experience finite rational numbers. Only finite numbers can exist in a computer. Thus to know that π is irrational is only to say that the algorithm for π is known, but is also known to be non-terminating. Any physical interpretation of π is a mapping obtained by arbitrarily terminating the program.

The number 1/3 does not exist in a representation scheme which is to a base n, which is relatively prime to 3. For example in decimal form, the number is 0.33333 recurring, and the total explicit representation of the number cannot be written! In this case, we all think we know what a third is, and moreover know it can be represented finitely in other numerical systems, e.g. duodecimally (0.4). This type of number is explicitly inexpressible, but totally and finitely predictable as a sequence of digits. Thus we can immediately say, with total confidence that the 10^{123} digit of decimal 1/3 is 3. All other rational fractions similarly have finite period, and are consequently predicable. The algorithmic information represented by the number can be defined as the length of the program needed to generate the sequence [8], [33].

Hence time complexity may be finite, periodic, aperiodic. The first represents algorithms which STOP, the second those which repeat cyclically, and the last algorithms which neither repeat, nor stop.

2. MATHEMATICS AS A PARADIGM FOR PHYSICS

Mathematics, is used as a paradigm for physics. Penrose [4], states that :
"the real number system is chosen in physics for its mathematical utility, simplicity and elegance."
Does this imply in a meaningful way that mathematics is true? Penrose in a discursive review of scientific philosophy asserts that,
"Mathematical truth is absolute, external, eternal and not based on man made criteria, and that mathematical objects have a timeless existence of their own, not dependent upon human society nor on particular physical objects."
The view is of an absolute Platonistic reality 'out there'. This is a common view amongst physicists, but less so by philosophers! Mathematics is true, but only

uninterpreted mathematics. Reviewing Godel's Theorem, (in which $P_k(k)$ represents a true but unprovable statement) Penrose further observes that:

> "The insight whereby we concluded that $P_k(k)$ is actually a true statement in arithmetic is an example of a general type of procedure known to logicians as the reflex principle: thus by reflecting upon the meaning of the axiom system and rules of procedure and convincing oneself that these indeed provide valid ways of arriving at mathematical truths, one may be able to code this insight into further true mathematical statements that were not deductible from these very axioms and rules." . . . . . ."The type of 'seeing' that is involved in a reflection principle requires mathematical insight that is not the result of purely algorithmic operations that could be coded into some mathematical formal system."

This argument admits one level of 'reality' upon which is another less explicit intuitive one (a reflexive level). The point here of course is that Penrose sees a way out of the Godel problem, but by the very nature of that escape it cannot be based upon a finite axiomatic method. The 'meaning' given to the reflex principle are very much part of the interpretative process, and this is a function of the real world.

C. THE INTERPRETIVE PROCESS

Symbolic systems require the application of sets and the relationships between sets and elements of sets. The mathematical manipulation and control of such structures, is *ab-initio* uninterpreted. That is they have no meaning in any real world context. In the domain of such symbols an expression is true if it satisfies the conditions defining that particular set. If we wish to use layered paradigms, we can take this as:

- layer 1: US = uninterpreted symbolic,
- layer 2: RE = explained by reflexive principle.

The next step in such a system is to apply a domain of interpretation to give:

- syntax (uninterpreted structure)
- reflexive viewpoint
- semantics (interpreted)

Semantics defines the way symbols are related to 'things' in the real world. For example in mathematics we interpret ideas such as line, planes, manifolds etc. and give them some degree of substance beyond the mere symbology, even though in a physical context they remain abstract. All of this appears to work well in mathematics, and gives birth to a common theme which seems intrinsic to human intellectual creativity - to use inductive inference to suggest that what works in one domain will work in another. Thus representing the world, or some aspect of it in symbolic terms, which have some interpretation, and then 'reasoning' about those objects by performing mathematics and logical operations on the symbols has become not just well entrenched in western thought, but a 'natural' way of approaching problems. Is it always correct to do this?

Although the mathematical distinctions between interpreted and uninterpreted systems is strongly maintained by mathematicians, without at least a background of mathematical intuition, the uninterpreted symbolism would be totally sterile. Even when an interpreted model is used, the question remains - does it always work?

IV. SETS AND LOGIC

A. THE PROBLEM WITH SETS

We have sketched the strong role of the mathematical approach to modelling the world. The idealized schema for this is:

Sets of objects with defined attributes,
Sets of relations between objects.

This perception can be interpreted as a language and its grammar.

In seeking a rational basis for defining natural categories, Lakoff [34] has shown that defining closed sets of objects - of categorizing the 'things' in the world - is so beset with difficulties as to be impossible in the mathematically formal sense for many real world cases. Here, following Lakoff, we examine some of the pitfalls in defining sets and in using rules of logic to define sets relationships.

The mathematical *set*, when used in the real world, can be shown to be frequently at odds with reality. A set from an objectivist viewpoint is a collection of objects whose set membership is determined by some rule. The objectivist viewpoint takes the existence of these sets as independent of human cognition and thought. The objectivist and mathematical viewpoints are almost synonymous. For many situations in the real world, including of course mathematics, this approach is satisfactory, but it is an incomplete model for categorization. Within this formulation is the strong intuition that things are of a 'natural kind', independent of the classification mechanisms of an observer. Sets whose boundaries are scalar can be included in the objectivist model.

The models which this type of classification does not admit include:

metynomic, where part of a category, or a single member stands for the class,
(e.g. use of the name of an institution for the people who are members of it).

radial categories, where many models are organized around a centre, but the links are not predictable, but motivated. Lakoff cites the concept of *mother* as an example of a radial category. Another example is *logics,* where no simple set of rules can classify all members as being of the same kind.

If members of a set are defined by a set of membership relations we should expect no one member of the set to be any different from others in terms of that set membership. Categorization is clearly more complex, usually depending on more complex criteria of membership. The pragmatic approach of early pattern recognition used feature bundles, or weighted sets of feature bundles to define categories. Although this approach often works for simple problems, it fails to allow any generalization where compounds occur. Lakof [34] cites the following examples:

Linguistic qualifiers such as:

- technically / strictly,

- doesn't have / lacks.

These qualifiers can very subtly change the emphasis of a sentence. Sometimes they are effectively synonymous, other times they are clearly different.

'Hedges' such as:

> Esther Williams is a fish = false,
> Esther Williams is a regular fish = true.

Set intersection failure,

> guppy = pet fish; poor example of pet; poor example of fish;

Intersection gives an even poorer example as pet fish. Other examples are:

> small galaxies (not the intersection of same things and galaxies,
> good thief (not the intersection of good things and thieves),
> heavy price (not the intersection of heavy things and prices).

These are known as non-compositional compounds.

Another basis for set definition is the prototypical member. This can usually be defined, but the definition does not fit set theoretic norms. There are always exceptions which defy any neat set of definitional conditions for set membership. Further such definitions cannot easily handle the difference between foreground and background examples.

Lakof cites several interesting examples where different races, nationalities often use quite different concepts of category, taking items together in one culture which seem at least strange to another, or even absurd.

The conclusions to be drawn from this type of review of categories is that whilst some forms of objects do appear to obey reasonable set theoretic rules, many others do not. The more categories are applied to man's abstractions the more extreme the failure. This critique is developed at some length by Lakoff, who proposes an answer to these ideas based on *Idealized Cognitive Models* (ICM's). This is a form of frame representation in which each idea, concept, or category is defined relative to a cognitive model, which can have a variety of structures according to the context.

This is reflected in linguistics by the difficulty of achieving any formulation for language which is not extensively context sensitive. Non mathematical classes, which I will call *categories*, fit uncomfortably within the mathematical paradigm for a class (defined by a set of logical conditionals).

Unlike sets, category membership is often sensitive to a wide range of environmental, viewpoint and temporal factors which condition any form of aggregation. Thus the set of fixed conditionals which we might attempt to define the class *bird* may run into difficulties with any of:

> trapped birds, dead birds, injured birds, walking birds, pictures of birds.

This line of inquiry leads to the conclusion that for many real world situations The proposition that categories are sets and are logically definable is often wrong. The factors which govern the admission of some real world thing to specific category membership is subtle, and not readily captured by mathematics.

B. PUTNAM'S THEOREM

It is well entrenched in western scientific culture that the only meaningful mode of rationality is logical thought. This may seem like a tautology but when rational thought is equated with logic, and logic to mathematical logic, we awaken the same dilemmas about mathematics, sets and symbolic logic interpreted to explain our world. Lakoff ([34], Ch. 14) rebuts it in the following form:

> " it is only by assuming the correctness of objectivist philosophy and by imposing such an understanding that mathematical logic can be viewed as the study of reason in general. Such an understanding has been imposed by objectivist philosophers. There is nothing inherent to mathematical logic that makes it the study of reason."

The unnatural element of this assumption is difficult to perceive. Van Wolferen, a western journalist who has lived for many years in Japan expressed this reservation thus [35]:

> "The occidental intellectual and moral traditions are so deeply rooted in the assumptions of the universal validity of certain beliefs that the possibility of a culture without such assumptions is hardly ever contemplated. Western child rearing practice inculcates suppositions that implicitly confirm the existence of an ultimate logic controlling the the universe independently of the desires and caprices of human beings."

The American philosopher Hilary Putnam has challenged the objectivist position in logic as a basis for understanding and reasoning about the world [36]. The objectivist position requires validity of two unsafe postulates:

> "P1: The meaning of a sentence is a function which assigns a truth value to to that sentence in each situation (or possible world);
> P2: The parts of the sentence cannot be changed without changing the meaning of the whole."

Putnam shows that this interpretation is logically flawed, which he demonstrates as a variant of the Lowenheim Skolem Theorem. This theorem shows that a set theoretic definition giving only non denumerable models can be shown to give denumerable models as well. Putnam goes on to illustrate the implication of this rather abstract paradox of mathematical sets with an example along the following lines ([36], Ch. 2).

Take the sentence The *cat* is on the *mat*, and define the three cases:

W1   *cat** = <some *cat* is on some *mat* AND some cherry is on some tree>
W2   <some *cat* is on some *mat* AND no cherry is on any tree>
W3   < neither W1 nor W2>
DEFINE *cat**: x is a *cat** IF and only if W1 holds AND x is a cherry OR
                        case W2 holds and x is a cat OR
                        case W3 holds and x is a cherry.
DEFINE *mat**: x is a *mat** IF and only if W1 holds AND x is a tree OR
                        case W2 holds AND x is a mat OR
                        case W3 holds AND x is a quark.

In any 'world' falling under cases W1 or W2, <a *cat* is on the *mat*> is true, and <a *cat** is on the *mat**> is true; in any world under case W3 both statements are false. So what? Well this contrived construction of *cat** and *mat** shows that by changing the definitions of *cat* and *mat*, the meaning of the sentence can remain unchanged. This style of construction can be extended. As Putnam comments, if a community of men and women defined a wide variety of *things* in this way with men using the basic *things* definition and women using the *things** definition there would be no way of telling, even though each might imply different intentions. The problem is acute because Putnam has shown that this postulate **P2** fails for every sentence in a theory of meaning.

By defining some rather convoluted criteria for class membership, Putnam has shown that a sentence can have the same meaning, even if constituent parts of that sentence are radically changed, i.e.

"no view which fixes the truth value of whole sentences can fix reference, even if it specifies truth values for sentences in every possible world."

What this implies is that a single objective meaning deriving from a theory of meaning of this sort is impossible. The reader is referred to Lakoff [34] for a more expansive demonstrative argument of the proof, and to Putnam for both demonstration and formal proof [36], [37].

Goodman examined the problem of reasoning about predicates such as *grue* (green before year 2000, blue thereafter) and *bleen* (blue before year 2000 and green thereafter) [38]. Applying conventional logic to these predicates can lead to paradoxical results . This and Putnam's example shows that for 'funny' predicates applying logic will lead to false or inadequate conclusions. What Putnam shows is a related result that not only can 'funny' predicates can lead to no unique interpretation, but can be used to demonstrate inconsistency in any scheme for implying a semantic interpretation of a formal symbol structure.

Both Goodman's and Putnam's problems have been criticized for using 'funny' predicates which should not be allowed (e.g [39]). This has a certain appeal, but what rules discriminate *'funny'* from normal? What logic applies to these rules, which must become part of a larger scheme R* in which we claim:

R* is the *real* relation of reference.

Unfortunately such a statement is also vulnerable to analysis by Putnam's funny predicate stratagem. The appeal to *natural* categories has been used by Watanabe [39], and by Lewis [40] and others in the argument about these paradoxes. Unfortunately it implies for a theory of meaning a form of constraint which is both mathematically arbitrary, and mathematically unevaluatable. Thus in Lakoff's words ([34], Ch. 15):

"Putman has shown that existing formal versions of objectivist epistemology are inconsistent: there can be no objectively current description of reality from a 'God's eye' view. This does not of course mean that there is no objective reality - only that we have no privileged access to it from an external viewpoint."

The symbolic theory of mathematics is claimed to be objectivist because it is independent of human values. It is perhaps not too surprising that as a model for human reasoning it is inadequate. What we next demonstrate by a brief overview of the logics largely deriving from the artificial intelligence research programme, is that there is little hope of a composite system of logic for reasoning about the world. For the isolated world of mathematics logic has a true sanctuary, but not 'outside'!

Much argument about reference and meaning is predicated on a degree of exactitude about the elements of reasoning - objects, rules, interpretation, which other than in mathematics there is little evidence for. Indeed in all but mathematics the evidence is overwhelmingly against such an interpretation. If understanding the action of a robot which can sense its environment, and reason about its (real world) environment (including itself) is a legitimate scientific objective, we must ask what knowledge and what constraints can guide such inquiry? Thus far mathematics is not sufficient.

C. THE LOGIC OF LOGICS

The search for a valid (under some common sense judgement) logic has resulted in numerous logic schemes intended to extend or supplement first order Classical Logic (CL). The objective is to model Common Sense Reasoning (CSR). A interesting feature of these new logics is the development of the semantics of implication and/or operators determining the interpretation of logical expressions. Some of the features leading to this diversity are:

Generalization and Quantification: the need to have effective constructs for the notions *in general* and *some.*

Modularity: ideally a knowledge base (KB) should be modular, whereby new knowledge adds to the KB rather than requiring its restructuring.

Non- Monotonicity: Many CSR problems require non-monotonic reasoning, where knowing $A \rightarrow B$ does the addition of the fact $A \cap C$ still permit $A \rightarrow B$. This is needed where knowledge is revised or falsified by new evidence. With non-monotonic reasoning new contradictory evidence does not necessarily imply that what was first believed is wrong. Consider the

'fact'   *<John is honest>*
new evidence: Fred says *<John stole her keys>*.
The following revisions must be considered:
-                         *<John is not honest >* OR
                          <John is still honest> AND *<Fred is mistaken>*, OR
                          <John is honest> AND *<Fred is lying>*.

Implication and Interpretation: The implication statement in CL, long a source of debate [3] has now become the subject of a series of extended definitions, and consequently the centerpiece of a variety of new logics. At the centre of these extensions is the need to better understand what any assertion means. Implication may be transitive or not according to the logic type:

$$(\text{ i.e. } A \rightarrow B; \ B \rightarrow C; \ C \rightarrow D \ \equiv A \rightarrow C). \tag{13}$$

A short resume of the distinguishing features of these some of these logics follows partly based on the fascinating review presented in Lea Sombe [41], where one simple problem is examined using these different logics.

Classical Logic is inadequate for many aspects of CSR: it cannot handle generalizations easily, thus $\forall(x) \ (student(x) = young(x))$ is violated by only one old student. Similarly an exception list $\exists(student(x) \neq young(x))$ does not allow the specific case that all students are young except one[2]. Exceptions can be handled explicitly but require the revision of the KB for each exception, i.e. CL fails the modularity requirement. Cases which are not a priori exceptional are usually difficult or impossible to prove.

1. OUTLINE OF NEW LOGICS

This outline is not intended to be tutorial. It is summarized here merely to illustrate the wide range of ideas that suffuse research on logic, and to dispel the thought that some may have that there is some immutable foundation logic. These logics are referred to by type and originator, and are only a small but interesting selection from the totality.

Reiter's Default Logic is non-monotonic [42]. The format for statements in this logic is

$$\frac{u(x) : v(x)}{w(x)} \qquad (14)$$

which reads:

IF $u(x)$ is known and IF $v(x)$ is consistent with $u(x)$ THEN $\rightarrow w(x)$.
It allows quantification of the forms:

$\exists(x)$ such that,

$\forall(x)$ there are,

x are, with exceptions.

The ordering of defaults in this logic can effect the result and is a significant problem.

Modal Logics weaken the blunt dichotomy of *true* and *false* in an attempt to better match CSR. There are many variants including:

McDermott and Doyle [43] in which a proposition is one of

true	P
false	$\neg$P
conceivable	$\lozenge$P

If a proposition is not provably false, it is conceivable: $(\neg P \rightarrow \lozenge P)$
Whatever is not conceivable is not true: $(\neg \lozenge P \rightarrow \neg P)$
This logic is not good for generalizations, allows no more quantification than CL, nor for information updating with data withdrawal.

Moore's Autoepistemic Logic [44] uses a belief operator *believe* P (denoted $\square$P) in which

---

2. $\forall(x)$ reads "for all x"; $\exists(x)$ reads "there exists an x such that"

$$\neg\Diamond\neg P \equiv \Box P$$

This logic does not allow belief without justification, whereas that of MacDermott and Doyle does.

Levesque's Logic [45] provides another nuance with

$\Box P$ = *what is known to be true is at least that*

ΔP = *what is known to be false is at most that*

Likelihood Logic of Halpern and Rabin [46]

$\otimes P$ = *it is likely that*

$\oslash P$ = *it is necessary that*

This provides a range of degrees of likelihood

$\neg[\otimes\otimes \ldots (\neg P)]$tends to P, and

$[\otimes\otimes \ldots (\Pi)]$ tends to $\neg\otimes P$

In this logic the basic relationships are:

$P \to \otimes P$

$\Box P \to \neg\otimes\neg P$

$\Box(P\to Q)\to(\otimes P\to\otimes Q)$

$\otimes(P \vee Q)\leftrightarrow(\otimes P \vee \otimes Q)$

Circumscription [47] is a non-monotonic logic founded on the rule that:

IF *every possible proof of C fails* THEN $\to\neg C$

This requires a closed world in which every possible proof can be enacted. A more appropriate 'open world ' alternative is to replace every possible proof with some resource bounded operation. In this type of alternative profound problems about the selection of such a subset need to be investigated   There is within circumscription the appeal to a 'normal' well behaved world by use of the operator *abnormal*, i.e.

$\forall(x)$ (scientist$(x) \wedge \neg$abnormal$(x) \to$ intelligent.

This logic admits only the classical quantifiers and is not easy to revise.

Conditional logics: here the implication statement is rephrased[3] by Stalnaker as [48]:

"A $\to$B is true in the actual state IFF in the state that mostly resembles the actual state and where A is true, B is true."

These logics have been developed to account for counterfactual statements [1], [38]. A later variant due to Delgrande is [49]:

" A B is true in the actual state IFF B is true in the most typical state where A is true, within the states more typical than the actual state."

 The question of determining *'that mostly resembles'* and *'the most typical state'* implies a strong context sensitivity, of unspecified form for this type of logic.

Possibilistic Logic defines possibility and necessity :

$\Pi(p)$ = possibility

$N(p)$ = necessity

$N(p) = 1 \to$ p true

_____

3.   IFF reads "IF and only IF"

$$\Pi(p) = 0 \; \rightarrow \text{impossible for p to be true}$$
$$\Pi(p) = 1 - N(\neg p)$$

This logic can indicate inconsistency when $\Pi(p)$, $N(p)$.> 0 occur, and dates from the early work of Finetti in 1937 [50] and includes the probabilistic belief functions of Shafer [51].

Numerical Quantifier Logic applies numerical values to predicates and set members and the reasoning is in terms of mathematical inequalities and numerical bounds.

Bayesian Reasoning uses Bayes' theorem to provide a form of quantified reasoning from apriori to posteriori probabilities as the consequence of an experiment. It is therefore causative. Problems with Bayes' theorem as a basis for reasoning include the assignment of priors is often arbitrary, and the need to make assumptions about the independence of variables. The view that most, if not all reasoning can be achieved by this means is strongly advocated by Cheeseman, and discussed by other contributers in [52].

Temporal logic: The lack of a temporal element is also a significant weakness of these logics. Our world is dominated by time, all our actions are governed by it, subject to it, and we run out of it, but the majority of logic is 'autopsy' logic. Each bit of evidence is laid before us static and invariant . Although there are a series of developments in temporal logics, again we lack any clear leaders in this field. The need for a <true now> variable is illustrated by the example:

    <I will take another breath after this one>
    = globally false,
    = true for planning tomorrow,
    = false for buying life insurance.

Temporal logics are being researched in AI (e.g. [53], [54]) and formalisms being considered in security systems and for communications protocol definition are also being investigated for the explicit representation of temporal features.

2. WHITHER LOGIC?

Lea Sombe shows that all these logics have various strengths and weaknesses. None can be proclaimed with any certitude as the logic. The authors comment that:

> "It would seem pointless to think that a single formalism would be capable of representing the multiple forms of common sense reasoning."

I suppose this itself to be a form of common sense reasoning about logic. Common sense reasoning only has any real expression in natural language. Unfortunately natural language has resolutely resisted attempts to encase it in a mathematically logical framework.

That none one of these logics are fully applicable for modelling reasoning poses the question *can they ever be?* The analysis of Putnam suggests that there will always be inadequacies in any attempt at a universal logic, and experimental evidence, which has resulted in this spate of logic development suggests there is no convergence to a single logical framework for reasoning, nor much experimental success in using these systems.

Can we invoke a meta logic which will hybridize some of the current developing logics? This is a form of Minsky's frame problem - how do we determine the set of rules?; how do we determine and categorize frames? This in turn is to admit that logic itself is not absolute, but is inevitably context dependent. Nillson, a strong proponent of logic based reasoning, argues against English as a means of explication because [55]:

> "Although humans understand English well enough, it is too ambiguous a presentational medium for present day computers - the meaning of English sentences depends too much upon the contexts in which they are uttered and understood "

This is to turn the problem around, and is based more on faith in logic than in evidence that it provides the basis we need to construct rationality. This belief is strongly voiced by Nillson in [55] and just as vigorously countered by Birnbaum [56]. The belief that there cannot be any alternative to logic as a foundation for AI is widespread. In a review of reasoning methods Post and Sage and assert that [57]:

> "Logic is descriptively universal. Thus it can be argued that classical logic can be made to handle real world domains in an effective manner; but this does not suggest that it can be done efficiently."

Lakoff's comment on this presumed inevitability of the role of logic is [34] :

> "It is only be assuming the correctness of objectivist philosophy, and by imposing such an understanding that mathematical logic can be viewed as the study of reason itself."

Analogical reasoning seems to be second order (that is reasoning about reasoning). Similarly the management of the ordering and creative process of taxonomy in human discourse must admit some metalogical features to manage this aspect. Lakoff's debates about classes would not be conducted at all without this.

McDermott, for long a strong (indeed leading) advocate of the logicist viewpoint has made a strong critique of the approach [58], and shows very frankly in his paper how any strong orthodoxy in science can inhibit dissenting views ([58] p 152, col. 2). The spectrum of his arguments against "pure reason" covers:

- paucity of significant real achievement,
- its inadequacy as a model for understanding planning,
- the seduction of presuming that to express a problem logically implies its solution,
- it fails as a model of abduction,
- it can only be sustained by a meta theory (what logic to use and how).

Of this McDermott complains that:

> "there are no constraints on such a theory from human intuition or anywhere else."

There is certainly no constraint that ensures that a meta theory can sustain deduction soundness.

Conventional first order logics are at least deterministically exponentially complex. Second order (meta) logics are of higher (worse) complexity. This appears to imply that whatever the merits of such logics they cannot formally be the mechanisms that are used by human brains. This is called the finitary predicament by Cherniak [59], who likewise argues that we cannot use CL because we haven't got the time. If fundamentally we are precluded from using CL, it might be claimed that we reason be some 'quick and dirty' technique, but surely the error is in ascribing reason to formal logics. This is after all the no more than the usual use of mathematical modelling - it is normally accepted as an approximation. The oddity with its use to represent reasoning is the conclusion that if we disagree with it, we are in some sense 'to a degree' wrong, whilst CL is right.

For us, evolution has adopted the set of deductive mechanisms which are both effective for survival and complexity contained. Also CSR *has generated* CL, and the rest of mathematics as well. Just as the human visual cortex (and the retina) have adapted to the range of special visual images and visual dynamics which we need to thrive (and ipso facto admits a wide range of visual phenomena we cannot perceive), so surely it must be for deduction?

V. COMMUNICATION

A. DEFINITION OF COMMUNICATION

Thus far we have examined limitations on the computational process, and the limitations in the definitions of sets, and the logic we use to reason about sets. However intelligent behaviour is crucially concerned with communication between agents. What is the communication process? We know we must communicate "information" which must be understood. Shannon has defined a statistical theory of communication in which the information transmitted can be equated to the recipients probability of anticipating that message [60]. There are inadequacies in this model, which relate to its dependence on probabilities, and what they mean in finite communications. Further the overall process of communication is not addressed by Shannon's theory; only the information outcome of a successful communication process.

A paradigm of the communications which has been developed for the definition of interoperability standards is the International Organisation for Standardisation's Open Systems Seven Layer Interconnection Reference Model [61]. This is a layered paradigm and is shown schematically in figure 4. This provides a useful checklist for showing the range of processes that underlie any effective communication. These include the need to recognize the protocols for establishing and terminating communication, and for recognizing that the process becomes increasingly (and more semantically) defined form as the layer level increases.

Layer Number	Layer Name	Short Explanation of Function
7	APPLICATION	What the user wants
6	PRESENTATION	How it is presented to the user
5	SESSION	Defines user to user communication
4	TRANSPORT	Defines the qualities needed by layer 5
3	NETWORK	Determines and executes routing over the available network
2	LINK	The communication between nodes of the network
1	PHYSICAL	The physical propagation process - wires, EMfibre optic etc.

Fig. 4. ISO OSI Seven Layer Model of Communication

B. WHAT IS INFORMATION?

1. EXTENDING THE DEFINITION

Shannon equated information in a message to the degree of surprise it causes [60], and I assume most readers are familiar with his theory. What information does a computation provide? We are engaged in the computing task on a massive scale internationally, and the demand is for more, and faster machines. Yet Bennett, in a review of fundamental limits in computing [19], notes that:

"since the input is implicit in the output no computation ever generates information."

In an earlier examination of the role of the computer Brillouin states [16]:

"It uses all the information of the input data, and translates it into the output data, which can at best contain all the information of the input data, but no more."

Similarly, but more powerfully, Chaitin in developing algorithmic information theory, asserts that no theorem can develop more information than its axioms [33], [62].

I call this viewpoint of information *the paradox of zero utility*. There must be more to computing than is implied by these statements or else there would be very little demand for our computers! This concept of zero utility is challenged on the following grounds. Firstly the output is a function not just of the input, but of the transforming programme as well. In general therefore the output of a computation is a function of the length of the input string $/ I_i /$ and the length of the program string $/ I_p /$. This point is covered explicitly in Chaitin's work.

I take the view that a computation is to fulfil two roles:

- it performs a task for an *External Agent* (EA), which is only understood by the EA at its topmost functional level,
- it performs a task which the EA cannot perform without computational support.

This second view point reinforces the concept of information only having any meaning in the context of a transaction between two entities. In the case of Shannon's theory it is between two entities with a statistical model of the occurrence of members of a finite symbol set. In the case of a computation it is a transaction between the computer and the EA, in which the computer performs information transformation which is of utility to the EA.

The probability of an event is meaningful only with respect to a predefined domain. The choice of domain of application is a feature of the EA, who observes the event so that a theory of the *computer*, its *program* and the *input*, in isolation cannot represent the totality of the information transfer role of which the computer is a part. That examination never answers the question 'what is the computation for?'

The examples of multiplication and primality testing illustrate this.

Multiplication: This is a well understood process, but for long strings may not be practical for the EA to undertake. The computation resolves two issues:
- it reduces the EA's uncertainty about the outcome of the operation.

The amount of information imparted by this is a function of his apriori knowledge about the answer. Thus for small trivial numbers the information provided is zero (he already 'knows' the answer). For large integers (e.g. a 200 digit decimal string) it provides an amount of information virtually equal to the symbol length. The fact that the output is implicit in the input in no way diminishes this information from the EA's viewpoint. Making information explicit, rather than implicit creates for an EA, information, which wasn't there before.

- secondly it performs the calculation economically from the EA's viewpoint.

He does very little work. This is basically the observation made long ago by Good [63], that any theory of inference should include some weighting to account for the cost of theorizing, which can be equated to time. For the EA to do the computation takes far more time than the computer. This point is a dominant one in the use of computers of course.

In this first example the computer is a cure for sloth on the part of the EA. There is no mystery, inasmuch as the EA understands multiplication, and could, in principle do it unaided. The relationship between the computer and the EA in this case is that of some monotonic relationship between the computer time to evaluate, and that for the EA's. In the case of the human EA, considerations of error, fatigue, and even mortality can be limitative factors in this comparison.

Primality: Consider next the problem of determining if a given (large) integer is prime. The EA will have a good understanding of what a prime number is, but perhaps very little understanding of the algorithm employed by the computer to make the determination of primality. In this case assume that the computer is again able to perform rapidly, and therefore assist the EA, who, for large integers, couldn't determine primality at all without the computer. So here both the speed aspect and the nature of the computer program are important.

What are his apriori expectations about the primality of the number under scrutiny ? We might argue that as the density of primes around a given value of x can be evaluated from a tabulated integral function (i.e $Li(x)$ [64]), a prior probability greater than $P_p(x)$ can be established, and $1 - P_p(x)$ is the probability that guessing primality is wrong. After running the computer with say Rabin's algorithm [65], it will reply (with a probability related to its run time) that:

I = prime (probability = p_r)

I = composite number (probability = 1).

Again this is a quite specific increase in information, which is a direct function of the observer's apriori knowledge.

This second example raises the additional question of the information content of the computer program. Although the programme can be sized in terms of a number of bits, it contains implicitly much information about the nature of prime numbers, and several important number theoretic relationships. There is the possibility of some communication of information between two EA's in this case: the one who created the program to test for primality, and the EA who uses the program. Note that if the problem was to determine factors of the input, the computed answer can be checked by multiplying the factors together. Indeed this method could be used to test guesses. So when defining a baseline of ignorance based on apriori probability, in some examples the guess can be tested, but in some others it cannot be. If a computer programme says that a 200 digit number is prime, you have to trust it. If a factoring program says a number is composite, this can be tested. The point here is that it is not sufficient to use a Shannon like model to define the base ignorance, because of the impracticality of meaningfully exploiting such a baseline schema.

If the answer to a computation question is required *exactly* then no guess with other than a very low probability of error is acceptable as a useful answer. (The result of a product with only one digit in error is WRONG!) For simple decision questions of the form <is x in class P> (e.g. is x a prime) it might seem that the answer resolves a simple ambiguity, and therefore constitutes only *1 bit* of information, whereas the answer to a product is $|I_a| + |I_b|$. However to retain the information in meaningful form we must note the input, I_i a descriptor of the program, I_{pd} and the output, I_o. The total information I_Σ therefore is:

$$I_\Sigma = |I_i| + |I_{pd}| + |I_o| \tag{15}$$

Note that although I_o is implicit in I_i it cannot be shown to be without explicitly noting I_{pd}. It is proposed that the information in this type of operation is called *transformational information*.

A higher level of information is *relational information* where the relation, $\Re$ is defined, but the algorithm for establishing it is not. The triples in this case are of the form of the following examples:

e.g.1:   [INTEGER], $\Re$(is prime), [ANSWER] e.g 2:   [INTEGER, INTEGER], $\Re$(product), [ANSWER]

e.g 3:   [INTEGER], $\Re$(divides by 3), [ANSWER]

In this case there must be the means to define a taxonomy of relational operations. Generally we would expect relational information to require less bits than transformational information because of the less explicit representation of the transformation. Note that both examples 1 and 3 require computation over the entire input string I, but the two problems are radically different in their complexity, i.e. the ease with which they can be answered. Does the former provide more information ? I suggest not so, but only that the (computational) cost of obtaining the information is greater.

2. THREE TYPES OF INFORMATION

Thus three levels of information are proposed:
> *algorithmic,*
> *transformational,*
> *relational.*

These lead naturally to the view that the function of a computer is to provide a transformation of the input which:
> i) changes the information as perceived by the recipient,
> ii) involves a 'cost of theorizing'.

The implications of point i) are described above. The second point concerns the differential cost of theorizing by: the EA unaided, the EA activating the computer.

Unaided the EA computing N operations will involve $\tau_{ea}N$ seconds, and for complex computation would need to take additional precautions to prevent errors. Our computer undertakes the task at a cost to the EA of entering the data, and reading the output. The cost to the computer are the resources of time, space and energy.

3. DISCUSSION

This informal discussion shows that the concept of information in bits must be viewed with care, and the statement that a computer does not generate information contradicts our intuitions about the computer's role. If we take the view that all agents are transformational machines, where does 'new' information come from?' I suggest that information does in a real sense grow, by the creative application of such transformations, which is directly a consequence of the computational process. If some agents are not transformation machines, in what sense are they defined, or does one take the metaphysical retreat to say they cannot be? I argue against the metaphysical excuse.

Clearly more needs to be done to define information types, and their relationships. The definitions of Shannon and Chaitin are not wrong, but do not cover the full spectrum of information types that we must understand in order to understand machines. The concept of layering may be a way forward here. For the higher levels of information Scarrott [66] has proposed defining organisation and information recursively thus:

> "An organized system (OS) is an interdependent assembly of elements and/or organized systems. Information is that which is exchanged between the components of the organized system to effect their interdependence."

Another important development is information based complexity, which is defined as the information (input and algorithm) needed to compute a function within an error bound ε [67]. Where a cost of computation is included (fixed cost per operation) the definition is with respect to that algorithm with the minimal cost among all algorithms with error at most ε.

Stonier [68] has proposed that information in physical structures is proportional to the order in that structure, where the highest degree of information is in the largest simplest structure, such as a crystal lattice. This definition does not accord with the viewpoint of this Chapter.

A more radical definition of information in Quantum theory has been proposed by Bohm et al [69], who assume that the the electron is a particle always accompanied by a a wave satisfying Shroedinger's equation. This leads to a formulation of a quantum potential which is independent of of the strength of the quantum field, but only depends upon its form. Since this potential controls the distribution of particles, Bohm et al interpret this as an information field. Although these emergent concepts do not appear to relate directly to the discussion of information here, any theory of information should be compatible with quantum effects, including quantum computation.

There is no cogent and widely accepted definition of information, but it is information which distinguishes much of our conscious activity from inanimate action. Much of the debate on logic, and on AI more generally takes some notion of information as implicitly self evident. I believe that any theory of machines must include this aspect of understanding our behaviour or that of a machine.

VI. BUILDING RATIONAL MACHINES

A. ALTERNATIVES TO LOGIC

In section III a range of objections to a purely logicist approach to understanding machines are raised and discussed. These objections are in my view serious ones, but leave the critic with the need to offer some competing paradigm which is better, or even to offer any other paradigm! There are some possibilities emerging and these are outlined in this last section.

B. MINIMAL RATIONALITY

It is clear that in human though and behaviour a complete deductive logic is not observed. Rather our rationality is restricted on the one hand by the contextual aspects of the environment and immediate goal, and on the other by the cost effectiveness of any 'deductive' or preparatory thought. Experience (dare I say common sense) shows that human deduction is rarely explicitly logical, and is often incomplete. It is this partial nature of human thought that is totally at variance with the view that conventional logics are in some sense 'reason'. Even if we accept that deductive thought occurs as a result of a *belief set, B* and a *desire set D* and a gestalt comprehension that reason can take us from *B* to *D*, the transfer process itself is far from logical.

Cherniak [59] has described this partial quality in our behaviour in terms of a minimal rationality needed to survive, in terms of the following fundamental propositions:

"If A has a particular belief set, A would undertake some, but not necessarily all of those actions that are apparently appropriate."

We can denote this by:

$$\underline{if}\, A(B,D)\ \underline{then}\ S_{\neg A}(Take\ Actions \in Apparently\ Appropriate).$$

A similar formulation is proposed for sound inferences, and inconsistencies:

$$\underline{if}\, A(B,D)\ \underline{then}\ S_{\neg A}(Make\ Sound\ Inferences \in Apparently\ Appropriate).$$

$$\underline{if}\, A(B,D)\ \underline{then}\ S_{\neg A}(Delete\ Inconsistencies \in All\ inconsistencies).$$

This thesis has appeal, but what determines the form of the operators *some but not all* $(S_{\neg A})$, *Apparently Appropriate*, and *Sound Inferences*. What Cherniak does here is to define a requirement for rationality that reflects human rationality, and which we know is adequate for our species to survive. Unfortunately the prescription is far from complete.

C. MOBOTICS

Brooks has developed a paradigm for AI, called *mobotics*, which does not seek, nor claim, to have explicit representation of concepts, actions or the like [70]. Further Brooks claims that

"if we are seeing to represent intelligent action, it must include the dynamic interaction "in the real world with real sensing and real action. Anything less provides a candidate with which we can delude ourselves."

In making this claim he is repeating a rather earlier assertion of Francis Bacon's [71]:

nec manus nisi intellectus, sibi permissus multant valent; instrumentis et auxilibus res perficitur (neither the hand nor the mind alone left to itself would amount to much).

This is important also for helping to understand the wide range of unavoidable context dependency of any reasoning system [1], and for ensuring that the finitary predicament is not only applied to 'reasoning' about static facts, but can apply realistically to the dynamic processing for sensing and motion and reaction. Brook's approach to achieving these ends is to produce an interacting family of processes acting at increasingly higher levels of operation/goal. This has similarities with the layering paradigm, and of neural networks, but a much higher level than the 'neuron'. Within this type of system, as with neural networks there is no explicit representation of concepts by symbols. The notion of a concept becomes blurred between the activities at different layers. Brooks calls this a subsumption architecture. Using this approach some reasonable success has been achieved with a wandering, exploring robot.

D. CONNECTIONISM

The theory of connectionist networks has provided as method of solving some machine problems without the need for explicit symbolic representation [72]. Whilst mobotics is aimed at constructing interacting groups of functions at a high level of functionality, the connectionist approach applies that model at the most primitive level. Symbolic representation only exists at the input and output, and Smolensky has referred to this style of operation as invoking the sub symbolic paradigm [73], in which symbol representation, although in principle possible, is of no particular value. A problem with connectionist approaches is that they are poor at executing symbolic representations, however results in the domain of speech recognition suggest that if the method works better than symbolic ones the only loss is some form of internal symbolic explanation of how it works [74]. A review of the scope for connectionist symbol processing is given in [75].

E. LANGUAGE

The medium that all this debate, in this Chapter and elsewhere is conducted is *natural language*. It is surely essential for any serious progress in understanding machines to understand language in its broadest setting, which must include the dynamics of speech and interaction physically with the environment. Vygotsky regarded language as the second signalling system, in contrast to the the first which is perception through the primary senses [76]. Bruner [71] has further asserted with some conviction that:

> "There are two modes of cognitive functioning, two modes of thought each providing distinctive ways of ordering experience, of constructing reality. The two (though complementary) are irreducible to one another. Efforts to reduce one mode to the other, or to ignore one at the expense of the other, inevitably fails to capture."

The two modes are "well formed argument" and "a good story", the one logical and conveying truth, the other narrative, and conveying versimmilitude.

These viewpoints support the belief that understanding must be a reactive process with the environment, and that language must similarly be strongly context sensitive. No theory of language can exist without including a theory of the world in which it is to be used and understood. No effective theory has yet emerged.

VII CONCLUSION

This review is only a few splashes of light in a still dark scene.. The logicist approach to AI has yielded limited achievement, mainly in analysing systems which were created with a hierarchical logical structure by man. AI achievement is difficult to assess objectively, and is not undertaken routinely by many of its practitioners, but some assessments have been made [77], [78], [79]. What does not emerge from these results is any clear confirmation of any fundamental principle which either underpins a 'theory of AI' or which defines limits on machine attempts at AI.

What fundamentally do we know about machine capability? At the algorithmic level the theory is now well developed, but in no way complete. The limits and capabilities we understand are still meagre, but the subject is only 60 years old. Physics after three centuries still has its problems. The delight is that there is still so much to discover.

REFERENCES

1. I. White, "The Limits and Capabilities of Machines", I.E.E.E. Transactions System Man and Cybernetics, Nov/Dec 1988.

2. A. Turing, "On Computable Numbers with an Application to the Entscheidungsproblem", Proc. London Math. Society, Ser. 2, vol. 42, 1936.

3 H. Delong, "A Profile of Mathematical Logic", Reading MA, Adison Wesley, 1970.

4. R. Penrose, "The Emperor's New Mind", Oxford University Press, 1989.

5. M. Minsky, "Finite and Infinite Machines", Prentice Hall, 1967.

6. M. R. Garey and D. S. Johnson, "Computers and Intractability", San Franscisco: W. H. Freeman, 1979.

7. A. L. Selman (ed.),"Complexity Theory - Retrospective", Springer Verlag, 1988.

8. G. J. Chaitin, "On the Difficulty of Computations", I.E.E.E. Trans. on Information Theory, vol. IT 16, no. 1, pp. 5 - 9, January 1970.

9. A. N. Kolmogorof, "Logical Basis for Information Theory and Probability Theory", I.E.E.E. Trans. on Information Theory, vol. IT 14, no. 5, pp. 662 -664, September 1968.

10. P. Martin-Lof, "On the Definition of a Random Sequence", Information and Control, vol. 9, pp. 602 - 619, 1966.

11. R. W. Landauer, "Wanted: a Physically Possible Theory of Physics", I.E.E.E. Spectrum, September, 1967.

12. R. W. Landauer, "Fundamental Physical Limitations of the Computational Process: an Informal Commentary", paper submitted to the Sackler Institute of Advanced Studies, Tel Aviv University, Israel, 1984.

13. R. W. Keyes, "Physical Limits in Digital Electronics", Proc. I.E.E.E., vol. 63, pp. 740-767, May, 1975.

14. J. von Neumann, "The Theory of Self Reproducing Automata", University of Illinois Press, Urbana, IL, 1966.

15. R. W. Landauer, "Irreversibility and Heat Generation in the Computing Process", IBM J. Res. Development., vol. 5, pp. 183 - 191, 1961.

16. L. Brillouin, "Science and Information Theory", Academic Press, New York, 1962.

17. E. Fredkin and T. Toffoli, "Conservative Logic", International J. of Theoretical Physics, vol. 21, 1982.

18. N. Margolus, "Physics Like Models of Computation", Physica, vol. 10D, no. 1-2, pp. 81 - 95, Jan. 1984.

19. C. H. Bennett, "The Thermodynamics of Computation - a Review", International Journal of Theoretical Physics, vol. 21, no. 12, 1982.

20. W. H. Zurek, "Reversibility and Stability of Information Processing Systems", Physical Review Letters, vol. 53, no. 4, 23 July 1984.

21. W. H Zurek, "Thermodynamic Cost of Computation, Algorithmic Complexity and the Information Metric", Nature, vol. 341, 14 September 1989.

22. J. Rothstein, "Physics of Selective Systems", Int. J. Theoretical Physics, vol. 21, nos. 3/4, 1982.

23. W. Porod et al, "Dissipation in Computing", Phys. Rev. Lett., vol. 52, no. 3, 1984.

24. C.H. Bennett et. al. " Thermodynamically Reversible Computing", Physical Review Letters, vol. 53, no. 12, 17 September 1984.

25. T. Toffoli, "Physics and Computation", International Journal of Theoretical Physics, vol. 21, no.'s 3/4 , 1982.

26. R. P. Feynman, "Quantum Mechanical Computers", Foundations of Physics, vol. 16, no. 6, 1988.

27. D. Deutsch, "Quantum theory, the Church - Turing Principle, and the Universal Quantum Machine", Proc. of the Royal Society, London, vol. A400, 1985, pp. 97 - 117.

28. A. I. M. Rae, "Quantum Physics: Illusion or Reality", Cambridge University Press, Cambridge UK, 1986.

29. R. P. Feynman, "Simulating Physics with Computers", International Journal of Theoretical Physics, vol. 21, no.'s 6/7 1982.

30. P. V. Coveney, "The Second Law of Thermodynamics: Entropy, Irreversibility and Dynamics", Nature, Vol. 333, No. 6172, 2 June 1988.

31. S. C. Pohlig and M. E. Hellman, "An Improved Algorithm for Computing Logarithms over GF(p) and its Cryptographic Significance", I.E.E.E. Trans. Information Theory, vol. IT-17, no. 24, pp. 106 - 110, Jan 1978.

32. K. R. Popper, "The Logic of Scientific Discovery", Century Hutchison, London, 1959.

33. G. Chaitin, "Algorithmic Information Theory", Cambridge University Press.

34. G. Lakoff, "Women Fire and Dangerous Things", Chicago University Press, 1987.

35. K. van Wolferen, "The Enigma of Japanese Power", MacMillan, 1989.

36. H. Putnam, "Reason, Truth and History", Cambridge University Press.

37. H. Putnam, "Models and Reality", J. Symbolic Logic, vol. 45, no. 3, Sept 1980.

38. N. Goodman, "Fact, Fiction and Forecast", Harvard University Press, Cambridge MA ,1983.

39. S. Watanabe,"Knowing and Guessing", John Wiley and Sons, New York, 1969.

40. D. Lewis, ""Putnam's Paradox", Australasian J. Philosophy, vol. 62, no. 3, pp. 221 - 236, 1984.

41. Lea Sombe, "Reasoning Under Incomplete Information in Artificial Intelligence", John Wiley

and Sons, September 1990.

42. R. Reiter, "A Logic for Default Reasoning", Artificial Intelligence, 13, pp. 81 - 132, 1980.

43. D. MacDermott and J. Doyle, "Non-monotonic Logic I", Artificial Intelligence, 13, pp. 41 - 72, 1980.

44. R. Moore, "Autoepistemic Logic", in "Non Standard Logic for Automated Reasoning", ed. P. Smuts et al, Academic Press, London UK, 1988.

45. H. J. Levesque, "All I Know: an Abridged Report", proc. Sixth National Conference on AI (AAAI-87), Seattle, WA, USA 1987.

46. J. Y Halpern and M. O. Rabin,, "A Logic to Reason about Likelihood", Artificial Intelligence, 32, 1987.

47 J. McCarthy,"Circumscription - a Form of Non-monotonic Reasoning", Artificial Intelligence, 13, 1980.

48. R. F. Stalnaker, "A Theory of Conditionals", in "Studies in Logical Theory", ed. N. Rescher, Blackwell, Oxford, 1968.

49. J. P. Delgrande, "A First Order Logic for Prototypical Properties", Artificial Intelligence, 33, pp. 105 - 130, 1986.

50. B. deFinetti, "La prevision ses lois logiques et ses sources subjectives" in H. Kyburg Jr. and H. Snokler (eds.) "Studies in Subjective Probability", Wiley 1937.

51. G. Shafer, "A Mathematical Theory of Evidence", Princeton University Press, Princeton, NJ, 1976.

52. P. Cheeseman (and others), Computational Intelligence, vol. 4, no. 1, Feb 1988. (An interesting debate about the role of probabilistic reasoning.)

53. M. L. Ginsberg and D. E.. Smith "Reasoning About Action I: A Possible Worlds Approach", Artificial Intelligence, 35 no. 2, June 1988.

54. M. L. Ginsberg and D. E,. Smith, "Reasoning about action II: the qualification problem", Artificial Intelligence, 35 no. 3, July 1988.

55. N. J. Nilsson, "Logic and Artificial Intelligence", Artificial Intelligence, vol. 47, pp. 31 - 56, 1991.

56. L. Birnbaum, "Rigor Mortis: a Response to Nilsson's "Logic in Artificial Intelligence"", vol. 47, pp. 57 - 77, 1991.

57. S. Post and A. P. Sage, "An Overview of Automated Reasoning", IEEE Transactions System Man and Cybernetics, vol. 20. no. 1, January/February 1990.

58. D. McDermott, "A Critique of Pure Reason", Computational Intelligence, 3, pp. 151 - 160, 1987.

59. C. Cherniak, "Minimal Rationality", MIT Press Cambridge Mass., USA, 1986.

60. C. E. Shannon and W. Weaver, "The Mathematical Theory of Communication", University of Illinois Press, Urbana, 1949.

61. U. D. Black, "OSI - A Model for Computer Communications Standards", Prentice Hall, Englewood Cliffs, NJ, 1991.

62. G. J. Chaitin, "Information Theoretic Limitations of Formal Systems", Journal of the A.C.M., vol. 21, no. 3, pp. 403 - 424, July 1974.

63. I. J. Good, "Probability and the Weighting of Evidence", in "Foundations of Probability Theory, Statistical Inference, and Statistical Theories of Science", vol. 2. Reidal Publishing Co., Dordrecht, Holland, 1976.

64. M. Abramowitz and I. A. Stegun, "Handbook of Mathematical Functions", Dover, 1970.

65. M. D. Rabin, "Probabilistic algorithms" in J.F. Traub (ed.), "Algorithms and Complexity". Academic Press, New York, 1976.

66. G. Scarrott, "The Nature of Information", Computer Journal, vol. 32, No. 3   1989.

67. J. F. Traub, "Information Based Complexity", Academic Press, 1988.

68. T. Stonier, "Towards a General Theory of Information II: Information and Entropy", ASLIB Proceedings 41(2), pp. 41 - 51, February 1989

69. D. Bohm, B. J. Hiley and P. N. Kaloyerov, "An Ontological Basis for the Quantum Theory", North Holland, Amsterdam.

70. R. A. Brooks, "Intelligence Without Representation", Artificial Intelligence, 47, 1991, pp. 139 - 159.

71. J. Bruner, "Actual Minds, Possible Worlds", Harvard University Press, Cambridge, Massachusetts, 1986.

72 D. E. Rummelhart, J. L. McClelland et al, "Parallel Distributed Processing", MIT Press, Cambridge, Mass., 1986.

73. P. Smolensky, "Information Processing in Dynamic Systems: Foundations of Harmony Theory", in D. E. Rummelhart, J. L. McClelland et al, "Parallel Distributed Processing", MIT Press, Cambridge, Mass., 1986.

74. J. Bridle, "Alpha-Nets: A Recurrent 'Neural' Network Architecture with a Hidden Markov Model Interpretation", Tech. Report SP 104, Royal Radar and Signal Establishment, UK, 1989.

75. Artificial Intelligence, 46, 1990, (Special Issue on Connectionist Signal Processing).

76. Vygostky, "Thought and Language", MIT Press Cambridge, Mass. 1962.

77. I. White, "W(h)ither Expert Systems", AI and Society, Spring, 1988.

78. J. W. Sutherland, "Assessing the Artificial Intelligence Contribution to Decision Technology", I.E.E.E. Trans. Systems Man and Cybernetics, vol. SMC16, no. 1, pp. 3 -20, January/February, 1986.

79. T. Niblett, "Machine Learning Applications", Turing Institute Note, December 1990, Turing Institute, 36 North Hanover St., Glasgow, Scotland.

Neural Network Techniques in Manufacturing and Automation Systems

Joseph C. Giarratano

University of Houston-Clear Lake

2700 Bay Area Blvd.

Houston, TX 77058

*David M. Skapura** *

Loral Space Information Systems

1322 Space Park Drive

Houston, TX 77058

INTRODUCTION

The 1980s saw a number of new computing paradigms applied
to manufacturing and automation systems. These paradigms
are artificial neural systems [1], expert systems [2], fuzzy logic
[3], and genetic algorithms [4]. The study of artificial neural
systems (ANS), or neural nets for short, is also sometimes
called *connectionism* or *parallel distributed processing*.

A neural net is typically composed of many simple
processing elements arranged in a massively interconnected
parallel network. Depending on the neural net design, the
artificial neurons may be sparsely, moderately, or fully
interconnected with other neurons. Two common
characteristics of many popular neural net designs are, (1) nets
are trained to produce a specified output when a specified
input is presented rather than being explicitly programmed,
and (2) their massive parallelism makes nets very fault
tolerant if part of the net is becomes destroyed or damaged.

Neural nets produce solutions to problems by training
artificial neurons connected in a network structure, which can
be represented in the form of a directed graph, such as the one
illustrated in Figure 1. Using this schematic representation,
each processing element is represented by the nodes and the

---

* Formerly with Ford Aerospace Corp.

CONTROL AND DYNAMIC SYSTEMS, VOL. 49

connections are modeled by the arcs. The normal direction of information flow through the network is indicated by the arrowheads on the arcs.

Neural nets are trained by automatically adjusting the strengths of their interconnections according to a learning rule. A neural net architecture or design is characterized by (1) the topology which defines the number of neurons, their interconnections, and layers, and (2) the learning rule such as one which trains the net to produce a specified output given a specified input. Figure 1 illustrates a common type of neural net architecture used with learning rules such as backpropagation and counterpropagation. In other architectures, such as the Hopfield, the output neurons are connected back to the input neurons. Many different types of architectures have been designed, as will be discussed shortly.

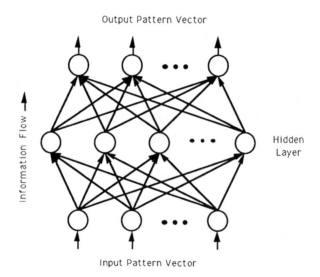

Figure 1 A Neural Net Architecture

A General Survey of Applications

After training, the neural net can efficiently provide the solution to a problem or a "best guess" to a set of inputs that it has not been explicitly trained to solve.

Some general applications of neural nets technology are the following.

- An explosive detection system for checked airline baggage [5]
- Automobile engine diagnosis [6]
- A typewriter that uses speech recognition [7]
- Inspection of solder points on electronic circuit boards [8]
- Cancelling echoes in long-distance telephone circuits [9]
- A machine vision system to verify that bottle caps of pharmaceuticals on a conveyor belt are properly sealed [10]
- Predicting the rating of corporate bonds [11]
- Economic prediction of the stock market [12]
- Mortgage underwriting judgments [13]
- Defense applications in processing sensor data to classify military targets [14]
- Medical diagnosis of back pain [15]
- Robot control [16]

One significant advantage of neural nets over conventional algorithmic solutions is that the neural net can give an answer in constant time and so avoids the combinatorial explosion in computing time that often occurs as the size of a problem increases.

A second advantage is that an algorithm does not have to be explicitly known and programmed by a human. In many cases, no algorithm may be known for a problem, or the algorithms may not be efficient enough for practical use. A neural net can learn to program its own solution to a problem, which may be adequate for a practical application even though it's not the best solution. Neural nets have found many practical applications in control, industrial inspection, robotics, target classification, and many other applications.

Neural nets have been implemented as electronic hardware using discrete components such as operational amplifiers, as integrated circuit chips, as optical components, as special purpose computers, and most commonly by software.

Neural nets is based on observed behavior of biological nervous systems. ANS is like the other paradigms of expert systems and genetic algorithms in that it is not intended to solve problems for which efficient algorithmic solutions exist. Rather, these paradigms are heuristic since they are not guaranteed to provide the correct answer. However, they have proven to be very useful in a wide number of applications that are not easily solvable.

While expert systems deals with IF THEN type of reasoning or inference, neural nets commonly operates at a lower level of pattern recognition. In many common types of neural nets, the neuron network is trained to generate a specified output when a specified input is given. The training is done by the network itself. In contrast, the expert systems generally require a knowledge engineer to explicitly code the rules used by the expert system in solving a problem.

Neural nets have also been used as an expert system [17]. The matrix controlled inference engine (MACIE) represents knowledge of diseases and symptoms in a neural net to diagnosis disease. MACIE also overcomes one of the basic limitations of neural nets which is the inability to explain its

knowledge. Most neural nets cannot explain why their internal
parameters are set to certain values after training. In contrast,
MACIE can interpret the net and generate IF THEN type rules
to explain the knowledge represented in the neural nets.

Advantages of Neural Nets

Neural nets offer a number of advantages compared to
algorithmic solutions.

> (1) *Fault tolerant*. Since the information in a net is
> distributed over its processing elements, a reduction in
> neurons only decreases the quality of the nets
> response. In contrast, changing a single memory
> location of an algorithm usually has a disastrous effect.

> (2) *Graceful degradation*. As more of a net is damaged or
> removed, the performance of a net degrades
> gracefully, rather than catastrophically as with an
> algorithm. This is important for computer systems that
> must operate unattended for long periods of time in
> hostile environments without the availability of spare
> parts.

> (3) *Associative recall*. A partial or noisy input may still
> elicit the complete original information.

> (4) *Interpolation from trained data*. A neural net may
> discover features in the original training data that it
> was not trained on. In one experiment [18], an neural
> learned family relationships of twenty-four

hypothetical people. Later, the net suggested new relationships between people that were correct but for which it had not been trained.

(S) *Plasticity.* If some neurons are removed or damaged, the net may be retrained to its original skill level if sufficient neurons remain.

If an algorithmic solution does not exist or is not practical, a neural net may be suitable. However, neural nets are generally not good for applications that require number crunching, the best solution, or where an explanation of the neural nets behavior is required. For example, a neural net would not be a good choice to balance your checkbook, although it would be a good choice to verify that the signature on your checks was really yours and not a forgery.

Commercial Vendors

A number of companies sell neural net hardware and software [19]. An excellent yet inexpensive way of learning about neural nets is volume 3 of Rumelhart's books called *Parallel Distributed Processing* available from the MIT Press. Vol. 3, titled *Explorations in Parallel Distributed Processing,* describes a half dozen neural nets simulators and includes diskettes for an IBM PC compatible machine.

An annual survey of AI applications, including neural nets products, is now available including information on neural nets software simulators, hardware neurocomputers and special purpose boards for image recognition [20].

One vendor of a neural net simulator called NeuroShell™ lists

101 applications of its product by users. The applications relating to automation and robotics are as follows.

- Analytical chemistry applications
- Circuit board diagnosis
- Disease diagnosis
- Factory and shop problem analysis
- Fault tracing systems
- Glass design
- Ground water quality control
- Inventory analysis
- Molding machine operation
- Oil exploration
- Optimizing operation of fusion furnace
- Optimizing raw material orders
- Optimizing scheduled machine maintenance
- Polymer identification
- Process control
- Product identification
- Pyroanalysis
- Quality control
- Response instructions for alarm activity reception operator
- Screening of estimates for time and material
- Selection of on-line sensors for chemical process
- Sensor interpretation
- Spectral analysis and interpretation
- Temperature and force predictions in mills and factories
- Troubleshooting of scientific instruments
- Water resources management

As a more detailed example, NeuroShell was used to primarily control a debutanizer, which is a type of distillation column, at Texaco's Puget Sound Refinery. The neural net was trained using 1440 historical data sets consisting of seven inputs (control and disturbance variables) and two outputs of steam to a reboiler for heating, and overhead reflux to an exchanger. A feedback mechanism is used in conjunction with the neural net since it was assumed that no process model, even a neural net, is 100% correct. Analysis shows that the neural net is correct a minimum of 80% of the time.

Another application of NeuroShell was to predict the number of crew members from Brooklyn Union Gas Company needed for service calls as a function of the month, predicted temperature, and day of week. It only took two weeks to develop the net since all the old service records were on file. The company no longer needs to rely on experienced supervisors for the crew predictions, and is more accurate.

A Brief History of Neural Nets

The study of neural nets is both new and old. The origin of neural nets is as old as artificial intelligence (AI) itself, dating back to Frank Rosenblatt's perceptrons of the 1950s and 60s. His work culminated in an influential book, *Principles of Neurodynamics*, dealing with a type of single-layer artificial neuron system he called a perceptron [22]. The perceptron consisted of two layers of neurons and a simple learning algorithm that could recognize certain patterns of inputs. However, the great significance of the perceptron was that it could learn new patterns and more importantly, recognize *similar* patterns to those it had learned.

The ability of the perceptron to recognize similar patterns, rather than exactly the same patterns it had been trained, is a

milestone in creating intelligent machines. The ability to
recognize similar patterns is very important because it greatly
reduces the amount of explicit programming involved. As an
analogy, once a person learns that this is a letter

A

a person has no trouble recognizing the following as A's in
Figure 2 even if you have not explicitly learned each font and
size.

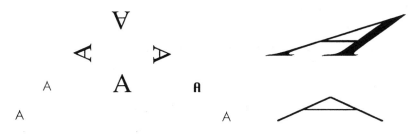

Figure 2 Patterns of A

In Figure 2, notice that there are eleven A's, not ten. While
people have no trouble seeing the eleventh A as a pattern
formed by some of the others, it is very difficult for algorithmic
programs (if possible at all) to recognize it.

The foundations of the mathematical modeling of biological
neurons had been developed in the 1940s by McCulloch and
Pitts [22]. The first mathematical theory of learning by neurons
was given by Donald O. Hebb in 1949 [23]. In this theory of
learning, a neuron's efficiency in contributing to the firing of
another neuron increases the more that the other neuron fires.
This models observed biological behavior in which the
conductivity of connections between neurons at their synapses
increases with use. In neural nets, the weight of connections
between neurons is changed analogously to the changing

conductance of natural neurons synapses. Thus, the neural net internal parameter called weight is directly dependent on the learning that the neural net experiences just as the conductivity of biological synapses changes as a biological neural net learns.

In addition to Rosenblatt, another prominent worker in neural nets was (and still is) Bernard Widrow, who developed an adaptive linear element model called the Adaline. This model and its successor, the Madaline were used for weather forecasting and adaptive control. Widrow achieved great success using adaptive filters to eliminate echoes in telecommunications and received the prestigious Alexander Graham Bell award. Widrow and Hoff developed the delta learning rule in 1960.

The early work on neural nets caused tremendous excitement in the popular press of the 1950s and 60s because it appeared that scientists would soon be capable of creating electronic devices that could learn rather than relying on programming. It appeared that robots and intelligent computers were just around the corner.

Many AI researchers worked on the perceptron during the 1960s. However, this early era of neural nets came to a shattering halt in 1969 when Marvin Minsky and Seymour Papert published a book called *Perceptrons* that showed the theoretical limitations of the perceptron as a general computing machine [24]. Their work showed that perceptrons could only do 14 of the 16 basic logical operations. In particular, a perceptron could not do the exclusive-OR logical function and thus could not solve interesting non-linear problems. Their book ended with the pessimistic view that their intuitive judgement was that multilayer neural nets would offer no advantage. Because of this book, research funding for non-symbolic work in AI essentially vanished. Instead, during the 1970s, AI research concentrated on symbolic computing, especially using LISP and PROLOG.

In spite of the lack of research funding, some significant work on neural nets continued in the 1970s. James Anderson developed the Brain-State-in-a-Box model. Teuvo Kohonen did work on self-organizing nets and competitive learning. Steven Grossberg work in the 1970s eventually culminated in his Adaptive Resonance Theory (ART) models.

The new period of neural nets originated with the work of Hopfield in 1982 [25]. Although some significant work had been done in the 1970s, it was Hopfield's work which made neural nets respectable again and a fit subject for research – and especially research dollars. By the early 80s, it was obvious that there were many interesting AI problems that were too difficult to solve by the symbolic method championed by Minsky. A new approach was needed and people began to investigate again the neural net approaches.

Hopfield developed a type of net that could solve a wide variety of problems. For example, a Hopfield net can be used as an associative memory or content-addressable memory in which a noisy or incomplete portion of an item suffices to retrieve the entire stored item [26]. In particular, his Hopfield net could solve the important Traveling Salesman Problem in constant time. The basic problem here is to visit a given number of cities exactly once so as to minimize the total distance traveled. This is an important problem that occurs in many delivery and telecommunications problems such as packet switching. In automation, the problem may be for a robot to deliver parts or materials to specified workcells in the minimum amount of time or to minimize some other utility function.

Although an algorithmic approach will yield the best answer, it may take too long for the answer to be of any good. In many real-world cases, a good answer rather than the best is adequate. The Hopfield net provides a good solution in constant time as the number of cities to be visited increased linearly. In contrast, the optimal algorithmic solution time grows by combinatorial explosion with a linear increase in cities and becomes useless after awhile. Other combinatorial optimization problems can easily be done by neural nets such as the four-color map [27] [28], the Euclidean-match [27], and the transposition code [29].

Minsky and Papert were unable to figure out how to update the weights in a multi-layer neural net so that it could do the exclusive-OR problem, and expressed doubt it could ever be done. Their reputation and prestige discouraged the majority of researchers in neural nets at the time. The lack of an algorithm to decide which weights should change and by how much became known as the Credit Assignment problem. In general, the Credit Assignment problem occurs in any case where it is necessary to identify specific actions that contribute to a undesirable outcome in a sequence of actions.

A solution to this problem was reported as early as 1974 [30] and independently rediscovered by a number of others [31] [32]. Rumelhart, Hinton, and Williams popularized this learning rule algorithm, which they referred to as the generalized-delta rule after Widrow and Hoff's delta rule for learning. The generalized-delta rule is also commonly referred to as backpropagation [33]. Backpropagation is essentially an implementation of gradient descent for multistage optimization [34]. The term *backpropagation* describes how the error signal is fed backwards from the output layers to the input layers. Appendix A provides a detailed mathematical discussion of backpropagation.

Neural nets such as the Hopfield nets which use feedback during operation are referred to as recurrent networks. Neural nets such as backpropagation that do not employ feedback after training are referred to as feedforward or layered nets.

The 1980s saw a renaissance for neural networks. The International Neural Network Society was formed and every year publishes hundreds of papers at its huge conference, the International Joint Conference on Neural Networks. A new journal, the IEEE *Transactions on Neural Networks* was published. Special issues on neural nets were published in the April 1988, April 1989, and April 1990 issues of the IEEE *Control Systems Magazine*.

Neural Net Architectures and Learning

The backpropagation algorithm was the first of many new neural net architectures developed in the 1980s. Over 50 different architectures have been developed so far. Some of the more popular ones are listed following [35].

- Adaptive Resonance Theory
- Avalanche
- Backpropagation
- Bidirectional Associative Memory (BAM)
- Boltzmann and Cauchy Machines
- Brain State in a Box
- Cerebellatron
- Counterpropagation
- Hopfield
- Madaline

- Neocognitron
- Perceptron
- Self-organizing Map

In the case of supervised learning, such as backpropagation, during training a teacher provides inputs and an error signal to the net if the output is not correct. This type of learning requires labeled training data because the teacher must know what the input is so that it can tell the net if the output is correct.

Not all neural nets are trained by example. in the case of unsupervised learning, as illustrated by the BAM, Adaptive Resonance Theory models, and Hopfield nets, there is no labeled training data and no teacher. For example, an application to identify sonar signatures would be a good candidate for a supervised learning net since a teacher could tell the net if it correctly identified the sonar signals. An application to help a mobile robot avoid obstacles would be a candidate for an unsupervised learning net.

The basic idea of unsupervised learning is to provide *no* feedback of its outputs. Instead, the network experiments to find interesting features or clusters that may illustrate important regularities and relationships of the inputs. Letting the neural net experiment to form clusters of related data is important in pattern recognition since otherwise all data would be in distinct categories. Unsupervised learning is sometimes called vector quantization when a net is used to convert analog signals into binary.

Several different unsupervised learning schemes have been developed such as winner-take-all, competitive learning, and leaky learning [36]. One type of unsupervised training called vector quantization converts analog inputs into a binary form.

Such nets have been used to compress and decompress signals for efficiency in telecommunications.

In addition to supervised and unsupervised learning, a third type is called self-supervised learning or reinforcement learning. In this type, the neural net monitors its internal performance all the time and no teacher is required. This type of learning is sometimes called learning-by-doing or learning-by-experimentation. A distinction between this and the other types is that an error signal is generated by the net that is proportional to the magnitude of the error. In the supervised and unsupervised learning, only the existence of the error is known.

Generic Neurocontroller Architectures

Although many different types of neural net architectures have been, and still are, being developed, five generic architectures have been identified for direct neural net control of actuators and effectors [37]. These five generic neurocontroller architectures comprise the following.

• *Supervised control.* The net learns how to control by being taught the correct outputs for specified inputs. The training may come from any source, including humans. One application of this has been to training robots by the Fuji Corp.

• *Direct inverse control.* Here, the net learns the system inverse dynamics so that the system tracks a specified trajectory. For example, given various input positions of a robot arm, the net learns the actuator signals which effected the corresponding arm positions. After training, the net can move the arm to a specified position when the actuator signals are supplied as input.

• *Neural adaptive control.* In this architecture, a neural net replaces conventional adaptive control designs such as the Self-Tuning Regulator and Model-Reference Adaptive Control. The neural net has the advantage of coping with nonlinearities and greater robustness. Biological control systems, i.e., living organisms, are very successful in dealing with the following.

- Uncertainty in unstructured environments (missing sensors and incomplete information, changing goals and resources, sensor overload)
- Multiple sensory input (data fusion)
- False sensory input (lies, mirages, virtual images, echoes)
- Strong nonlinearities
- Complex planning involving multiple degrees of freedom
- Real-time response

However, standard control systems for automation and robotics are generally not very good in these areas. In particular, the problem of dealing with uncertainty is becoming more critical as the need to develop intelligent control systems grows. The events of Three-Mile-Island and Chernobyl vividly illustrate the need for intelligent control systems which can replace failure-prone human beings who make mistakes that cost human lives and property.

• *Backpropagation-through-time.* This can be derived from the ordinary backpropagation algorithm [38]. It is also called backpropagation of utility, since it involves information flow backward through time. Applications include a truck backer-upper [39], and an official model of the natural gas industry [40].

• *Adaptive critic methods.* This can be derived as approximations to the Bellman equation of dynamic programming, and can be used to maximize any utility function or reinforcement parameter as a function of time [41]. For example, an adaptive critic neural net may be designed to minimize the energy used by a robot arm in following a specified trajectory. In this example, the utility function is the energy expended along the trajectory. Such considerations are important in self powered robots which cannot easily be recharged.

Process Controllers

Process control is essential to manufacturing. As we have seen so well in the 20th Century, automation of manufacturing is a critical factor in the competitiveness of industry and entire nations. Those nations which keep their automation current keep their industries current and therefore remain profitable. While simple process control is adequate for simple products, people today are demanding much better quality for their money, as American automakers learned the hard way in the 1980s. More complex products and those of higher quality require better process control.

The introduction of low-cost microcomputers in the 1980s was a significant advance in providing economical distributed process controllers that could be programmed with complex algorithms. This contrasts with the traditional hard-wired controllers dating back to the 1940s (and still often used today) such as Ziegler-Nichols tuning of feedback controllers.)

A number of process characteristics may be required for proper control, such as the following.

- Time delays
- Nonminimum phase
- Disturbances
- Unmeasured variables
- Noise
- Time-varying parameters
- Nonlinearities
- Constraints
- Multivariable interactions

While microcomputer-based controllers are adequate hardware to theoretically overcome these problems, the remaining – but hard – problem is to come up with the appropriate software to direct the hardware. Neural nets is an innovative new paradigm that is beginning to be applied to these areas with excellent results.

During the 1980s, neural nets began to be applied in this area because of the limitations of the classic single loop PID (proportional-integral-derivative) controllers. These limitations include the following [42].

- Simple empirical rules such as the Ziegler-Nichols dating from the 1940s
- Poor tuning due to neglect or lack of time
- Problems with processes having long time delays and high oscillations

One solution to the manual tuning of PID controllers has been autotuned controllers such as the pattern recognition autotuner and the relay autotuner. However, autotuning has not proven to be adequate for all problems.

Another approach is adaptive control, which attempts to

maintain good robustness even if unpredictable changes occur in the process, sensors, and perhaps the controller itself due to damage. Classic linear controllers provide robustness only when there is little uncertainty while adaptive controllers are designed to perform better over a larger range of uncertainty. Adaptive control is successful if a model exists with sufficiently accurate parameters, while robust control is applicable if constant parameters are acceptable in the controller. An intelligent system is designed for robustness under maximum uncertainty. Intelligent systems have been devised using a variety of approaches such as expert systems, fuzzy logic, and neural nets [43].

Adaptive control has complementary properties to autotuning. For simpler cases, an autotuner is satisfactory since it requires little prior information about the process. If better control is required, an adaptive controller may be used at the cost of more prior information. Hybrid approaches may also be used. For example, a commercial PID regulator with autotuning, gain scheduling, adaptation, and adaptive feedforward was announced by SattControl in 1988.

One problem with adaptive controllers is inherent in their design since classic adaptive algorithms are based on local gradient descent. For optimum performance, the adaptive algorithm requires prior knowledge of the sampling period, the signal range, the dead time, and the model structure. Another problem is that the calculations required for the desired precision and accuracy may be too computationally expensive, particularly if real-time performance is a strong constraint. Lastly, there is the scaling problem which occurs as the process size increases. There may be a combinatorial explosion such as in the Traveling Salesman Problem which prohibits a practical solution and sets an upper limit to the plant size. Neural nets overcome these disadvantages.

In one application, a neural net is used as a real-time autonomous adaptive controller for a robot which uses stereo vision to enable it to grasp objects in 3D space using a five-degree-of-freedom arm [44]. The robot can adapt to unexpected changes in its motor system due to wear and faults, and to the location, orientation, size, and shape of objects. This neurocontroller can learn and maintain its own calibration, and is expandable as more robot joints are added. In addition, the neurocontroller was designed for portability to other plants and actuators. No special knowledge was included in the general design. Such a system would be enormously difficult, if indeed possible, to program with classic algorithms.

An expert system controller represents another control strategy in which the emphasis is on heuristic knowledge for process control rather than mathematical theory. The expert system can be used for direct process control or to select the optimum algorithm depending on the process state. Thus, instead of using only one algorithm or paradigm for all time, the expert system can make an intelligent choice for best performance. For example, one expert system controller includes the following components [45].

- PID regulator
- Relay tuner
- Least squares recursive estimator
- Pole-placement self-tuner
- Supervision and analysis
- Signal generation to improve identifiability

One advantage of the expert controller compared to other types is its explanation facility. This is a general characteristics of expert systems by which they can explain their knowledge. Such explanations can be useful for process operation, for debugging and enhancements, and for training.

Nonlinearities in Process Control

Classic control theory has primarily dealt with linear systems because these were the only systems in which the mathematics were well understood. As the maxim says, "If all you have is a hammer, everything looks like a nail." While the problems of nonlinearities have been recognized, the tendency has been to try and consider them as approximations to linear systems.The traditional solution is to linearize the process model and then apply linear control techniques. However, this solution is only satisfactory if deviations from the steady-state are small [46].

Neural nets appear to have important advantages in dealing with the problem of nonlinearities. This is a difficult but important area, especially in the chemical industry since many chemical processes exhibit nonlinearities.

One application of neural nets to nonlinear systems has been the use of a backpropagation net for sensor failure detection in a chemical process system [47]. The neural net approach has three significant advantages compared to the traditional FISE (finite integral squared error) criteria.

- The neural net can detect significantly smaller failures because the signal frequency spectrum is used instead of the gross FISE.

- The neural net can be applied to nonlinear systems which cannot currently be analyzed because too many calculations are required.

- The neural net can learn new failure modes as it is being used on-line.

Another application is the classic inverted pendulum. Although it is an inherently unstable system, it is basic to many real-world tasks such as walking, rocket control, and ship propellors. Traditional controllers require detailed knowledge of the system dynamics and an analytic objective function. A much harder inverted pendulum problem occurs when neither the system knowledge nor the objective function is available. This general problem corresponds to the real world in which systems do not maintain their known, ideal characteristics for all time. An adaptive controller based on neural nets has been used to solve the general problem [48].

One difficulty that was solved in this application was the Credit Assignment problem. Since no objective function was provided to evaluate states and actions, the controller had to rely on the occurrence of failure signals, which were the only performance measure supplied. The Credit Assignment problem was to determine which of a long sequence of actions generated the failure signal.

Supervised learning could not be applied since the correct action for most states is not well-defined. In addition, there are infinitely many possible trajectories that indefinitely avoid failure. The appropriate type of learning is reinforcement.

ROBOTICS

Robotics is a field that is rapidly growing in importance due to the need of industry to produce higher quality products at lower cost. Industrial robots are now an indispensable element of automation for mass production. In a sense, factory automation using robots is easy since the trajectory of the robot arm is fixed in repetitive assembly. However, major problems can occur if the components to be assembled are not in exactly the right position and orientation, or if supplies run short. In other words, nothing is perfect in the real world and the robot should be able to deal with some uncertainty to prevent shutting down the assembly line.

In addition to their use in factory automation, robots are used in special applications. For example, they are used in environments where it would be hazardous for humans, such as bomb disposal and nuclear reactor inspection.

Current industrial robots are still generally explicitly programmed. However, many workers are investigating neural nets to provide robots with learning ability so that they can adapt to a changing environment and unforeseen conditions. Another problem is sensor fusion so that robots can relate inputs from sensors such as machine vision and tactile to the appropriate actuator/effector response.These requirements are particularly strict if we wish to create autonomous, adaptive, mobile robots.

The general idea of using neural net models for robot control is to let the a net learn how the robot works rather than the classical method of explicitly programming the robot dynamics. This neural net approach has several advantages. First, it allows easy changes as the robot configuration is intentionally altered, such as adding additional joints. Secondly, wear in the system is automatically compensated. Third, the robot can learn to adapt to unexpected changes in itself due to faults such as sensor failure, and also the environment, such as components for assembly being out of place.

A number of neural network models have been developed for robot control [49]. One neurocontroller has been used with a Kawasaki-Unimate PUMA 260 industrial robot [50]. Another application of neural nets has been made to a mobile robot which has ultrasonic, infrared, tactile, and limit sensors [51]. This robot has two different types of neural nets for control. The reason neural net guides the robot based directly on short-term sensory input, such as moving toward a source of light. The instinct net provides long-term control such as directing the robot to execute a sequence of actions in the hope of encountering a suitable stimulus. The two nets are arranged in a hierarchical manner with the reason net at the lower level and its output affecting the input of the instinct net. This application illustrates that multiple neural nets may be structured to achieve complex compound behavior having short- and long-term components.

A popular group of neurocontroller models is based on the CMAC (cerebellar model articulation controller) design. These models were based on the work of David Marr [52] and James Albus, who gave the model its name [53]. In the past decades, much work on the cerebellum has shown that it is essential for adaptation, learning, and reflexive and motor actions. The CMAC is essentially a table-driven model that uses supervised learning to set the weights stored in the table. A number of workers have used CMAC or modified versions. For example, a CMAC model was used to control a five-axis industrial robot (General Electric P-5) [54], and another robot interacted with randomly oriented objects on a moving conveyor belt during both repetitive and non-repetitive operations using machine vision [55]. The CMAC neurocontroller has also been integrated with an expert system to provide intelligent robotic control

[56].

One challenging problem of modern robotics is to create multiple-degree-of-freedom robot hands. These hands should be based on the prehensile model of human hands because of their increased dexterity compared to simple grasp-and-release effectors. It is desirable to have a generic controller that can easily be used with a wide variety of robotic hand designs.

An adaptive neural net controller has been designed that meets these criteria [57]. This approach uses a grasp table to map the geometric object primitives derived from machine vision, and the prehensile hand behaviors so that the robot hand can grasp various objects. The table approach offers device independence rather than a conventional dynamic systems approach.

Although a grasp table could be implemented by a case statement or in an expert system, the size would be enormous because of the number of possible grasp modes for each value of each dimension for every primitive. Instead of explicitly building the table, a neural net is used to build the grasp table using supervised learning and backpropagation.

Detailed Applications of Neural Networks in Industry

The preceding sections have provided a general background and overview of neural networks. In the following sections, we will describe in detail the implementation of several systems that integrate neural network technology into their architecture. In all of the applications described, the neural network is used to perform a complex transformation of the input data into a meaningful output for the application; that is, each of these networks are utilized in their production mode, with training occurring off-line prior to the actual deployment of the system. Furthermore, in each of the systems discussed, the neural network performs a function for the system that might not have been achievable using conventional computer programming methods.

In many cases, the complexity of the transformation formed an **N-P** complete problem. Thus, any attempt to algorithmically solve the problem would have resulted in a non-polynomial consumption of computer resources. As you will see, the use of neural network technology offered the application programmers an alternative method for solving their specific problem in unit time.

In each of the sections that follow, we will describe the implementation of a system designed to address a specific problem in industry, including details on the nature of the problem the system was designed to address. Finally, we will show how the neural network was utilized to solve the problem.

Automotive Systems

The automotive industry is always seeking new methods for increasing automation in their assembly-line and product maintenance operations. The reason for this push to increase automation is obvious: The more work that can be done by machines, the lower the cost of the product to the consumer, whether in the initial price of the vehicle, or in the cost of ownership. In either case, lower costs imply more sales, and higher profits, for the automobile manufacturer. Thus, quite a bit of time and research money is spent on developing new methods for increasing automation.

Paint Quality Assurance.

Inspection of the paint on the body panels of a car coming down the assembly line is currently a labor-intensive, statistical sampling process. Specifically, a random selection of cars are pulled off the assembly line, and each of these is then visually inspected by a paint quality assurance expert, whose job it is to not only ensure a high-quality paint job on each of the sample vehicles, but also determine what might be wrong in the painting process when problems are detected.

To understand the complexity of this task, imagine yourself standing in front of a car with a reasonably good paint job. How might you go about determining whether or not the paint process is operating at its highest level? One immediately observable clue is in the quality of the paint finish on the car you are examining. Does the painted surface exhibit a high-luster, glass-like finish, or is it more appropriately described as "orange-peel?" Are there visible "runs" in the painted surface? An even more interesting question is: If there are "smudges" on the painted panel, are they permanent (i.e. induced by some anomaly in the paint process itself), or are they easily removed (i.e. a result of someone touching the painted surface prior to inspection)?

In each of the paint quality scenarios described above, a human "observer" is expected to examine the painted surfaces, and judge the overall quality of the paint job. Moreover, if the paint job is deemed "unacceptable," the human expert is then expected to recommend changes to the paint process control foreman about how to correct the deficiencies. Let us now consider how we might go about automating the paint inspection process. Obviously, if we can determine how to automate the "visual" inspection of the painted surface, we can have a computer perform the analysis of the image.

One technique, utilized by Ford Motor Company, used to help the human expert perform the visual inspection provides the basis for defining how the inspection process can be automated. This technique simply requires that a laser beam be projected onto the paint panel under inspection. Having done that, the quality of the paint job can be determined by examining the reflected image of the laser. Using this method, a "good" quality paint job minimizes the amount of scatter in the coherent laser beam, and reflects an image that exhibits a very tight clustering of photons. On the other hand, a poor quality paint job severely scatters the coherent laser, resulting in a reflected image that bears little resemblance to the original laser beam.

To acquire the image data for the computer analysis, all we need to do is utilize existing video processing technology to "frame-grab" an image of the reflected laser beam. The resulting pixel (picture element) matrix can then be analyzed by the computer to determine the quality of the paint job based on the amount of scatter exhibited by the video frame. However, this proposed analysis is more complicated than it might initially seem.

Consider that what we are proposing to do is (algorithmically) examine each pixel in the input image, correlate the intensity of the pixel with a subset of its neighboring pixels, determine where the "center" of the image is, and then determine the amount of scatter exhibited (on average) over the entire image. For even a relatively small input image (100 rows of 900 pixels), we are talking about examining 90,000 pixels *sequentially,* further examining 30-50 neighboring pixels at each pixel. This exhaustive examination is necessary, because, under the rigors of the assembly line, we cannot guarantee that the laser image will always be centered in the frame, nor can we guarantee the paint job on the panel under inspection will be good enough to ensure a very low scattering of the reflected image.

While it is possible to write an algorithmic computer program to perform this analysis, researchers at Ford Motor Company (and its former subsidiary, Ford Aerospace Corporation) have developed a prototype system that utilizes a backpropagation neural network to perform the image analysis function. The architecture of the system is shown in Figure 3 below.

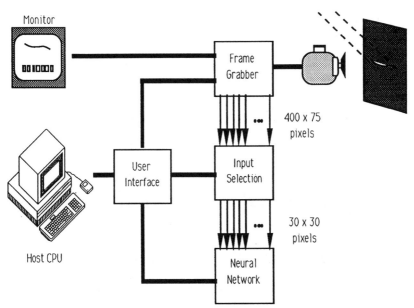

Figure 3 Architecture for the Paint Quality Assessment Application.

In this application, the backpropagation network was

constructed to accept as input the digitized image of 900 pixels
taken randomly from the reflected laser image, and trained to
produce a "score" of paint quality (based on the amount of
scatter in the image) that corresponded to the human experts
assessment of the same paint panel. One of the benefits of this
approach, as shown in Figure 4, is the fact that the network
assesses all 900 input pixels *in parallel*, as an inherent part of
its structure. This means that limiting the pixel neighborhood
around any particular point under examination to a subset of
the total number of neighboring pixels is no longer required.
Thus, this system approach provides a more comprehensive
assessment of the paint quality, since each pixel in the input
image is automatically correlated to all neighboring pixels as
part of the network information processing.

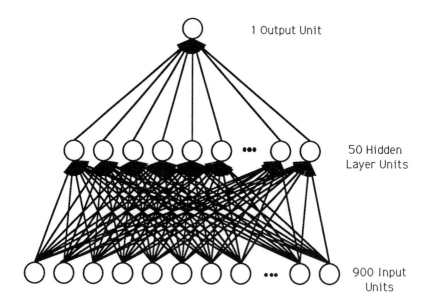

*Figure 4 Structure of the Backpropagation Network Used in the
Paint Quality Assessment Application.*

The network utilized in this application required 50 sigmoidal hidden layer units (used as feature detectors in the input pattern), and one linear output layer unit. The output unit was defined to produce a linear output so that it could be more easily interpreted as an analog "score" of paint quality. To illustrate how this network behaves, consider the forward propagation mode of the backpropagation network. Assume that some external source (i.e. another subroutine in the system software) has acquired the digitized video image of the reflected laser beam, and it has also selected the limited sample to be assessed. Each pixel in the 900 pixel sample image is then applied to one of the input units in the network. Each input unit in the network simply generates an output signal that exactly corresponds to the magnitude of the single input pixel it receives (each input unit is thus a fan-out only unit). Then, each of the 50 hidden layer units either saturate, or remain off, based on the total input stimulation they receive from the input layer units through the connection weights that have been trained to identify certain input features. Thus, each hidden layer unit, in activating its output signal, can be thought of as providing an indication to the single output layer unit of the presence of a certain feature in the input image.

Finally, the single output layer unit generates an output based on the number (and type) of features identified in the input by the hidden layer units. This output, since it is a linear transformation of the total input stimulation it receives, can then be interpreted by the application program as the score for the paint panel image. All that remains to do is perform a statistical sampling of the entire image (in this case, ten samples selected sequentially across the length of the reflected laser image), and average the scores produced by the trained backpropagation network. The resulting average is then provided to other traditional software modules that utilize the score to inform human quality inspectors (via the user interface) and provide feedback to the paint application process system (through as yet undefined interfaces).

In summary, the benefits of using the backpropagation network to perform the paint quality assessment included:

- The ability to produce a computer application that solved the problem in unit time, without requiring expensive hardware accelerators. The final application was hosted on a standard '386 class PC/AT computer system.

- A robust system that could deal well with minor anomalies in the input image. Relatively small perturbations in the input were either rejected by the network as insignificant, or if large enough to effect the score of a limited number of sample images, were damped by the averaging technique used to produce the aggregate score for the panel.

- A system architecture that could evolve to suit other applications, if the processing requirements were similar. Since the "knowledge" in this system was stored in the form of network connection weights, other networks could be trained off-line, perhaps using different scoring criteria, and then "implanted" in the application by simply moving a connection weight data file from the off-line system to the on-line system.

Engine Diagnostics

As anyone who has ever worked on a car knows, engines are becoming increasingly complex. Fuel economy subsystems, air pollution control equipment, and an increasing number of options all contribute to the increase in the number of hoses, belts, and special-purpose equipment found under the hood. By extension, as the complexity of the power-plants increase, car manufacturers are becoming increasingly interested in automating the diagnosis of the engines when problems occur.

To accomplish this automatic diagnosis, we find an ever increasing number of system sensors and microprocessor based control systems that monitor the sensors and make adjustments to the operation of the engine as it is running. However, there remains a large number of problems that cannot be fixed automatically; thus, we must occasionally take our cars into the shop for repairs.

Unfortunately, with the increasing complexity of the engines, the cost (to the public and the car-makers) of training mechanics is growing very quickly. Even mechanics that have received specialized training in the diagnosis and repair of certain systems in a single manufacturers car-line must be continuously retrained, since many automotive systems change (hopefully improving) on an annual basis. Therefore, most automotive manufacturers believe the cost-savings associated with a faster, and more reliable, diagnosis of the customers problem.will more than offset the cost of developing a diagnostic system that can assist a service-bay mechanic by detecting specific problems and identifying failed components.

Ford Motor Company has recently developed another neural network based application that performs exactly that diagnostic function. Originally based on a backpropagation neural network, and eventually a reduced-coulomb energy (RCE) model network, the service-bay diagnostic system prototype illustrates the power of using a neural network to analyze a relatively large number of data streams that are not necessarily highly correlated.

In this application, data was collected from twelve on-board sensors, providing a mix of analog and digital data, as illustrated in Figure 5. Samples were collected sixteen times per engine revolution, yielding 192 data parameters that varied according to the state of the engine (i.e. how well it was performing) and the point in the engine cycle at which the data was sampled. To construct a training set, a mechanic was employed to "break" the engine by deliberately faulting an engine component.

After the engine was broken, the car was run for a period of time, and data from the sensors was collected and stored in a training data file. Six faults were introduced in this manner, each fault being initiated individually after verifying that the engine was operating correctly prior to inducing the fault. Furthermore, when applicable, each of the faults were induced in all possible cases. For example, a failure in the operation of one of the spark plugs was replicated for all spark plugs in the power plant.

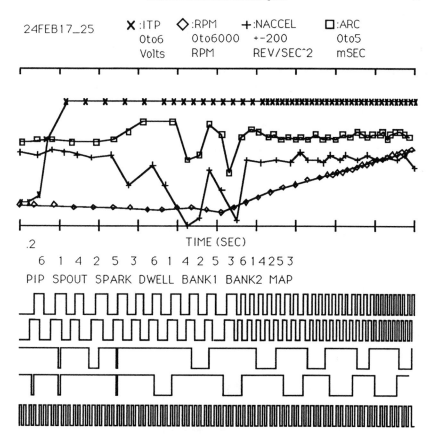

Figure 5 Typical Plot of Engine Sensor Data Collected for the Neural Network Diagnostic Application.

In this manner, data sets indicating the characteristics of the engine behavior at nominal operation, and with six known faults were collected and used to train a neural network. The outputs desired from the network were indications of the specific faults, as well as the location of the fault if it was determinable. For example, a "spark plug short" could occur in any of the six cylinders in the engine under test. Thus, it was desirable to have the network not only indicate the presence of

a spark plug "short" fault, but also have it identify the
particular cylinder that was experiencing the problem. In this
manner, the network was expected to significantly reduce the
amount of time needed to diagnose the failed component.

The architecture of the system constructed to address this
problem is illustrated in Figure 6. As you can see from this
diagram, all aspects of the system were expected to be
contained within a low-cost PC/AT computer system, so there
were really no opportunities for exploiting hardware
techniques to accelerate the neural network simulation. The
data acquisition hardware was designed to interface directly to
the sensor output channels available from connections directly
to the engine. In this manner, the state of the engine could be
ascertained unobtrusively, which was desirable since often the
symptoms of a problem are changed when the engine must be
modified in order to make measurements.

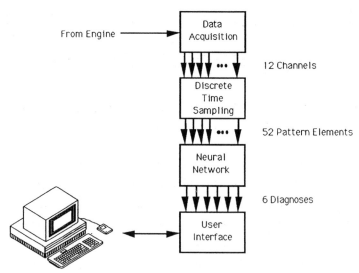

Figure 6 Block Diagram for the Engine Diagnostic Application

Some of the results of the tests described in the paper that originally detailed this work include the following findings.

- Using the backpropagation network, 24 hours of computer time were required to train the network. When the RCE network was employed, training time was reduced to 10 minutes.

- In both network models, the accuracy of the single-problem diagnosis after training was completed was greater than 99%. Moreover, both networks did reasonably well when presented with multiple faults, a condition not encountered during training. On average, the network would accurately diagnose two distinct failures accurately 78% of the time.

- After training was complete, the amount of time required to diagnose the problem was well within the acceptable criteria for real-time operation: less than 500 milliseconds for the backpropagation network, and less than 100 milliseconds for the RCE model.

Scheduling Systems

Scheduling is an extremely difficult application to automate for a variety of reasons. For example, it is difficult to specify an "ideal" scheduling system for all applications, because many applications have different types of inputs and priorities for schedule determination. To illustrate this we will now consider two examples of "expert" scheduling systems that are interesting not only because they employ a neural network solution, but also because they utilize a variety of different technologies to solve a complex pattern matching (or optimization) problem.

Workstation Maintenance Expert System

The Workstation Maintenance Expert System (WMES) was developed by Loral Space Information Systems in 1986 to address a complicated manpower scheduling problem at the NASA Johnson Space Center. To understand the nature of this application, consider that prior to 1986, NASA/JSC had more than a dozen major aerospace contractors designated as "prime" for various aspects of the operation of the center.

For example, IBM was the primary contractor for all the software that ran the mission control center mainframe computer systems, providing the foundation for the operation of the mission control center itself. Singer-Link was the primary contractor responsible for the development and maintenance of the Shuttle Simulator. Loral had been the primary contractor for all of the custom data processing systems that interfaced to the IBM mainframes in the mission control center. Several other contractors had various other support roles.

In the early 1980's, with the advent of the IBM PC/AT desktop computer, NASA undertook a project to upgrade its facilities to include the desktop computer system as part of its office automation programs. However, there could be no "standard" configuration for the desktop systems, because each was used in a different manner, depending on which organization within NASA used the computer. Different software packages might be required, or different third-party expansion boards might be needed, depending on the types of applications that the machine had to run.

Compounding this configuration management problem for the desktop systems was the previously mentioned fact that NASA/JSC had a conglomeration of contractors responsible for providing different types of support. For example, IBM was responsible for the PC/AT hardware, while Loral was responsible for configuring the various systems per NASA's requirements. If something broke, another contractor was called in to make repairs, and so on. The only method NASA managers had for understanding how long certain "change requests" instituted to effect the reconfiguration of any of the desktop systems was based on verbal status obtained in the proverbial smoke-filled room.

Depending on who was scheduled for vacation in the upcoming weeks, the arbitrary priority assigned to certain change requests by NASA managers, and the ability of twelve independent contractors to work together cohesively, the only insight NASA had into how long each change request would take to implement was a "shot in the dark," at best. To correct this situation, Loral, under NASA's direction, undertook the development of the WMES package to project not only the manpower levels needed to address the requirements of any specific change request, but also to detect potential bottlenecks in the change-management system, and to forecast eventual manpower shortages.

To accomplish these goals, Loral interviewed a number of NASA and contractor management officials to ascertain the role and scope of each of the players in the change-management system. Furthermore, the activities that had to be performed to implement the requirements of any arbitrary change-request were determined and broken into low-level work units. Any interdependencies were defined, and each work unit was allowed to be assigned one of three priorities, depending on the importance of a particular task.

Once this initial systems engineering activity had concluded, a rule-based expert system was constructed that could decompose the high level requirements of a change request into the associated work units (with associated priority levels) was constructed. As shown in Figure 7, this implementation solved the first of the scheduling requirements: any arbitrary change request could now be expressed in terms of who needed to do what. What was missing was an ability to cohesively schedule these work units based on any kind of heuristic, because there was not a good method for dynamically redefining the ability of the contractors to perform their assigned functions. It was truely a dynamic situation, in that there was no way to take into account other factors, such as how non-NASA contract related work took time away from low- and medium-priority NASA activities.

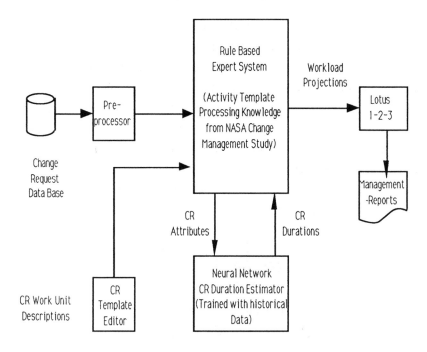

Figure 7 Block Diagram of the Workstation Maintenance Expert System.

A work-around technique for this aspect of the scheduling application was constructed using a backpropagation network. Using the previous three months of actual times charged to the work-unit tasks, we trained a network to indicate the amount of time (in days) required by the various contractors to perform a set of work-units. Using this historical perspective, it was then possible to project how much time would be required to satisfy the requirements of a new change request that consisted of a similar set of work-units.

Once the amount of time needed to complete all of the tasks associated with the change request was known (as projected by the neural network), it was then an easy task to program the expert system to back-schedule the work-unit activities, and provide a "projected schedule" time chart for management.The backpropagation network utilized for this application had 120 input units, one for each of the three priorities for the 40 work-unit activities defined. The output consisted of three sigmoidal units that were used to represent a binary encoding of the number of days associated with the set of input tasks. The single hidden layer contained 60 sigmoidal units, again utilized for feature detection in the input pattern. This network is illustrated in Figure 8.

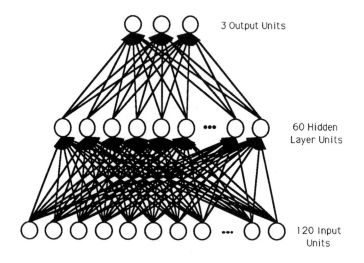

3 Output Units

60 Hidden Layer Units

120 Input Units

Figure 8 Neural Net for the Workstation Maintenance Expert System

Among the benefits of using a neural network to solve this scheduling problem are the following.

• The backpropagation network offered a means for allowing the WMES system to evolve with improvements in the NASA change management system. As the contractors ability to satisfy the requirements of any of their assigned work-units varied (depending on personnel availability, etc.), the WMES could adapt to provide a better assessment of the responsiveness of the contractors. This was achieved by simply replacing the data files that described the configuration of the trained backpropagation network with new files that were developed off-line, by training another network using more recent data. Thus, the WMES could always provide reasonably accurate estimates of the amount of time required to implement any change request.

• The use of the neural network solved an N-P complete combinatorial problem that could not have been solved any other way. It was virtually impossible to write enough expert system rules to address all of the possible combinations of work-units in various priorities to predict time requirements.

• From a practical viewpoint, while it might have been possible to implement a statistical analysis of the work-unit requirements and construct a program that embodied the results of that analysis, it would have been impractical to consider redoing the analysis and modifying the program every month to update the ability of the system to track current contractor performances.

Flexible Manufacturing Systems

As we have already seen, it is often desirable to have an automatic system that can easily deal with the combinatorial explosion of alternatives that must be considered when

implementing a scheduling system. The neural networks
utilized in all the applications described up until now provided
the system with a solution (in a fixed amount of time) to a
complex pattern-matching problem that might otherwise have
consumed more computer time than practically available for
the application. Let us now consider another method for
utilizing a neural network to achieve the same type of results.

In this application, the complete system is again an
assembly-line type of scheduling system, where the users have
a fixed number of tools, a target number of products to
generate, and many alternative methods for achieving the
production goal. There are a number of contraints in the
scheduler that must be accounted for, including.

- It is desirable to keep all active tools busy whenever
  possible.

- No two jobs can be scheduled on the same tool at the same
  time.

- There are several tool-related interdependencies that must
  be satisfied (i.e. a product must be processed by tool "A"
  before it can be processed on tools "B", "C", or "D."

To satisfy all the users constraints, yet still maintain a
flexible solution for the practical scheduling of resources on an
assembly-line operation, the developers of the flexible
manufacturing system (FMS) combined several neural
networks with a knowledge-based expert system (KBES)
application written in PROLOG [58]. However, unlike the WMES
application described previously, the developers of the FMS
utilized the neural networks as dynamic pattern matchers

trained to select rules for execution within the KBES. Using this approach, the design of the FMS allows the application developers to dynamically reconfigure the scheduling strategies employed by developing new neural networks that in turn determine which rules are most applicable for execution within the application.

To understand how this approach is utilized, let us digress for a moment and consider how a traditional, forward-chaining expert system typically operates. The system is "knowledge"-based, in the sense that expertise about the application can be captured and expressed in the form of modular "if-then" rules. Each rule contains a set of activation patterns following the "if" clause, forming the left-hand side (LHS) of the rule. Each of the patterns specified on the LHS must be satisfied by one or more instance patterns contained in the current situation memory. Each LHS pattern must match its corresponding instance pattern by either exact match, or by replacing pattern variables with specific instance values in order to validate the rule.

Once validated, the rule becomes available for execution. A higher level priority mechanism selects one rule from the set of currently valid rules, and executes the selected rule by sequentially performing the set of instructions encoded in the right-hand side of the rule. Typically, the execution of these instructions change the patterns in the instance memory, and validate another set of rules. The process then repeats until no valid rules can be found, or until the user deliberately stops the process.

The pattern-matching function employed in many expert systems is based on an algorithmic technique that relies on the temporal-persistence of many patterns in the instance memory; that is, it is expected that most instance patterns do not change

during a rule execution. However, there are many drawbacks to
the use of this algorithmic approach to finding the current set
of valid rules. For instance, if an application employed only a
few activation patterns in its rule set, the computer would
spend a disproportionate amount of time trying to discriminate
between different instances of similar patterns. Also,.the
priority scheme employed to determine which of the currently
valid rules is most applicable to the current situation is usually
fixed by the expert system "inference engine." At best, an
application developer can usually only choose between several
different rule-selection strategies offered by the expert system
language.

The developers of the FMS application took another approach
to deciding which of the systems "rules" was most applicable at
any given time. Specifically, they developed a set of three layer
backpropagation networks that were trained to recognize when
each of the KBES rules were most applicable to the current
situation. Then, using another set of neural networks trained to
decide which one of the rule selection networks was most
appropriate under a set of existing conditions, the KBES utilized
the selected network to select and prioritize the set of rules
available for execution. Given the results of this selection, the
KBES was then employed to execute the selected rule(s), and
the process was repeated until the schedule was developed or
the set of selected rules was exhausted.

In Figure 9, we illustrate the system architecture of the FMS
as documented by the developers of the system. At this
writing, the system is still in the prototype stages, but the
developers report that by using a neural network approach to
select the appropriate rule, the system selected the "best" rule
92% of the time, with the second "best" rule being selected the
remainder of the time. They also report that the response time

of the system is vastly improved over a traditional rule-selection technique, in that the system can select and execute more than 200 rules per second, whereas the traditional KBES method (restricted to a rather small subset of possible rules) can only execute 4 rules per second. The reason given for this disparity in performance is attributed to the fixed amount of time required to perform the neural network simulation as opposed to performing an algorithmic search for the "best" rule when the possible combinations of patterns that can validate the rule grows geometrically, thus requiring a geometric growth in computer time to perform the search.

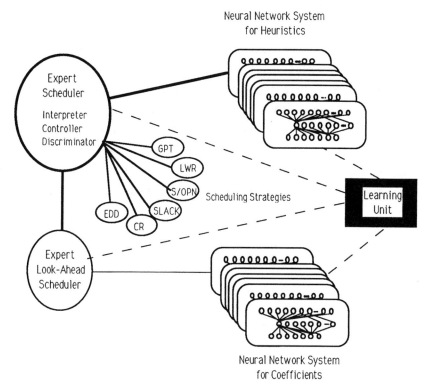

Figure 9 The Flexible Manufacturing System (FMS) Architecture

First Break Picking

When one thinks of "industrial" applications, the first thought that might come to mind is that of the typical Detroit assembly-line environment. The image of people interacting with robotic systems to produce a close-tolerance end-product, where the environment is full of noise, dirt, and other significant impediments for both the people and robotic systems. However, there are other industrial applications outside of the close-quarters environment of the assembly line. Seismic analysis and the performance of geophysical surveys is one such application [59].

In this environment, the application is to determine the structure of the rock below the "weathering layer," which is the top-most layer of relatively loose earth. The reason these structures are of interest is because they reveal much about the potential of the earth to contain oil, natural gas, and other hydrocarbon resources. The methods employed for determining these structures vary, but usually consist of performing some type of acoustic imaging based on the sound propagation characteristics of the different layers of earth, as shown in Figure 10.

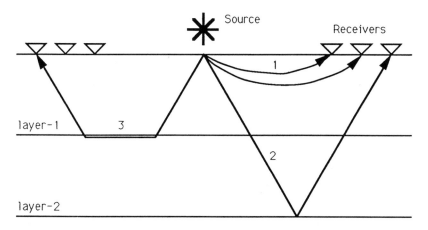

1 – Direct Arrivals
2 – Reflected Arrivals
3 – Refracted Arrivals

Figure 10 Behavior of Acoustic Signals During Seismic Imaging.

The weathering layer is the most troublesome layer to characterize, because sound travels through this layer slowly with respect to the other rocky layers (due to the lower density of the surface material with respect to the subterranean layers). Moreover, since the thickness of the weathering layer varies greatly from location to location, the movement of sound energy through this layer can introduce errors in the analysis of the structure of the earth if the sound travel-time anomalies introduced by the weathering layer are not removed prior to seismic image processing. Thus, the identification of the "first break" signals (critically refracted signals that have travelled along the boundary between subsurface layers and leak energy back to the receivers) is critical to being able to obtain an accurate seismic section.

Currently, the problem of identifying first breaks in a typical seismic survey is a time consuming, labor intensive process.

Algorithmic approaches to performing the first break identification have not met with much success, due to the variability of the weathering layer. Manual determination of first breaks is an extremely time consuming process, since a typical three-dimensional survey will contain more than a half a million seismic traces, each one of which must be examined by hand. Figure 11 illustrates the form of the data in a seismic trace, and provides an indication of the manual complexity of the task. In an attempt to automate this application, a system that would automatically locate the first break in a seismic survey was developed, using a neural network to perform the pattern-matching function needed to identify the first arrival (FAR) signals in a survey.

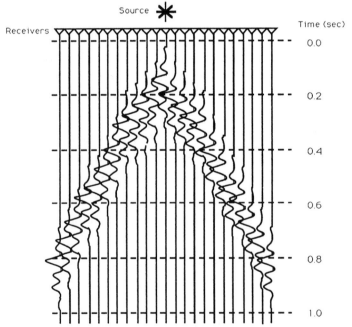

Figure 11 Synthetic Data Representing a Single Seismic Trace.

The process of identifying the arrival of a FAR consists of placing a window over five samples centered at the peak to be classified, and a set of attributes is extracted from each of the five samples. The network utilized to perform this function is a very small, three-layer backpropagation network, containing twelve input units, five hidden units, and one output unit. The twelve inputs consisted of the following parameters for each of three peaks (the peak to be classified, and one peak on either side of the peak to be classified to provide a contextual reference).

- Amplitude value for the peak under consideration
- Mean power level in the window
- Power ratio between the forward and reverse sliding window
- Envelope slope for the peak

The output unit was used to simply indicate whether or not the peak under consideration was a FAR: if the output unit value exceeded a 0.8 value threshold, the peak was classified as a FAR.

In an interesting twist to the typical approach to utilizing a neural network, the developers of this system decided to train the network each time a new survey was conducted, due to the wide variation in geophysical characteristics. Specifically, since the amount of data to be analyzed was extremely voluminous, the user preselects a set of representative seismic traces for the survey. The FAR arrival peaks on those selected traces are then manually determined and interactively entered into the computer system. The system then performs the remainder of the analysis by first training the backpropagation network using the data specified by the user at the beginning of the run.

Once training completes (taking about a minute of compute time on the Sun-4 host), the system uses the trained network to identify the FAR peak on the remainder of the seismic traces.

The developers report that by using this approach, the identification of first breaks on two representative (a marine and vibroseis) surveys indicated that the neural network based system could accurately identify the first break 95% of the time. Moreover, the turn-around time for the survey was improved by 88%. Both of these measures are reported to be well above any other automated techniques available at the time of this writing. Based on these successes, the developers are planning further expansion of the application to include the development of a generalized network that does not require training prior to conducting a survey.

Summary

The preceding sections have shown that neural networks have a growing place in industry by providing solutions to difficult and intractable problems in automation and robotics. This growth will increase now that commercial neural net chips have been introduced by vendors such as Intel Corp. Neural net chips will find many applications in embedded systems so that the technology will spread outside the factory.

Already, neural networks have been employed to solve problems related to assembly-line resource scheduling, automotive diagnostics, paint quality assessment, and analysis of seismic imaging data. We believe that these applications represent only the beginning. As neural network technology flourishes, many more successful applications will be developed. While not all of them will utilize a neural network to solve a previously intractable problem, many of them will provide solutions to problems for which a conventional algorithmic approach is not cost-effective. Based on the success of these applications, we look forward to the development of future applications.

REFERENCES

(1) Giarratano, J., Villarreal, J. and Savely, R., "Future Impacts of Artificial Neural Systems on Industry," ISA *Transactions*, 29 (1), Jan. 1990, pp. 9-14.

(2) Giarratano, J., Culbert, C. and Savely, R., "The State of the Art for Current and Future Expert Systems Tools," ISA *Transactions*, 29 (1), Jan. 1990, pp. 17-25.

(3) Giarratano, J., Riley, G., *Expert Systems Principles and Programming*, 1989, pp. 291-341.

(4) Goldberg, D., *Genetic Algorithms in Search, Optimization, and Machine Learning*, Addison-Wesley, 1989.

(5) Shea, P. M., "Detection of Explosives in Checked Airline Baggage Using an Artificial Neural System," *IJCNN-89*, pp. II-3 1 to II-34.

(6) Marko, K. A., et. al., "Automotive Control System Diagnostics Using Neural Nets for Rapid Pattern Classification of Large Data Sets," *IJCNN-89*, pp. II-13 to II-16.

(7) Kohonen, T., "The Neural Phonetic Typewriter," IEEE *Computer*, March 1988, pp. 11-22.

(8) Simmons, A., "Neural Nets in Aerospace: Visions and Problems," *Proceedings of AIAA Computers in Aerospace VII* (1989), pp. 804-810.

(9) Widrow, B., and Stearns, S. D., "Cancelling Echoes in Long

Distance Telephone Circuits," in *Adaptive Signal Processing*, Prentice-Hall Pub. Co., 1985, pp. 339-347.

(10) Glover, D., "A Hybrid Optical Fourier/Electronic Neurocomputer Machine Vision Inspection System," *Proceedings of Vision '88 Conference*, June 1988.

(11) Dutta, S., and Shekhar, S., "Bond Rating: A Non-Conservative Application of Neural Networks," *IJCNN89*, pp. II-43 to II-450.

(12) White, H., "Economic Prediction Using Neural Networks: The Case of IBM Daily Stock Returns," *IJCNN-89*, pp. II-451 to II-458.

(13) Collins, E., *et. al.*, "An Application of a Multiple Neural Network Learning System to Emulation of Mortgage Underwriting Judgments," *IJCNN-89*, pp. II-459 to II-466.

(14) Castelaz, P., "Neural Networks in Defense Applications," *IJCNN-89*, pp. II-473 to II-480.

(15) Bounds, D., and Lloyd, P., "A Multi-Layer Perceptron Network for the Diagnosis of Low Back Pain, " *IJCNN89*, pp. II-481 to II-489.

(16) Sobajic, D., *et. al.*, "Intelligent Control of the Intelledex 605T Robot Manipulator," *IJCNN-89*, pp. II-633 to II-640.

(17) Gallant, S., "Connectionist Expert Systems," *Comm. of the ACM*, 31, (2), Feb. 1988, pp. 152-169.

(18) Hinton, Geoffrey, E., "Learning Distributed Representations of Concepts," *Proceedings, 8th Annual Conference of the Cognitive Science Society*, Lawrence Eribaum Assoc., Hillsdale, NJ, 1986.

(19) "PC-Based Development Products Move AI Out of the Classroom, into the Workplace, " *EDN*, August 18, 1988, pp. 71-80.

(20) *PC AI Magazine*, May 1989, pp. 62-68.

(21) Rosenblatt, Frank, *Principles of Neurodynamics:*

Perceptrons and the Theory of Brain Mechanisms, Spartan Books, 1962.

(22) McCulloch, Warren S., and Pitts, Walter, "A Logical Calculus of the Ideas Imminent in Nervous Activity," *Bulletin of Mathematical Biophysics,* 5, 1943, pp. 115-137.

(23) Hebb, D. O., *Organization of Behavior,* John Wiley, 1949.

(24) Minsky and Papert, *Perceptrons - An Introduction to Computational Geometry,* MIT Press, 1969. (Note: a second edition was published in 1988 containing some handwritten corrections and an additional chapter criticizing the modern neural nets algorithms such as backpropagation.)

(25) Hopfield, J. J., "Neural Networks and Physical Systems with Emergent Collective Computational Abilities," *Proceedings of the National Academy of Sciences,* USA, 79, No. 8, 1982, pp. 2554-2558.

(26) DARPA Neural Network Study, Oct. 1987-Feb. 1988, pp. 81-85.

(27) Hopfield, John J., and Tank, David, W., *Disordered Systems and Biological Organization,* Springer-Verlag, 1986.

(28) Dahl, E. D., "Neural Network Algorithm for an NP Complete Problem: Map and Graph Coloring," Lawrence Livermore National Laboratory Report.

(29) Hopfield, John J., and Tank, David W., "Neural Computations of Decisions in Optimization Problems," *Biol. Cybern.* 52,1985, pp. 141-152.

(30) Werbos, Paul, *Beyond regression: New Tools for Prediction and Analysis in the Behaviorial Sciences,* Ph.D. Dissertation, Harvard Univ., 1974.

(31) Parker, D. B., "Learning Logic," *Tech. Report TR-47,* Center for Computational Research in Economics and Management Science, MIT, April 1985.

(32) le Cun, Y., "Une Procedure d'Apprentissage pour reseau a

Sequil Assymetrique (A Learning Procedure for Assymetric Threshold Networks)," *Proceedings of Cognitiva*, 85, 1985, pp. 599-604.

(33) Rumelhart, D. E., Hinton, G. E., and Williams, R. J., "Learning Internal Representations by Error Propagation," *Parallel Distributed Processing, Volume 1: Foundations*, The MIT Press, 1986, pp. 318-362.

(34) Drefyus, S. E., "Artificial Neural Networks, Back Propagation, and the Kelly Bryson Gradient Procedure," *J. Guidance Control Dynamics*, 13, 1990, pp. 926-928.

(35) Hecht-Nielsen, Robert, "Neurocomputing: Picking the Human Brain," IEEE *Spectrum*, March 1988, pp. 36-41.

(36) Knight, Kevin, "Connectionist Ideas and Algorithms," *Comm. of the ACM*, Nov. 1990, 33 (11), pp. 59-74.

(37) Werbos, Paul J., "Overview of Designs and Capabilities," in *Neural Networks for Control*, edited by Miller, W., Sutton, R. and Werbos, Paul, MIT Press, 1990, pp. 59-65.

(38) Werbos, Paul, "An Overview of Neural Networks for Control," IEEE *Control Systems Magazine*, Jan. 1991, pp. 40-42.

(39) Nguyen, Derrick, and Widrow, Bernard, "The Truck Backer-Upper: An Example of Self-Learning in Neural Networks," in *Neural Networks for Control*, edited by Miller, W., Sutton, R. and Werbos, Paul, MIT Press, 1990, pp. 287-299.

(40) Werbos, Paul, "Generalization of Backpropagation with Applications to a Recurrent Gas Market Model," *Neural Networks*, 1, pp. 339-356.

(41) Werbos, Paul J., "A Menu of Designs for Reinforcement Learning Over Time," in *Neural Networks for Control*, edited by Miller, W., Sutton, R. and Werbos, Paul, MIT Press, 1990, pp. 67-95.

(42) Astrom, Karl Johan, "Toward Intelligent Control," IEEE *Control Systems Magazine*, April 1989, pp. 60-64.

(43) Bavarian, Behnam, "Introduction to Neural Networks for Intelligent Control," IEEE *Control Systems Magazine*, April 1988, pp. 3-7.

(44) Kuperstein, Michael, and Rubinstein, Jorge, "Implementation of an Adaptive Neural Controller for Sensory-Motor Coordination," IEEE *Control Systems Magazine*, April 1989, pp. 25-30.

(45) Astrom, Karl Johan, Anton, J. J., and Arzen, K. E., "Expert Control," *Automatica*, 22 (3), 1986, pp. 277-286.

(46) Edgar, Thomas F., "Current Problems in Process Control," IEEE *Control Systems Magazine*, April 1987, pp. 13-15.

(47) Naidu, Sinnasamy R. *et al.*, "Use of Neural Networks for Sensor Failure Detection in a Control System," IEEE *Control Systems Magazine*, April 1990, pp. 49-55.

(48) Anderson, Charles R., "Learning to Control an Inverted Pendulum Using Neural Networks," IEEE *Control Systems Magazine*, April 1989, pp. 31-37.

(49) DARPA Neural Network Study, Oct. 1987-Feb. 1988, pp. 123-129.

(50) Miyamoto, H., *et al.*, "Hierarchical Learning Control of an Industrial Manipulator Using a Model of the Central Nervous System," *Japan IEICE Technical Report*, Vol. MBE86-81, 1987, pp. 25-32.

(51) Nagata, S., *et al.*, "Mobile Robot Control by a Structured Hierarchical Neural Network," IEEE *Control Systems Magazine*, April 1990, pp. 69-76.

(52) Marr, David, "A Theory of Cerebellar Cortex," J. *Physiology*, London, 202, 1969, pp. 437-470.

(53) Albus, James, "A New Approach to Manipulator Control: The Cerebellar Model Articulation Control (CMAC)," *Trans. ASME, J. Dyn. Syst., Meas. Contr.*, 97, Sept. 1975, pp. 220-227.

(54) Miller, W. T. *et al.*, "Real-Time Dynamic Control of an

Industrial Manipulator Using a Neural-Network-Based Learning Controller," IEEE *Transactions on Robotics and Automation*, 6 (1), Feb. 1990, pp. 1-9.

(55) Miller, W. T., and Hewes, R. P., "Real-Time Experiments in Neural-Network-Based Learning Control During High-Speed Nonrepetitive Robot Operations," *Proc. 3rd IEEE Int. Symp. on Intelligent Control*, Washington, D.C., Aug. 1988, pp. 24-26.

(56) Handleman, *et al.*, "Integrating Neural Networks and Knowledge-Based Systems for Intelligent Robotic Control," IEEE *Control Systems Magazine*, April 1990, pp. 77-86.

(57) Liu, Huan, *et al.*, "Neural Network Architecture for Robot Hand Control," IEEE *Control Systems Magazine*, April 1989, pp. 38-43.

(58) Rabelo, Luis Carlos, *et al.*, "Synergy of Artificial Neural Networks and Knowledge-Based Expert Systems for Intelligent FMS Scheduling," *IJCNN-89*, pp. 1/359 - 1/366.

(59) Veezhinathan, Jay, and Wagner, Don, "A Neural Network Approach to First Break Picking," *IJCNN-89*, pp. 1/235 - 1/238.

Appendix A

Derivation of the LMS error signal
from the Generalized Delta Equation.

Given:

$$\Delta_p W_{ji} = \eta \delta_{pj} O_{pi} + \alpha \Delta_{pi} W_{ji} \tag{1}$$

$$net_{pj} = \sum_i W_{ji} O_{pi} \tag{2}$$

$$O_{pj} = f_j(net_{pj}) \tag{3}$$

$$E_p = \frac{1}{2} \sum_j (T_{pj} - O_{pj})^2 \tag{4}$$

where (1) is the generalized delta rule as defined in Chapter 8, Volume I of the PDP[33], (2) is the input calculation performed at each unit, (3) is the output activation function, (4) defines the least-mean-square (LMS) error minimization function, and η and α represent the learning rate and momentum terms respectively, the goal is to derive

$$\delta_{pj}^o = (T_{pj} - O_{pj}) O_{pj}(1 - O_{pj}) \tag{5}$$

$$\delta_{pj}^{\neg o} = O_{pj}(1 - O_{pj}) \sum_k \delta_{pk}^* W_{ki} \tag{6}$$

Prove gradient descent in E. To get the correct generalization of the delta rule, let

$$\Delta_p W_{ji} \propto -\frac{\partial E_p}{\partial W_{ji}} \tag{7}$$

Now, expand the right hand side of equation (7) by the chain rule to view this as a product of two parts; the first reflecting the change in error as a function of the change in net input to the unit, and the second representing the effect of changing a particular weight in the input. Then,

$$\frac{\partial E_p}{\partial W_{ji}} = \frac{\partial E_p}{\partial net_{pj}} \frac{\partial net_{pj}}{\partial W_{ji}} \tag{8}$$

Next, we expand the second term in the right hand side of equation (8) by substituting the definition of net_{pj} as an input sum of products.

$$\frac{\partial net_{pj}}{\partial W_{ji}} = \frac{\partial}{\partial W_{ji}}(\sum_k W_{jk}O_{pk}) \qquad (9)$$

This partial reduces to O_{pi}, because the derivative with respect to W_{ji} of $W_{jk}O_{pk}$ is O_{pk} when $k = i$ and 0 otherwise.

Then, by analogy to the derivation of the equations for the delta rule for units with a linear activation function (as described in [33]), let

$$\delta_{pj} = -\frac{\partial E_p}{\partial net_{pj}} \qquad (10)$$

Note that this is consistent with the definition of δ_{pj} used in the delta rule for linear units since $O_{pj} = net_{pj}$ when the unit is linear. Thus, by the delta rule, weight changes should be made according to the equation

$$\Delta_p W_{ji} = \eta \delta_{pj} O_{pi} \qquad (11)$$

in order to implement a gradient descent in E.

The concern at this point is the δ_{pj} term, because it will differ for units in the network according to their relative contribution to the total error (which is a function of the position of the unit in the network). We must therefore derive a generalized delta rule that can be applied to any unit in the network. We shall accomplish this in two parts: the first will represent the delta computation for an output unit, and the second will show the delta function for a non-output unit.

For output units: Apply the chain rule to equation (10) to expand the partial derivative to reflect the change in error as a function of the output of the unit and the change in output as a function of the change in the input.

$$\delta_{pj} = -\frac{\partial E_p}{\partial net_{pj}} = -\frac{\partial E_p}{\partial O_{pj}} \frac{\partial O_{pj}}{\partial net_{pj}} \qquad (12)$$

Now we can determine the second factor in the expansion by substituting the definition of O_{pj} from the given equations. Thus,

$$\frac{\partial O_{pj}}{\partial net_{pj}} = \frac{\partial}{\partial net_{pj}}(\frac{1}{1 + e^{-net_{pj}}}) \qquad (13)$$

$$= -1(1 + e^{-net_{pj}})^{-2}(-e^{-net_{pj}}) \tag{14}$$

$$= \frac{1 + e^{-net_{pj}} - 1}{(1 + e^{-net_{pj}})^2} \tag{15}$$

$$= \frac{1}{1 + e^{-net_{pj}}} - \frac{1}{(1 + e^{-net_{pj}})^2} \tag{16}$$

$$= O_{pj} - O_{pj}^2 \tag{17}$$

$$= O_{pj}(1 - O_{pj}) \tag{18}$$

Then, for output unit U_j, it follows from the definition of E_p that

$$\frac{\partial E_p}{\partial O_{pj}} = \frac{\partial}{\partial O_{pj}}(\frac{1}{2}\sum_j(T_{pj} - O_{pj})^2) \tag{19}$$

$$= \frac{1}{2}(2)\sum_j(T_{pj} - O_{pj})(-1) \tag{20}$$

$$= 0 + 0 + \ldots - (T_{pj} - O_{pj}) + 0 + \ldots \tag{21}$$

$$= -(T_{pj} - O_{pj}) \tag{22}$$

Finally, by substituting the results obtained in equations 22 and 18 into equation 12, we have

$$\delta_{pj}^o = (T_{pj} - O_{pj})O_{pj}(1 - Opj) \tag{23}$$

which is the desired result when U_j is a network output unit.

For non-output units: When U_j is not an output unit, the error contributed to the output units by the U_j unit is given by equation 12:

$$\delta_{pj} = -\frac{\partial E_p}{\partial net_{pj}} = -\frac{\partial E_p}{\partial O_{pj}}\frac{\partial O_{pj}}{\partial net_{pj}} \tag{24}$$

The second term in this equation is the same as derived in equation 18, as this partial contributes nothing to the error term. However, the first term in equation 24 relates the unit error to the output produced by the unit. We can solve for the first term by using the chain rule to expand the error term into a sum of partial errors as seen by the output units. Thus,

$$\frac{\partial E_p}{\partial O_{pj}} = \sum_k \frac{-\partial E_p}{\partial net_{pk}}\frac{\partial net_{pk}}{\partial O_{pj}} \tag{25}$$

This expansion allows us to determine the total error at the non-output unit as the sum of the partial errors contributed by the non-output unit to all the output units in the subsequent layer. It follows, then, that the error signal at any non-output unit U_j is proportional to the output produced by the unit and the current value of the connection weights between the U_j and the output units it feeds. We shall therefore decompose equation 25 into its component parts and find the minimal solution for the error at any non-output unit.

$$\frac{\partial E_p}{\partial O_{pj}} = \sum_k \frac{-\partial E_p}{\partial net_{pk}} \frac{\partial net_{pk}}{\partial O_{pj}} \tag{26}$$

$$= \sum_k \frac{\partial E_p}{\partial net_{pk}} \frac{\partial}{\partial O_{pj}} (\sum_i W_{ki} O_{pi}) \tag{27}$$

$$= \sum_k \frac{\partial E_p}{\partial net_{pk}} \sum_i 0 + 0 + \ldots + W_{kj} + 0 + \ldots \tag{28}$$

$$= \sum_k \frac{\partial E_p}{\partial net_{pk}} W_{kj} \tag{29}$$

$$= \sum_k \delta_{pk}^* W_{kj} \tag{30}$$

In this case, by substituting the results found in equations 18 and 30 into equation 12, we obtain

$$\delta_{pj}^{\neg o} = O_{pj}(1 - O_{pj}) \sum_k \delta_{pk}^* W_{kj} \tag{31}$$

which is the desired result when U_j is not an output layer unit.

TECHNIQUES FOR AUTOMATION SYSTEMS IN THE AGRICULTURE INDUSTRY

Frederick E. Sistler

Louisiana State University Agricultural Center
Department of Agricultural Engineering
Baton Rouge, LA 70803-4505

I. INTRODUCTION

The agriculture industry encompasses the growing, distributing, and processing of food and fiber, and their suppliers of goods and services. This chapter concentrates on techniques and control systems used in on-farm agriculture. It is applications oriented rather than mathematically oriented because the primary contribution is seen to be in the unique applications of existing sensors, systems, and techniques to biological systems. The properties and behavior of plants and animals vary greatly both among and within species. The response of a biological system is greatly dependent upon its environment (moisture, temperature, relative humidity, soil, solar radiation, etc.), which itself can be highly variable and difficult to model. All of this makes biological systems more difficult to model than inorganic systems and materials.

Automation is used in agriculture for machine control, environmental (building) control, water management, sorting and grading, and food processing. Farming has traditionally been associated with a very low level of automation. Tractors, tillage, and harvesting equipment were developed to reduce the manual labor involved, but most functions were performed with an open loop control. The operator steered the tractor, regulated its speed, and adjusted the implements. He continued to re-adjust and observe until the proper response was obtained. Environmental control for the livestock consisted of opening or closing the windows and doors in the barn, and turning on the fans or heat lamps. Crops were irrigated by pumping water until the farmer thought they had enough. In almost every situation the farmer performed the functions of feedback and control.

II. Automation of Tractor Operations

A. Guidance Systems

Misalignment with the rows of a few centimeters during cultivating and harvesting can be enough to severely damage or even destroy a crop. Automated or partially-automated machine guidance can be very useful to the operator because of the other operations requiring constant monitoring. Tillage implements must be checked for plugging and proper tilling depth. Planters must place seeds, fertilizer, insecticides, herbicides, and fungicides in the proper locations at the desired rates. Grain harvesters require adjustments as moisture levels change during the day.

Extensive research has been done on automatic guidance systems for tractors. Automatic steering systems use contact, optical, buried cable, or spatial position sensing.

Tasks such as mowing have the potential to be completely automated since guidance is often the only function which needs to be performed by the operator. Surbrook *et al.* [1] used an automatic steering system for an electric-powered robotic lawn tractor. The steering feedback was sensed by the position of the edge of the cut grass. Two microswitches located at the right front edge of the mower were activated as the unit moved through the grass. Cut grass was too low to activate them. When both switches indicated the cut grass area, a linear actuator connected to the steering linkage moved the front wheels to the left. The actuator steered the tractor to the right when both microswitches were in the uncut grass. The mower was properly aligned when the left switch was in the uncut grass and the right switch was in the cut grass.

The operator was required to mow the initial swath around the perimeter of the area to be cut. This furnished an on-board computer with the shape of the lawn, and also created the initial cut edge for the edge sensor to follow. A proximity switch was used to detect magnets attached to the rim of one of the rear wheels to measure the distance travelled. By monitoring the distance travelled, the computer determined when the edge of the field was reached, and turned the tractor to begin its next traverse. The process continued until any dimension of the uncut lawn was less than one meter. At this point the system was automatically stopped. The cut-edge sensors worked satisfactorily except when there were bare spots in the lawn. The sensors could not distinguish between bare spots and cut grass. Signal averaging helped to alleviate the problem. The mower cut an average of 95% of the area with the swath overlap set to 8%.

The use of vision for guiding agricultural vehicles has thus far not proved to be a very robust guidance method, but progress is being made. Fehr and Gerrish [2] mounted a video camera over the right rear axle of an electric-powered tractor. The camera was positioned to focus on the ground at various distances in front of the tractor. Problems were encountered when the lighting conditions changed rapidly in a corn field such as when a cloud passed in front of the sun.

Stability of the system depended upon how often the video guidance images were updated and how far ahead of the tractor the camera was aimed. Higher ground speeds necessitated both more rapid image updating and looking further ahead to allow an adequate response time for the steering system. In bright sunlight the system had difficulty distinguishing between plants and brown dirt. The problem was reduced by the use of color filters which increased the contrast between the plants and the soil. The system was able to follow rows of young corn and stayed within 5 cm of the intended path without destroying any plants.

B. Apple Harvester Guidance

McMahon *et al.* [3] used non-contact sensing for an auto-assist steering-control system on an apple harvester. The harvester used a 4 wheel hydrostatic drive system with front and rear wheel steering. It was an over-the-row vehicle that straddled the rows and harvested semi-dwarf trees. The operator sat on top of the machine above the row of trees. Shaker bars on the machine shook the tree trunks and dislodged the apples onto a catching frame.

To minimize damage to the trees or the harvester, the centerline of the harvester had to be within ± 22 cm of the tree trunk centerline. Maintaining this distance was sometimes very difficult because the foliage prevented the operator from precisely locating the center of the tree trunk as the harvester passed over it. Five ultrasonic sensors were located on the harvester ahead of the front wheel centerline to determine the location of the tree trunk as it moved through the harvester. Each sensor made four measurements of the trunk's location. A shaft encoder sensed the steering angle of the front wheels. A proportional control algorithm was used to set the steering angle of the front wheels at an angle proportional to the position error of the tree trunk. The steering was executed with two hydraulic solenoid valves controlling two linear actuators on the front wheels. After the harvester was properly aligned, the front wheels were set to a straight-ahead position (zero steering angle) until the next tree was sensed. The system was able to keep the harvester centerline within ± 22.5 cm of the tree trunk centerline for speeds up to 3.2 km/hr.

C. Combine Control

A combine for small grains cuts the stalks with a cutterbar, gathers the material with a rotating reel and auger, threshes the grain with a concave and rotating cylinder, and separates and cleans the grain with straw walkers, sieves, and a fan. The clean grain is stored in a tank until it is unloaded from the combine. Because harvesting delays result in rapidly increasing losses, the combining must be performed as quickly as possible. The combine operator must drive the machine at the maximum speed which will still allow the greatest harvesting efficiency. Harvesting efficiency is a measure of how much of the crop is harvested, threshed, and cleaned without excessive damage to the grain.

During combining the operator must maintain the header at the proper height for the cutter bar to cut the stalks with a minimum of excess material. The height can change with changes in soil type, drainage, and plant variety. The speed of the gathering reel must be synchronized with the ground speed to avoid excessive forces on the plants which could result in shattering of the pods and grain lost in the field. The cylinder speed must be adjusted as the moisture content changes to maximize the threshing and minimize damage to the grain. The straw walkers and fan must be set to rapidly discharge the chaff and straw without losing any of the grain yet slow enough to allow adequate cleaning. All of these functions must be performed while the operator is steering the machine and running it fast enough to keep it at its maximum operating capacity.

Several techniques have been adopted to assist in these combining operations. Kruse *et al.* [4] installed an automatic system to control the ground speed based upon the material load in the combine. The material load was the flow rate of the harvested material through the combine. If it was too large, some of the grain was lost through the rear of the combine. It is was too low, the timeliness of harvesting was reduced.

A pressure sensor was installed on the hydraulic motor which drove the feeder conveyor. The feeder conveyor delivered the cut material to the concave and rotor. As the amount of material increased, more power was required to drive the conveyor and the hydraulic pressure increased. A magnetic pickup was used to monitor the engine speed. The engine speed had to be kept constant to keep all of the various threshing and cleaning components operating at their optimal speed. Ground speed was controlled by changing the effective drive ratio of the hydrostatic transmission. A computer monitored the engine speed and the feeder conveyor motor pressure. When the engine speed was too low, the controller electronics for the hydrostatic transmission decreased the ground speed. When the engine speed was too high, the ground speed was increased. If the pressure transducer indicated a plugged condition (too much material entering the combine) the combine was stopped until the plug worked its way through the harvester. Maximum acceleration and deceleration rates were controlled by the computer. The maximum ground speed was set by the operator.

Mailander *et al.* [5] worked on the same combine to control the rotor speed based upon the material flow rate and the moisture content of the grain. The threshing efficiency of the cylinder and concave was very dependent upon the cylinder's rotational speed. The optimal speed varied as the grain moisture content varied. A hydraulic motor and an electro-hydraulic flow control servo valve were used to drive the rotor. A tachometer sensed the rotor speed. The servo valve was used to keep the rotor speed constant with changes in flow rate. A moisture sensor was placed in the auger to the grain storage tank to sense the grain moisture content by measuring its dielectric constant. The moisture content was used to determine the desired rotor speed which would minimize grain damage.

D. Location Sensing

Self-propelled, linear-moving irrigation systems are used to irrigate large areas of land. They are very wide (up to 500 meters), multi-jointed, mobile trusses with drive wheels located at each end and at the truss joints. Sprinkler heads are mounted along the truss and water the crops as the structure moves across the field in a straight path. A single, flexible hose supplies water from a well to the irrigation system. Because the trusses are so large and the wheel travel may not be the same for each wheel, it is important to know the position of each wheel. The driving wheels can be individually controlled to maintain proper alignment of the truss.

Shmulevich *et al.* [6] used triangular positioning and a single laser source to determine the position of a linear-move multi-jointed truss. The goal was to use the continuous position information to steer the wheels and move the system in a straight path. The positioning system contained five modules: a laser source, 2 scanners, a detection module, a computer module, and a retroreflector module. Four retroreflectors were mounted on the field machine. The other modules were mounted at stationary locations near the edge of the field.

A 1.7 mW He-Ne laser beam was used as the light source. The beam was enlarged with a beam expander, then split into beams which were directed at two mirrors. The adjustable, rotating mirrors were driven by stepping motors at an angular velocity of approximately 0.18 radians/second.

The position of each mirror was monitored with an incremental angle encoder with a resolution of 0.018 degrees. The laser beam was reflected off each mirror to one of the four retroreflectors on the gantry system. The beam bounced back from the retroreflector to the mirror and then to the beam splitter where it was split again and focused on a photodetector surface by means of a condensing lens. Two detectors were used: one was exposed to ambient light and the other to ambient light plus the laser light. They were used in a differential mode to detect the laser beam. As the mirrors rotated, the laser was consecutively bounced off each of the four corner-cube type retroreflectors.

Given the distance between the mirrors and the beam splitter, the position of each truss joint on the field machine could be determined by measuring the angular position of each mirror when the return laser signal was received. The system was effective up to 1000 m from the source, and the positional accuracy in a 400 m by 400 m field was within 15 cm.

E. Ballast Position Control

A large number of the farm tractors in the United States are of the two-wheel drive type with two large driven wheels at the rear of the vehicle and two smaller, non-driven wheels in the front. Steering is accomplished through turning the front wheels and with individually-applied braking on the rear wheels. Weight distribution between the rear and front axles is very important because it affects the amount of soil compaction, fuel efficiency, tractive efficiency, steering ability (too little weight on the front axle when it is on soft soil makes the steering ineffective), and the resistive forces caused by the soil-tire interactions. Large weights, called "ballast," are often mounted on the front end of the tractor to counterbalance rear-mounted implements when they are raised for turning and for transport to and from the field. When the implement is used in the field in its lowered position, the tractive and fuel efficiencies could often be improved if the ballast was shifted towards the rear axle to transfer more weight to the drive wheels.

Zhang and Chancellor [7] controlled the weight transfer by sensing the vertical reaction on the front axle with a strain gage bridge. The minimum front axle force was set at a level which would allow for safe operation. A 500 kg ballast was located on outboard rails on each side of a 67 kW tractor, and was moved fore-and-aft by a hydraulic motor through a worm gear box and attached roller-chain loops. Adjustments were made by turning on a hydraulic solenoid valve for a period of time (0.1s increments) proportional to the difference between the desired front axle reaction and the measured value. Adjustments were made every two seconds. The maximum on-time for the valve in any two second period was 1.5 seconds. The total travel time for the ballast to traverse the total fore-aft distance was approximately 13 seconds.

Tests for plowing and harrowing showed an average fuel savings of 22.3% and a time savings of 24.5% when the front axle vertical reaction was reduced from 11.12 kN to 7.12 kN. Greater savings were realized with harrowing than with plowing because the harrow was semi-mounted and it exerted very little vertical force on the drawbar. The fully-mounted plow created large vertical forces on the drawbar and created its own weight transfer from the front to the rear axle.

F. Planter Depth Control

Planting depth is critical to the survival of the seeds, and is dependent upon the size of the seeds. Until a young plant emerges from the soil surface and begins photosynthesizing, it is dependent upon the energy reserves within the seed. If the seed is planted too deep, it will not contain sufficient energy to reach the soil surface, and dies before emergence. Seeds which are planted too shallow may be eaten or may not be able to extract sufficient moisture from the soil to germinate. If the soil is level and of a uniform type, it is not difficult to maintain the proper depth control for planting. However, if the surface has rapid undulations or if the soil type changes rapidly across a field, proper seed placement becomes more difficult to maintain. Different methods have been tested to automatically control planting depth and some commercial systems incorporating these principles have been produced and marketed.

Dyck *et al.* [8] reported on a depth sensing device marketed by Sakundiak (Saskatoon, Saskatchewan, Canada). It used gage wheels as the depth sensor and a proportional electro-hydraulic servo valve to adjust the depth. The gage wheels were attached to the cultivator seeder frame with mounting brackets and rolled along the soil surface to limit the planting depth. As the gage wheels were forced down by the hydraulic cylinder(s), the planter frame was forced up. This raised the furrow openers for the seeds and decreased the planting depth. Potentiometers located on the mounting bracket sensed the position of the gage wheels. Planting depth was defined as the distance between the bottom of the cultivator sweep (furrow opener) and the gage wheel. The desired planting depth could be set on the control panel. The summed and averaged signal from the gage wheels was compared with the set point signal, and the error signal was fed to the controller circuit. The solenoid-operated servo valve activated the hydraulic cylinder to move the gage wheels up and down. The cylinder's spool travel was proportional to the error signal. Different rates of travel had to be used for the up and down travel of the gage wheels because the dynamics were different for the two operations.

Another approach to depth control has been to use an ultrasonic depth sensor [8]. The sensor was mounted on the planter frame and sensed the distance to the soil surface. Each sensor was activated approximately 18 times per second. A temperature sensor monitored the air temperature so compensation could be made for changes in sound velocity with changes in air temperature. The processing unit averaged the signals and compared the measured depth with the desired depth. The microprocessor energized a double-solenoid pilot-operated three-position valve for a time which was dependent upon the magnitude of the position error and recent control actions. The greatest benefits of the system compared to conventional depth control methods were obtained on varying soil types and uneven land. They were of the least benefit on level land and with uniform soil types.

Morrison [9] used a pneumatic down-pressure system for depth control on a no-till planter. An automotive pneumatic shock absorbing unit was mounted on the depth control wheel in each row. The shock absorbers were equipped with individual ON-OFF-EXHAUST pneumatic switches. An air compressor on the tractor supplied air for all of the shock absorbers. Planting depth was monitored by a linked arrangement of the front and rear planter wheels which sensed the average soil surface elevations. A compression spring in the link between the front and rear wheel was compressed when the furrow opener went deeper than the desired depth.

A pneumatic switch connected in parallel with the spring routed air to the shock absorber until the pressure on the depth control system link decompressed the spring and moved the switch to the off position. When the link spring was further compressed, the switch vented air from the shock to the atmosphere until the planter returned to the desired depth. An advantage of this kind of depth control system was that no adjustments were needed when going from one field to another as long as the desired planting depth remained constant. The limits of the system were determined by the maximum air pressure and the weight of the planter. The system performed well on the no-tillage planter with corn, grain sorghum, and cotton.

Morrison [10] used hydraulic cylinders to control planting depth on a 20-row grain drill modified for controlled-traffic, no-till operation. Each row of the drill had a single-acting cylinder to supply down pressure on the furrow opener. The operator set the pressure of the closed-center hydraulic system which was connected to all of the cylinders through a single constant pressure controller. The system eliminated the need for manually adjusting down pressure springs on each row and reduced the need for extra ballast requirements on the planter units.

The main problem occurred when a furrow opener encountered an elevated area of ground or heavier soil which caused a high pressure surge to be transmitted throughout the hydraulic system. The pressure surges in the hydraulic system were considerably higher than in the pneumatic system, but they were reduced by using pneumatic dampers. In spite of this, the hydraulic cylinder control system performed well in field use over several seasons.

III. Automation of Chemical Application

Chemicals are applied to crops to supply nutrients, control insects, diseases, and fungi, and to stimulate defoliation for harvesting. Controlling the rate of chemical applications is highly desirable because excessive applications result in extra cost and environmental hazards, and can sometimes damage the crop. Inadequate application rates provide insufficient protection against crop damage. The application rate of granular chemicals is usually controlled by the size of the openings on the spreader and/or the speed of a metering auger which is mechanically connected to a ground wheel. The auger delivers the material at a rate which is directly proportional to the ground speed. A more demanding situation arises when applying liquids. A pump usually delivers the material at either a constant pressure or a constant flow rate to a set of spray nozzles. The application rate is constant for a given set of nozzles and can be varied by changing the pressure, the size of the nozzles, or the ground speed of the sprayer. This works satisfactorily as long as the proper nozzles are used and the operator maintains a constant ground speed and the crop growth is uniform. If the ground speed changes, the chemical application becomes too large or too small. A non-uniform crop coverage can also be a problem because the chemicals need only be applied on or near the plants - not wasted on areas between plants.

A. Orchard spraying

Orchard spraying is a prime example of non-continuous plant coverage. Giles *et al.* [11] used ultrasonic sensors to control a towed-type air-carrier orchard sprayer using a large fan to blow the chemical mist into the tree canopy. The chemicals were injected into the airstream by means of three, equal-capacity nozzle manifolds equally spaced around the blower outlet. There were two sets of nozzle manifolds so the rows on both sides of the sprayer could be sprayed with a single pass through the orchard. The ultrasonic sensors were positioned at three different heights on the spray rig ahead of the fan with one set of three directed to the left and the other set aimed to the right of the sprayer. Sprayer movement was monitored by the displacement of a ground wheel. Every 1/16th of a revolution, the ultrasonic sensors measured the distance from the sprayer to the tree at three vertical positions in the canopy. Based upon initial tree measurements, tree canopy extensions were determined for various tree sizes. Larger trees had more foliage and extended further into the rows. The manifolds to be activated were determined by the sensed distance to the tree foliage. The control algorithm selected the nozzle manifold(s) to activate at a spray rate of 0, 1/3, 2/3, or full output. For peach and apple orchards, the spraying system using the ultrasonic sensors had a 31% spray savings compared to a conventional sprayer in peaches, and a 44% spray savings in an apple orchard. There was, however, a significant reduction in coverage at some sample locations.

IV. Automation of Irrigation and Drainage

For crops to reach their maximum yields it is necessary to maintain the proper soil moisture level. This involves modifying the soil water content through irrigation and/or drainage, and is dependent upon the geographical location, time of year, stage of the crop, and the weather. Irrigation is used to maintain an optimal soil moisture content by adding the difference between the precipitation and the soil water losses due to drainage and evapotranspiration (evaporation + crop transpiration).

A. Lysimeter Control

Computer models have been developed to predict the evapotranspiration (ET) based upon crop type and stage of development, soil type, temperature, humidity, and wind speed. However, a direct measurement of the soil moisture with the crop growing on it is more accurate. A lysimeter is a device developed for this purpose. It consists of a large container of soil set in a field and subjected to the same soil, crop, and weather conditions as the fields around it. The lysimeter is set on a buried scale to allow continuous weight measurements, and it has a drain for excessive water to emulate sub-surface drainage or percolation.

Phene, *et al.* [12] developed a real time drip irrigation controller with a lysimeter used for the feedback control. A water reservoir for irrigating the lysimeter was attached to the lysimeter and weighed with it so the amount of irrigation water could be accounted for in the water-balance equation. Weight measurements were made on an hourly basis. When the water deficit (ET + drainage) reached 1 mm, a subsurface drip irrigation system was turned on to add 1 mm of water to the lysimeter and to the same crop growing in the surrounding field.

The lysimeter was successfully used to control and maintain the desired soil moisture level in surrounding fields for both tomatoes and grass. The system was used in a low-rainfall area, so water drainage through the soil was not a factor. In areas of significant precipitation it would also be necessary to measure the amount of rainfall, subsurface drainage, and the amount of surface runoff in order to control the soil moisture at the desired level.

B. Center-Pivot and Surface Irrigation Control

Cahoon *et al.* [13] used a computer-based scheduler for center pivot and surface irrigation systems for soybeans, cotton, corn, and grain sorghum. Inputs included daily maximum temperature, precipitation, irrigation dates, and, for center pivot systems, depths of irrigation applications. The system was designed for use in the humid Mid-South region of the United States.

The scheduler was a system scheduler rather than a field scheduler. A field-based scheduling system only considers the growing conditions for one crop grown in a single field. It does not consider adjacent fields and crops which may be irrigated with the system. The goal of the system scheduler was to minimize travel of the irrigation system while maintaining adequate soil moisture in each field.

The irrigation scheduler used a soil moisture balance to determine irrigation times. The soil moisture balance had units of depth of water, which was how much water needed to be added to the crop. The daily soil moisture balance was the balance from the previous day plus the amount of evapotranspiration during the day, minus the day's precipitation and irrigation. Irrigation was recommended when the soil moisture balance reached a pre-determined maximum depth.

The predicted evapotranspiration was a function of the daily pan evaporation and a modified crop coefficient. The daily pan evaporation is a measure of how much water evaporates from a standard class A weather station pan in one day. Predicted ET was calculated using maximum daily temperature and the number of hours of daylight. The modified crop coefficient was dependent on the type of crop, stage of growth, and the wetness of the soil surface.

Because the scheduler was set to irrigate multiple fields with a single system, the water requirements for each crop were computed for the next 14 days. Using this information the scheduler was designed to position the irrigation system to irrigate the critical fields by the time they were expected to exceed their critical soil moisture deficit. Irrigation predictions could be based either on the long-term average weather for the area or on local weather forecasts. The system was successful in tests on both cotton and soybeans.

C. Solid-Set Irrigation Control

Lamb *et al.* [14] used weather data to improve the efficiency of a remotely-operated irrigation system and implemented an automation shutdown for the system if the irrigation equipment malfunctioned. They used a stationary irrigation system with underground laterals, risers, and sprinklers to independently irrigate 24 plots. The system was located at a remote site and controlled via a telephone modem and computer. An anemometer at the remote site measured wind speed and a tipping bucket rain gage measured the rainfall amount and its intensity. Pressure gages were used to monitor the water system pressure, and a flow meter measured the total flow rate. Each plot had its own solenoid to control the sprinkler operation. There was also a solenoid control on the main water supply. All of the gages and solenoids were connected to the irrigation control computer.

The operator called the computer and instructed it as to which plots needed irrigating and the amount that needed to be applied. With this information the computer would apply the water at the appropriate times if the wind velocity was not too high. If the wind speed was excessive, irrigation was delayed until the wind subsided. While the plots were being irrigated, the total water flow rate was compared to the expected value for the number of plots being irrigated. If the flow rate was excessive, the main water supply solenoid closed the line because there was a malfunction such as a broken pipe. The control computer notified the operator of the problem. The system performed satisfactorily during a three-year test.

V. Automation of Grading and Sorting

Most agricultural products must be graded and/or sorted before they are sold because the price is related to size and quality. Small grains are graded according to the number of cracked and broken kernels, size, foreign matter (trash and undesirable seeds), diseases, and moisture content. Eggs, potatoes, apples, pears, tomatoes, and many other types of produce are sorted by size, shape, and color or ripeness. Grading and sorting become much more difficult when the product shape is non-uniform, as is often the case with agricultural products. Grading is also very specific to the product, so the machinery and criteria are often unique to one product or a single type of products.

A. Peach Grading

Peach grading is a manual operation. Inspectors view the peaches passing on a conveyor belt in front of them and sort them according to size, color, and defects. Color is a very important component of the grade. Peach color is classified according to the amount of overcolor (pink or red) and the ground color (yellow or green). Ground color is indicative of the maturity level of the peach. The highest peach grade requires the proper maturity level and one-third of the surface to show pink or red color.

Miller and Delwiche [15] developed a machine vision system to perform a color analysis for inspecting and grading fresh market peaches. The peaches were transported on a conveyor belt beneath a solid-state RGB video camera. Diffused tungsten halogen lighting was used. A conveyor belt presented the peaches individually to the camera mounted above the belt. The imaging system usually digitized images at 512 x 512 pixels per frame which was more resolution than was required to inspect the peaches. In order to reduce the processing time, an interface was developed to multiplex the RGB camera inputs into a single image, dc restore the video signals, and generate timing signals for the frame grabber. The resulting image had a pixel resolution of 61 rows and 128 columns with full RGB information for each pixel stored in a single frame for each image.

Segmentation - Peaches were detected and segmented in the images by using the red histogram which was distinctly bimodal.

Feature extraction - To establish a procedure unaffected by variations in illumination intensity, a pair of normalized color coordinates were defined in the RGB color space as:

$$\text{and}$$

$$r = \frac{R}{R + G + B} \qquad\qquad g = \frac{G}{R + G + B}$$

These normalized colors were sensitive to changes in chrominance but insensitive to changes in luminance. Standard color levels were defined by taking video measurements of the six official ground color standards and converting them to chromaticity coordinates. A blush standard color was added to the ground color standards to identify blush pixels. To classify a peach, chromaticity values for each pixel were assigned to the nearest standard color level using a minimum distance quantization technique. The peach was classified to the maturity class for which not more than 15% of the surface area was assigned to greener color standards.

When the system was tested on several varieties of peaches, the vision classification was within one color standard of the manual maturity classification for 88% of the tests. The correlation coefficient between machine vision and manual estimates of blush surface area was 0.92.

B. Small Fruit Sorting

McClure and Rohrbach [16] developed an asynchronous method for maturity sorting of blueberries and grapes. Maturity was based on the amount of anthocyanin in the fruit. The level of anthocyanin could be determined by measuring the light transmitted through the fruit. The differential optical density at 740 nm and 800 nm was used to measure the maturity of blueberries and black grapes. For bronze grapes it was 540 nm and 610 nm.

The sorter used parallel wires with M-shaped grooves to convey the fruit asynchronously and in a single file. A photo-diode was used to detect the presence of the fruit. When the fruit was detected, a 21 ms time delay was activated to allow the center of the fruit to be located at the differential optical density sensor (maturity discriminator). This constant delay time worked well with all sizes of fruit. Light from a tungsten lamp was projected up through the fruit and measured by the maturity discriminator. A pulse height was generated which was proportional to the anthocyanin concentration in the fruit. The level of anthocyanin increased with maturity, so the pulse height was proportional to the maturity of the fruit.

This "maturity pulse" was fed to two comparators. One was set at the low level of maturity which indicated the minimum acceptable ripeness. If the signal was less than this value, the fruit was considered to be green (not ripe). The high level comparator was set to the maximum acceptable maturity level. Discriminator values exceeding this threshold were considered to be over-ripe. Either condition resulted in rejecting the fruit. Two air valves controlled by the maturity discriminators were used to remove undesirable fruit. One was for green fruit and the other was for over-ripe fruit. Ripe fruit passed through the system undisturbed.

C. Tomato Grading

Tomatoes are graded according to defects, color, and soluble solids. Defects are classified as worms, mold, green, materials other than tomatoes, and limited use. A load of tomatoes can be rejected if any of the following limits are exceeded: 3% worms, 10% mold, 5% green, 4% material other than tomatoes, 8% limited use, or a low color reading indicated by an Agtron E-5-M color intensity meter reading greater than 39.

O'Brien and Garrett [17] helped to semi-automate tomato grading and significantly reduce the grading time. An automatic sampler was developed which could draw a core of fruit from top to bottom from pallet bins or bulk trucks. The core samples were delivered to a mechanized grading table where they were mechanically divided in half, automatically weighed, and manually inspected. Sample weights were averaged and automatically recorded by the central computer. Inspection for defects was done by inspectors as the tomatoes passed by on belts in front of them. Sub-samples were tested for color and soluble solids. After the samples were sorted by defect, the weight of each defect group was recorded in the computer, as were the results of the soluble solids and color measurements.

The semi-automated grading resulted in a reduction in sampling and grading time from 9.2 minutes to 2.2 minutes per double trailer truckload of tomatoes. An added benefit was an overall reduction in the number of loads rejected because the grading results were produced and transmitted more quickly to the crews who could respond and modify their harvesting quality control on a more timely basis.

VI. Automation of Greenhouse and Nursery Operations

A. Container Handling

Handling and movement of container-grown plants in plant nurseries is a labor-intensive operation. Young and Dunlap [18] used a self-propelled machine with a series of front-mounted, open-flight screw conveyors for loading, transporting, and unloading the containers. The containers were oriented in rows on the ground with the row spacing equal to the screw conveyor spacing. A large-pitch screw was mounted above each conveyor. As the unit was driven into the rows of containers, the screws turned and the containers were drawn onto the conveyor. One container fit between each flight of the screw conveyor. The turning rate of the screws had to be synchronized with the forward speed for the system to function properly. The most desirable situation was for the containers to experience a zero horizontal (ground) speed as they were loaded onto the transporter unit.

Control of the screw conveyors - The screw conveyors were driven with a hydraulic servomotor. An optical shaft encoder on one of the non-powered ground wheels was used to measure the forward speed of the machine. A second encoder was attached to one of the screw conveyors through a bevel gear. A reference signal was provided by the encoder on the ground wheel and the feedback was provided by the conveyor shaft encoder. The two signals were fed to a digital summing junction. The error signal was used to control the hydraulic motor with a third-order, single-loop servomechanism which was conditionally stable based on Rough-Hurwitz and root locus analyses, depending upon the magnitude of the forward loop gain. The hydraulic servomotor kept the screw conveyors turning at a rate which maintained the relative container speed at zero for both loading and unloading.

B. Greenhouse Control

Gauthier and Gray [19] presented an object-oriented model for greenhouse climate control. The model was designed to meet the following criteria:

1.   It must respond in real time to the sensors' perceptions of a somewhat unpredictable and changing environment.

2.   Data from sensors must be interpreted and reasoned about. The system must identify sensor or time-based events by matching them with patterns stored in the knowledge base. The temporal and spatial properties of objects must also be conveyed.

3.   It must react safely and reasonably to inaccurate and/or uncertain data inputs (*e.g.*, faulty sensors or user mistakes).

4.   It must handle multiple zones within the greenhouse(s) having different configurations and possibly different crops or crop stages.

5.   It must interface with many process controllers and allow the transfer of information between controllers installed in a greenhouse complex and a central computer.

6.   It must allow the dynamic adjustment of set-points and parameters used by the regulation algorithms.

7.   It should log and maintain histories describing the evolution and performance of crops, climate, equipment, and resources.

8.   It must support the representation and employment of generic and knowledge-base production scenarios designed to respond to the evolving needs of a given crop and to goals set by production managers.

The model was written in Smalltalk V/286 (Digitalk, Inc., Los Angeles, CA). It provided a conceptual framework for knowledge intensive control strategies based either on heuristic or analytical knowledge. The entire system was built assuming a distributed processing system with a central computer communicating with several programmable process computers over a local network. Each process computer regulated one or more greenhouse zones. All data structures and controls were implemented through the definition of classes, instances, and methods. Examples of these include a ZONE class which represented an area of the greenhouse with a uniform environment. Areas for crop production were defined as sub-classes of ZONE. Others classes included AIR-TEMP (air temperature), CLIMATIC-PARAMETER (parameters describing a climate zone) and DEVICE (holds protocol for interpreting the data received from programmable process computers). Gauthier and Gray stated that, "The object-based environment used for the study greatly facilitates the prototyping and encoding of knowledge and appears adequate for a full-scale operation in a production environment."

VII. Automation of Environmental Control

A. Potato Storage

Constant monitoring and control of the storage conditions is necessary to preserve the quality of many agricultural products. In large storage facilities, several points must be monitored for specific gases released by various disease organisms in order to rapidly detect and halt the deteriorating effects of these organisms. One solution for controlling the growth of disease organisms in potatoes is to regulate the air flow through the storage bins. A high temperature, high humidity environment is conducive to the growth of these organisms, so a low humidity, increased ventilation rate can be used as a control mechanism. However, if the potatoes are healthy, the increased ventilation will result in excessive drying of the potatoes.

A system to control ventilation flow rate, temperature, and humidity requires the use of an automated system to measure the gas concentrations at several points within the storage area. Schaper *et al.* [20] developed an automated sampling system to monitor the CO_2 concentration at several places in a potato storage facility. The gas concentration measurements could be used to provide the feedback for the control loop and also served to measure the efficiency of the ventilation system throughout the storage area.

B. Poultry Housing

Poultry houses are one of the most critical animal housing environments. There have been cases where several thousand birds have died in a confined facility because of a power failure lasting only a few hours. To maintain bird health and maximum growth rate, the temperature, humidity, air flow, lighting, and maximum allowable ammonia concentration must be carefully monitored and controlled. Ammonia is produced by growing the birds on reused litter - a common practice in some areas. Maintaining the proper environment requires changes with the weather and with the number and age of the birds in the house. The temperature needs to be gradually lowered from 35°C at placement to 21.1°C at 28 days of age, and then held constant for the rest of the grow-out period [21,22] while maintaining a relative humidity between 50 and 70%. Humidity is usually controlled by the ventilation rate. The ventilation rate must also be sufficient to maintain an adequate oxygen supply and keep the ammonia concentration to a safe level.

Allison *et al.* [23,24] used a computer system to control the environment in a broiler house. It controlled the width of the air inlet slots, operation of the heaters, and the ventilation fans. There were multiple heaters and ventilation fans, and the computer controlled each one separately. Thermocouples, humidity sensors, ammonia sensors, and pressure gages were used to measure inside and outside temperatures and relative humidities. Ammonia sensors measured the ammonia concentration within the house. The computer also maintained data on the size of the building and its thermal properties. It monitored the age of the birds and was programmed to account for changes in their specific and latent heat production levels with age, and their changing temperature requirements for optimal health and growth rates.

Compared to conventional control systems for broiler houses, the computer system was able to maintain the air temperature closer to the desired level when the outside air temperature was lower than the desired air temperature. When the outside air temperature was higher than desired, the computer maintained an inside air temperature lower than or equal to the air temperature in the conventionally-controlled house. The relative humidity was also kept closer to the desired level. The overall ventilation rate was lower with the computer system which meant the energy usage would be lower than with a conventional system using only thermostats and timers for fan control.

VIII. Automation of Electrical Energy Management

A. Electrical Demand Control

Electrical energy suppliers use different methods for energy charges. Some charge higher rates for energy consumption during peak demand hours (time-of-day rates), some have a minimum charge which is a function of the peak demand over a long period of time (ratchets), and others may include charges to cover the capital investment needed to meet the peak demands (demand charges).

Heber *et al.* [25] demonstrated a system which could schedule the electrical demands on the farmstead in a way to reduce these various energy charges. Because electrical demands vary with the types of crops and livestock grown, season, weather, types of equipment, usual working hours, and management style, the demand controller must be configured for each individual situation. The various electrical demands must be identified and priorities assigned to them. Ventilation of total-confinement livestock systems, for example, would have a very high priority because of the danger of suffocation if it was turned off. Grain drying demands will be the highest in the fall and winter. Grain drying must be accomplished in a timely manner to prevent spoilage. However, it is sometimes possible to delay the drying for short periods (hours) during a day without causing problems. Feed grinding and delivery power requirements will be relatively constant throughout the year, but the time of day for performing the operations may be very flexible. Milking-related operations such as milk cooling and equipment cleaning must be performed at the actual milking times.

Heber classified the electrical loads into nine operating and management categories. They were as follows:

1.   Programmable - operator presence not required.

2.   Independent - performs a task independent of other loads.

3.   Dependent - dependent on one or more other loads.

4.   Monitored - control is external to the electrical load management system (e.g., lights and grain truck unloading augers).

5.   Controlled - under electrical load management system control.

6.   Non-interruptible - temporary shutdown is not allowed.

7.   Multiple - subsystem with more than one dependent load.

8.   Dynamic priority - priority is automatically adjusted.

9.   Static priority - priority is constant.

The load management system was used in a feed processing and grain handling center. The highest priority was assigned to the grain-transfer equipment for loading and unloading the trucks and storage bins. Lower priorities were assigned to feed processing and grain drying. The peak electrical demand was reduced more than 52%, and the daily load factor (ratio of actual energy use if the peak demand had been maintained throughout the specified time period) was increased from 53% to 97%. The efficacy of the electrical management system was very sensitive to the priorities and management schemes. Some changes actually resulted in increasing the peak demand.

The techniques presented in this chapter are representative of the types of control systems presently used in agriculture. Some of them have been incorporated into commercial systems, and others will require further refinement before they can become economically justifiable. One of the biggest barriers in further automation of agricultural operations is the development of reliable sensors for biological materials. Real-time moisture content evaluation is not yet possible for many products. Reliable defect detection is still difficult because of the many types of defects which fruits and vegetables can possess. Modeling of animal and plant behavior is in need of refinement. It is expected that significant progress will be made in the next decade as it has in the last decade. However, challenges in these areas will continue to exist because of the cost restrictions imposed by sensors which are used with low unit-value products for only short periods of the year.

References

1. Surbrook, T.C., J.B. Gerrish, R.E. Squires, and M.A. Schanblatt, "An Automatic Steering Controller for a Robotic Lawn Tractor," Paper 82-3039, ASAE, St. Joseph, MI (1982).

2. Fehr, B.W. and J.B. Gerrish, "Vision-Guided Off-Road Vehicle, Paper 87-7516, ASAE, St. Joseph, MI (1989).

3. McMahon, and B.R. Tennese, "Performance of Auto-Assist Steering System Using Ultrasonic Sensors," *Applied Engineering in Agriculture* (1):87-89 (1987).

4. Kruse, J., G.W. Krutz, and L.F. Huggins, "Computer Controls for the Combine," *Agricultural Engineering* 64(2):7-9 (1983).

5. Mailander, M.P., G.W. Krutz, and L.F. Huggins, "Computer Controlled Hydraulics - A Combine Application," Paper 79-1615, ASAE, St. Joseph, MI (1979).

6. Shmulevich, I., G. Zeltzer, and A. Brunfeld, "Laser Scanning Method for Guidance of Field Machinery," *Transactions of the ASAE* 32(2):425-430 (1989).

7. Zhang, N. and W. Chancellor, "Automatic Ballast Position Control for Tractors," *Transactions of the ASAE* 32(4):1159-1164 (1989).

8. Dyck, F.B., W.K. Wu, and R. Lesko, "Automatic Depth Control for Cultivators and Air Seeders Developed under the AERD Program," *Proceedings of the Agri-Mation 1 Conference & Exposition* 265-271, ASAE, St. Joseph, MI (1985).

9. Morrison, J.E., Jr., "Interactive Planter Depth Control and Pneumatic Downpressure System," *Transactions of the ASAE* 31(1):14-18 (1988).

10. Morrison, J.E., Jr., "Hydraulic Downpressure System Performance for Conservation Planting Machines," *Transactions of the ASAE*

31(1):19-23 (1988).

11.   Giles, D.K., M.J. Delwiche, and R.B. Dodd, "Control of Orchard Spraying Based on Electronic Sensing of Target Characteristics," *Transactions of the ASAE* 30(6):1624-1630, 1636 (1987).

12.   Phene, C.J., R.L. McCormick, K.R. Davis, J.D. Pierro, and D.W. Meek, "A Lysimeter Feedback Irrigation Controller System for Evapotranspiration Measurements and Real Time Irrigation Scheduling," *Transactions of the ASAE* 32(2):477-484 (1989).

13.   Cahoon, J., J. Ferguson, D. Edwards, and P. Tucker, "A Microcomputer-Based Irrigation Scheduler for the Humid Mid-South Region," *Applied Engineering in Agriculture* 6(3):289-295 (1990).

14.   Lamb, J.A., G.E. Meyer, and P.E. Fischbach, "Microprocessor Automation of a Solid-Set Irrigation Facility for Research and Demonstration," *Transactions of the ASAE* 28(3):943-948 (1985).

15.   Miller, B.K. and M.J. Delwiche, "A Color Vision System for Peach Grading," *Transactions of the ASAE* 32(4):1484-1490 (1989).

16.   McClure, W.F. and R.P. Rohrbach, "Asynchronous Sensing for Sorting Small Fruit," *Agricultural Engineering* 59(6):13-14 (1978).

17.   O'Brien, M. and R.E. Garrett, "Development of a Mechanical and Electronic Assisted Tomato Grading Station," *Transactions of the ASAE* 27(5):1614-1618 (1984).

18.   Young, R.E., and J.L. Dunlap, "Digital Controller for speed synchronization of Nursery can handler," *Transactions of the ASAE* 30(4):866-872 (1987).

19.   Gauthier, L. and R. Guay, "An Object-Oriented Design for a Greenhouse Climate Control System," *Transactions of the ASAE* 33(3):999-1004 (1990).

20. Schaper, L.A., J.L. Varns, and M.T. Glynn, "Computerized Gas Sampling and Analysis System for Potato Storages," Transactions of the ASAE 30(6):1807-1810 (1987).

21. Reece, F.N., "Ventilation of Livestock Confinement Structures," Paper 76-3530, ASAE, St. Joseph, MI (1976).

22. Reece, F.N., "The Effect of Ammonia on the Performance of Broiler Chickens," Paper 79-4004, ASAE, St. Joseph, MI (1979).

23. Allison, J.M., J.M. White, J.W. Worley, and F.W. Kay, "Algorithms for Microcomputer Control of the Environment of a Production Broiler House," *Transactions of the ASAE* 34(1):313-320 (1991).

24. Allison, J.M. and J.M. White, "Hardware for Microcomputer Control of the Environment of a Production Broiler House," *Applied Engineering in Agriculture* 7(1):119-123 (1991).

25. Heber, A.J., T.L. Thompson, and D.D. Schulte, "Development of an On-Farm Electrical Load Management System," *Transactions of the ASAE* 29(1):281-287 (1986).

MODELING AND SIMULATION
OF
MANUFACTURING SYSTEMS

NDY N. EKERE
University of Salford, Salford, UK

ROGER G. HANNAM
University of Manchester Institute of Science
and Technology, (UMIST),
Manchester, UK

I INTRODUCTION

The term 'manufacturing system' is a generic term which is
capable of many interpretations. Its generic meaning is a
system in which raw materials are processed from one form
into another, the product. A manufacturing system generally
includes many linked processes, the machines to carry out
those processes, handling equipment, control equipment and
various types of personnel. A manufacturing system for an
automobile could include all the presslines to produce the
body panels, the foundries to produce the engine blocks and
transmission housings, forge shops to produce highly stressed
parts such as suspension components and crankshafts, the
machine shops which convert the forgings, castings and other
raw material to accurately sized components and the subassembly and final assembly lines which result in the final
product being produced. Most writers would also call each of

these sub-sections a manufacturing system although each is also constituent of the larger manufacturing system. Thus, each of the transfer lines that machine parts for the engines in mass production quantities could be termed a manufacturing system, the treatment tanks imparting corrosion resistance and the painting lines could similarly be a manufacturing system and so on.

The machines and processes involved in manufacturing systems for mass production (such as just quoted) are dedicated to repetitive manufacture. The majority of products are, however, produced by batch manufacturing in which many different parts and products are produced on the same machines and the machines and processes are re-set at intervals to start producing a different part.

The techniques which will be discussed in this chapter apply to manufacturing systems which extend from a few machines (which are related,- generally because they are involved in processing the same components) up to systems which might comprise the machines in a complete machine shop or complete processing line. The characteristics of batch manufacturing mean it is often analysed by simulation, mass production systems are more analysed by mathematical analysis.

Like 'manufacturing system', the term 'model' is similarly a very broad term. Models are widely used to represent physical and engineering phenomena. They can take many forms but all permit real systems or structures to be investigated without the expense of having the real system available. This is naturally of significant benefit when designing something new or when planning to modify something which already exists. Models may be physical models, they may be expressed in mathematical terms such as a set of equations, they may be graphically expressed models, such as a circuit diagram or a flow chart, or they may be numerical or number based models. Simulation comes in this last category. Manufacturing systems have been modeled in all these ways. This chapter will concentrate on modeling techniques which are supportive of simulation but some mention will be made of the others.

Both large and small scale manufacturing facilities have been modeled through physical models,- scale models which show the layout of a complete facility. As with other forms of modeling, these models help system designers and eventual users to check on their ideas, on how parts of the facility relate and how accessible any parts of the facilities are from any other. The price of scale models for a complete plant can be over $2M but they are considered value for money for the insights they give. Realistic 3-D images of a new manufacturing system can now be generated by CADCAM software and these are now being used alongside physical models as a means of giving designers a 3-dimensional view of what they are creating.

Scale models are normally static,- the parts to be processed do not move about and machines do not run. While scale models allow spatial relationships to be checked, they do not permit a system's operation to be investigated. This is where simulation and other types of numerical and mathematical model are used. These techniques are used to investigate how to operate a system efficiently - and efficiency has many aspects in manufacturing systems.

A. Efficiency in manufacturing systems

There is no single definition of efficiency of manufacture because manufacture has to have many variables in control.

or example, it is often impossible to minimise the cost of production and maximise the speed of production simultaneously. To get high speed production can involve having specialised machines and to get quick production can involve having spare capacity to bring on-stream when demanded. These can both be costly. Thus manufacturing is about balancing demands against resources and against efficiency. Because manufacturers do not want to carry spare capacity, in many manufacturing systems there is 'competition' for resources such as machines, operators, inspectors, etc. This competition results in some items having to wait (ie queuing) until resources become available. This is particularly true when different items (or batches of such items) are to be produced. This requires more adaptable or flexible machinery to be used. The machines have to be reset and these change overs and new set-ups take time to carry out. This is time lost.

Batch production also generates inventory which is not all immediately required, so the inventory has to be stored and managed. Inventory storage costs money and this may necessitate investing in automatic warehousing and workhandling systems. Yet there are naturally pressures to produce parts as cheaply as possible. Cheap parts are not necessarily produced on high cost machines if volumes are not high. A further pressure is to respond to the demands of the market quickly. This needs low inventory and adaptable machinery. Thus manufacturing system engineers have to cope with many conflicting demands which have to be balanced against each other. Manufacturing systems are modeled to explore these competitive demands and to endeavour to maximise the efficiency of as many system elements as possible.

II INTRODUCTION TO SIMULATION

Simulation is one of the most powerful analysis tools available to those responsible for the design and operation of manufacturing systems. Simulation can be defined as the process of designing a model of a system and conducting experiments with this model for the purpose of either understanding the behaviour of the system or evaluating the various strategies for the operation of the system. Experiments are conducted because simulation is not directly used as a means of carrying out any form of optimization. (Some analysis tools which will be discussed later can produce an optimum.) In a simulation, a set of initial conditions and operating parameters are set, and then the behaviour under these conditions and parameters is determined by running the simulation. Other conditions and operating parameters are investigated through additional runs. Skill in analysing the results is required and in planning further experiments to identify any optimum.

As a method of analysis, simulation can be used in various ways, including:-
* investigating the operation of a real system,
* investigating proposed or hypothetical systems,
* investigating effects of organisational change,
* testing new policies and decisions before implementation,
* verifying analytical results.
Simulation models can also be used interactively as a means of training operating personnel. Thus the reason for building and using simulation models is to allow inferences to be drawn about a system (or systems); without building them if they are only proposed systems; without disturbing them if they are operating systems that are costly or unsafe

to experiment with; or without destroying them, if the object of the experiment is to determine their limits of performance.

In relation to manufacturing systems, the main purposes of a simulation study will generally be to:-
* assess the resource requirements and their utilisation, (ie machines, workhandling systems, operators)
* identify system bottlenecks,
* compare the performance of alternative designs, and to
* develop strategies for scheduling/sequencing jobs.
Simulation models have been used fairly widely since the early 1960s for the design, operation and performance assessment of manufacturing systems[1].

A. Simulation modeling fundamentals

A simulation model is a conceptualisation or representation of the elements and the behaviour (the operation) of a real system. The model can represent the behaviour very closely or it can have less detail in it and approximate to the behaviour of the real system. The model may first be represented by a diagram which may take the form of a network or a flow diagram or any activity cycle diagram. The forms of these diagrams will be clearer later once more of the detail of modeling is understood. The creation of the diagram requires that the modeler has a good understanding of how the proposed or actual system operates. The creation of the diagram is a first stage of system's analysis, which, by its nature, will require the modeler to address aspects of the logical operation of the system.

Some models may then be manipulated by hand rather like a board game. The passage of time is monitored and objects can be moved around the diagram to correspond to the behaviour of the resources in the real system. A close and careful record of all changes in the state of the objects has to be kept so that the behaviour of the model can be determined. While this approach is laborious, it can provide an initial check on a modeler's understanding of a system and it can be helpful in discussing a system's behaviour with others involved in its development.

For most simulations, the diagram is simply a convenient step on the way to converting the model into a computer coding which can then be run so that the results are automatically generated.

Any general purpose computer programming language such as FORTRAN, BASIC or PASCAL can be used for simulation modeling. However, because simulation is used for representing the dynamic behaviour of a system as it operates over time, features are needed which are not normally found in most general purpose programming languages. These features are provided by simulation software to give a natural framework for simulation modeling, and for making it easier to modify models and conduct experiments with them. The net result should be a reduction in programming time. The typical features needed are:-

*   a mechanism for managing time and advancing time,
*   a framework for model formulation and description,
*   means for representing variability and sampling,
*   means of inputting operating data,
*   means of collecting data as the simulation runs,
*   means of analysing the data collected and representing the analysed results back to the modeler,
*   means to monitor the program as it runs,
*   debugging and diagnostic facilities.

The availability and ease of use of these features can be used as the criteria for selecting a simulation language for a given application. Other user considerations in simulation language selection include the availability of vendor support, the ease of use for experimentation and execution efficiency.

Before proceeding further with considerations of languages and alternatives to languages, other elements of the terminology of the modeling framework need to be explained.

A basic step in developing a simulation model is the selection of a conceptual framework for describing the system to be modeled. A real system is conceptualised with a model and this terminology refers to the model rather than the real system. Models of systems are normally described as being continuous or discrete, and deterministic or stochastic.

1. CONTINUOUS and DISCRETE systems

The terms continuous and discrete describe the nature or behaviour of changes of state of the dependent variables in the model. Normally, time is the major independent variable and other variables are functions of time. When the dependent variables are modeled as changing discretely at specified points in time, the model is called a discrete change model. When they change continuously over time, the model is referred to as a continuous change model. Systems may contain both discrete and continuous variables, and can therefore be modeled effectively as a discrete or continuous system or as a combination of these. Most manufacturing systems producing individual products (ranging from aircraft to toys) can be modeled using discrete change models. The discrete changes in the state of the model typically occur at the start and end of all processes occurring in the model.

2. STOCHASTIC and DETERMINISTIC systems

A deterministic system is one in which the output of the system is completely determined by the input and initial values of system parameters or state of the system. In contrast, a stochastic system contains a certain amount of variability and randomness, and it may not even be possible to assign a positive probability to a specific output, given the input and the initial state of the system. The variability of the system behaviour is generally characterised by using a frequency (or probability) distribution and this distribution is sampled to obtain values to use in running the simulation. In this type of system therefore, the variability is typically reflected through to the output which can only be described in terms of a possible range of values. This may also be represented by a conditional probability distribution.

Most manufacturing systems of the metal cutting/batch manufacturing type can be modeled using the discrete change method but such systems also have various sources of randomness. They are therefore usually modeled using the stochastic approach. Examples of random processes in a manufacturing system are:-

* inter-arrival times of orders
* processing times of jobs on machines
* machine operating time between failures
* repair times for a machine
* outcome of an inspection, etc.

3. SIMULATION and EMULATION

Almost all models are representations of a real system. The accuracy of the representation varies however. One characteristic of systems which simulation is particularly

used to model is variability. Variability in the operation
of a manufacturing system can arise from the processes in the
system, from the human interactions with system, or from
external influences such as the non-arrival of raw material
when expected. The variability may be represented as a
histogram or as a frequency distribution. These may then be
used in the simulation and samples taken from them. For
example, a histogram or frequency distribution could be used
to characterise the range and form of variation of machining
cycle times for a number of different batches scheduled to
various machines. Each time a machining time is required, a
sample is taken from the distribution. Such a model is not
exact, it is representative,- this is simulation.

The same batches could have their exact cycle times at the
particular machines they visit known. Data on these
combinations could be stored and accessed when required so
that the model of the operation of the manufacturing system
is more precise. Models and simulations that are set up to
model systems in reasonably precise detail are known as
emulations. Not all workers draw the distinction between
simulations and emulations, but it is always important to
know the precision of a model. The distinction does not
typically occur in computer languages which are available to
transfer models on to computers. Most languages can be used
to model at the simulation or the emulation level.

III ELEMENTS OF A DISCRETE SIMULATION MODEL

In discrete change simulation, a system is viewed as
comprising various objects termed ENTITIES which have
ATTRIBUTES. The entities interact under certain conditions
in ACTIVITIES creating EVENTS that change the STATE of the

system.   The static structure of a system is described by
identifying the classes of objects that compose the system,
the number of objects in each class, and the relationships
among the objects within a class and among the classes of
objects.   The dynamic structure of a system is usually
described by the behaviour of the dynamic objects and their
activities in the system, and the conditions that are
required for the occurrence of events which change the state
of the system.

Entities: In a batch manufacturing system, typical entities
are the parts or the batches of parts, the machines, the
operators and workhandling and storage devices.   Entities are
either permanent or temporary.   Permanent entities are those
which are in the model for the entire duration of the
simulation run.   Temporary entities only exist for a while
(for instance, they are created or arrive, pass or flow
through the system and are then destroyed at some time).   In
a machine shop, the machines are examples of permanent
entities, while the batches are temporary entities.

Activities: Entities in a simulation model engage in
activities (eg batches and machines can be involved in a
machining activity).   In most activities, more than one type
of entity is involved.   An activity can therefore be
described as the coming together of two or more entities for
a period of time.   A fundamental assumption in simulation
modeling is that activities begin and end instantaneously.
These instantaneous happenings are also termed events.

Events: An event is an instantaneous occurrence which changes
the state of a system, for example when one or more
activities begin or end.   The state of the model as a whole,
and of the entities and queues may change at each event
time.   Simulation languages may characterise the operation of

a system as a series of activities or a series of events. A third characterisation is in terms of processes.

Processes: A series of related activities which refer to a particular entity is referred to as a process. For example, the sequence of machining operations (activities) such as facing, turning, drilling and tapping performed on a raw forging as it progresses through a machine shop are referred to collectively as a process.

Queues: Queues are the passive states of an entity, while it waits for conditions to change so that it can participate in another activity. Queues have a discipline governing how entities join and leave, for example, first in, first out (FIFO), last in first out (LIFO) etc.

Attributes: Attributes are the named properties or characteristics of entities such as the type of machine or the number of operations required on a job. They are usually expressed numerically. They are used to distinguish entities of the same type from another, and also to distinguish between entities in the same group. For example, to select a part from the input buffer of a machine tool, we can use the part's attribute 'type' or 'due date' as a criterion for selection.

Sets: In simulation, a set is used to refer to a group of entities, for example, parts in a queue, tools in a tool magazine or the machines in a cell. Sets are therefore used to group entities in any convenient way.

States: Generally, the term state is used to refer to the condition of the model or its entities at any instant. During simulation, the state of a system is updated at each event time, and it is normal to test the state of the system before deciding on a course of action. In activity based simulation, there are two main states,- the active and passive states. These refer to when entities are engaged in an activity or are waiting in a queue.

IV SIMULATION MODELING WORLD VIEWS

In discrete simulation, a system can be modeled in one of three ways, termed the world views or modeling orientation. These refer to the approach which discrete change simulation languages use to describe the static and dynamic structure of a system and to effect changes in the modeled system. These world views are referred to by terms just described.

Event: A system is modeled by describing the changes in state at event times. The modeler must define events, and develop and program the logic associated with their occurrences.

Activity: A system is modeled by describing the activities in which entities in the system engage. The modeler defines the conditions necessary to start and end each activity in the system. As with event orientation, the modeler must develop and program the logic associated with each activity.

Process: A system is modeled by describing the series of processes in the system through which entities in the system flow. For instance, the flow of a part through a manufacturing line consisting of several workstations may be represented by a process. The user describes the flow of parts through these processes which are then translated into equivalent sequence of events and activities by the language during a simulation.

Process oriented languages may be classified into three types:-

* transaction type, in which entities flow through stylised activity blocks,
* network type, which employs activity-on-branch, and nodes

for modelling queues and decision points,
* statement type, in which the modeler must define and
program each process.

All discrete change simulation languages are based on one
or a combination of these world views.

V. SIMULATION LANGUAGES and PACKAGES

So far, the only means identified to computer code a
system model has been a simulation language. These can be
described as special purpose programming languages with their
own vocabulary, syntax and programming structure for
performing the basic functions of a simulation. Simulation
languages can, however, take a long time to learn and the
user also needs a basic knowledge of computer programming and
simulation concepts. There are other approaches to providing
simulation tools and these, together with simulation
languages, are classified in Table 1 in terms of their world
views. Notice that some languages provide the framework for
combining two or more of these world views, and thus provide
greater modeling flexibility. The other types of simulation
software are now described.

Simulation packages

These are a collection of routines written in a computer
language (such as FORTRAN or PASCAL), which provides the
basic simulation functions. Most notable examples are GASP,
SLAM and SEE-WHY, which are primarily event oriented. With
these packages, the user develops a model by writing routines
to characterise the events and logical relationships of the
entities and calls the package routines to handle the
simulation functions.

Code generators

These are software tools which pair up directly with a

S/No.	WORLD VIEW	LANGUAGES/PACKAGES
1	EVENT ORIENTED	SLAM II (2), MAP/1 (3), TESS (4) GASP IV (5) SIMSCRIPT II.5 (6) SEE-WHY, WITNESS (7)
2	ACTIVITY ORIENTED	CSL/ECSL, CAPS (8) SIMON (9), DRAFT (10) GENETIK/OPTIK (11) FORSSIGHT/FORGE (12) HOCUS (13)
3	PROCESS ORIENTED	SIMULA (14) GPSS V, GPSS H (15) Q-GERT (16) SIMAN (17), CINEMA (18) SIMSCRIPT II.5 (6) SIMFACTORY (19)
4	COMBINED	SLAM II (2) GASP IV (5) SIMSCRIPT II.5 (6)

TABLE 1: SOME DISCRETE SIMULATION LANGUAGES, PACKAGES AND WORLD VIEWS

particular simulation language and generate code in that language. They accept a simple description of a system (usually in a graphical form such as activity cycle diagram), and will lead the user through a dialogue to learn of the logic of the system. The code generator will then check this for consistency before using the information to generate code in its related simulation language.

Simulators

The modeling process with simulation languages, packages and code generators can be very complex and time consuming. It

takes relatively long to develop a model, code it, debug it
and to verify that it works as intended. In contrast,
simulators already contain the logic defining the operation
of various components of a manufacturing system and their
interactions. The user simply has to tailor these pre-
packaged features to correspond to the system to be analysed.

Simulation and graphics

Simulation languages were initially developed in the late
1950s and 1960s and run on mainframes. In the late 1970s and
1980s, language developers responded to the increases in
computing power and the advent of medium priced color
graphics screens either by writing new languages for or by
adapting their languages to take advantage of these
facilities. Graphics are beneficial in three main areas of
manufacturing simulation. The first is in model building
where they help to facilitate model definition and
debugging. The second is in improving communication between
model builder and user through the use of graphic display of
the physical system layout or the logical model during model
validation. And the final area is for displaying and
presenting results to user and management. In general the
use of interactive graphics in simulation greatly improves
the user friendliness and helps in the understanding of
simulation results.

Simulation workbench

The 1980s saw the coming of age of artificial intelligence
and in particular expert systems and knowledge bases. The
simulation workbench was conceived as a means to make all the
software tools and supporting information necessary for
carrying out a complete simulation study available to the
modeler in the form of an integrated environment. In such a
system, there is the potential for using an expert system

shell as a simulation consultant to a non-expert user for model building and experimentation. An example of a commercial implementation is the ISI [20], an intelligent simulation environment in which simulation models are constructed using interactive menu-driven graphics and a database of information related to model building. ISI is a SIMAN/SLAM language implementation. Another example of a similar approach has been reported by Schroer et al [21].

Having reviewed these simulation alternatives, attention can now be turned to the classification of simulation languages.

A very widely used classification scheme for simulation languages and simulation techniques in general, has been given by Shannon[22]. The scheme encompasses the analog, hybrid and digital (discrete and continuous) simulation techniques. The scheme presented here is for digital simulation software which is particularly relevant to manufacturing systems simulation and was adapted from Shannon's work. Digital simulation software can be classified at three different levels: the system, application and structural levels. Figure 1 shows the relationship between these levels and representative software.

At the application level, discrete simulation languages can be classified as special or general purpose. The special purpose packages are designed to model specific environments and they use terms which reflect the jargon of that environment. These packages are either code generators or pre-packaged simulation models which the user adapts to his particular requirements. The sophistication of the model and of the simulation which can be created depends on facilities provided by the package writer. On the other hand, general purpose simulation languages allow any system to be modeled,

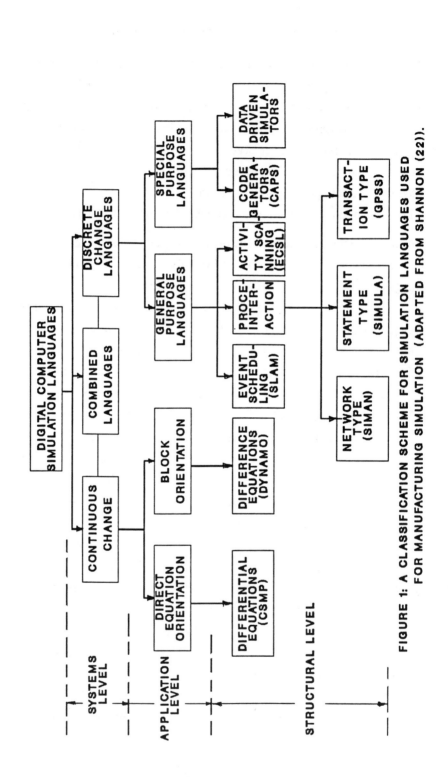

FIGURE 1: A CLASSIFICATION SCHEME FOR SIMULATION LANGUAGES USED FOR MANUFACTURING SIMULATION (ADAPTED FROM SHANNON (22)).

for systems to be simulated or emulated and for practically any type of analysis to be performed. It should be noted that the classification of languages into general purpose and special purpose is based on subjective judgement. Some code generators and simulators have wide applicability whilst some general purpose simulation languages provide modules that may be configured and used for specific applications.

At the structural level, simulation software products can be classified on the basis of the nature of the input and modeling orientation. While the nature of the input applies to special purpose packages which are classified as code generators or simulators (as they are data driven), the modeling orientation applies to the general purpose simulation languages. The three modeling orientations correspond to the world views just described.

VI ROLE OF SIMULATION IN MANUFACTURING

In a survey of the industrial applications of simulation in 1980, Christy et al [23] found that of the total applications of simulation in the USA, 59% were in manufacturing. The abundance of references to successful manufacturing simulation studies in the literature is a good indicator of the wide applicability and acceptance of simulation modeling techniques in manufacturing industry. As the complexity of manufacturing systems increases and new concepts evolve, the need to test design ideas and potential operational strategies before their implementation has also increased. The use of simulation allows the designer or system manager to obtain a system-wide view of the effects of changes in an existing or proposed manufacturing system by providing answers to "what if" questions.

When introducing new equipment to a new or existing manufacturing facility, and the balance of equipment and personnel to operate it must be assessed. The system hardware components must be selected to meet current or anticipated production requirements. The system configuration and the number and type of workstations and machines may be varied to determine the resources required for production. The material handling and transportation equipment, involving equipment such as AGVs, carts, conveyors and other support equipment such as pallets and fixtures must be selected to suit the part types and the processing requirements. In general, the type of equipment used will influence the nature of the operational control procedures. The type of workhandling and transportation equipment, and the production rates of the workstations influence the location and size of in-process storage.

As most systems are designed to manufacture a family of parts, the determination of appropriate balanced part mixes and schedules is critical to system performance. For example, the number of parts in the system at any instance must be compatible with the in-process storage capacity, as very small numbers may led to under utilisation of resources, while too many parts may cause congestion and blockage, increasing throughput times and work-in progress levels. Similarly the effect of different order entries and job arrival patterns, changes in part mix, and the impact of new orders can be evaluated through simulation experiments.

When operating existing manufacturing systems, greater efficiencies in production can often only be attained by using improved operational and control policies within constraints imposed by the fixed system hardware. The objective is usually to obtain better system performance levels using the same resources. Simulation allows the investigation of system behaviour under alternative loading and control strategies to ensure that the most effective alternative is chosen for implementation.

It must be emphasised that each change in system hardware or operating strategy will often need investigating with separate simulation runs. Simulation does not optimise, it allows the behaviour of a system to be analysed for a given set of resources and particular operating strategy.

Some of the performance evaluation measures that may be investigated in a manufacturing systems simulation study are:

i)   throughput or production per period,

ii)   throughput time or makespan,

iii) in-process storage or queuing times,

iv)   transportation times,

v)   in-process inventory size (queue lengths),

vi)   utilisation of resources (equipment/personnel),

vii) idle time of resources,

viii) rework and scrap level.

A brief look at the performance evaluation of a manufacturing system will conclude the case study on features of a simulation language which is presented in the next section. This is followed by a case study on a simulation package.

VII A SIMULATION LANGUAGE CASE STUDY

To demonstrate the typical features of simulation languages, a case study will be presented. This will be contrasted later with a short case study on a simulation package which also uses different world views. This study uses a representative simulation language to model a real manufacturing system typical of those for which simulation is used. The Extended Control and Simulation Language or ECSL is used for this study and the Leyland Vehicles FMS Facility at Farington is the system modeled. An overview of ECSL language features and the philosophy of its activity orientation is presented first to aid understanding of subsequent material. This is then followed by a description of the system modeled, the specification and details of the model developed, and the experiments used to analyse the typical problems.

ECSL is an activity activity-oriented language which was developed by Clementson [24]. With this activity orientation, Activity Cycle Diagrams (or ACDs) are used to describe the behaviour of a system. ECSL statements are mainly English-like, and the "simplified english" syntax makes the ECSL programs easily readable. It is also an aid in obtaining the understanding and commitment needed from management to make any simulation study a success.

The model of a system in ECSL is described in terms of classes of entities, which belong to sets and have specific attributes. The modeler must specifically declare the classes of entities, the size of each class, the possible states (ie queues or sets) of entities in each class, and the list of attributes associated with each class. The state of a system modeled in ECSL at any time is then given by a listing of entities belonging to each set in the system and the numerical values of the attributes of the entities.

The behaviour of a class of entities may be represented by a series of alternating active and passive states (ie activities and queues). A graphical representation of these states is known as an Activity Cycle Diagram (ACD) and one is shown in Fig.2. In these diagrams, queues are represented by circles and activities by rectangles. Connecting arrows are used to show the flow of entities in the system.

ACDs provide the modeler with the means for setting out the systematic outline of the simulation model. Its use can be considered complementary to the thought and narrative processes normally used by designers, engineers and managers when describing a system. The importance of ACDs for manufacturing systems simulation and other related applications have been summarised by Hutchinson et al [8] as follows:

i)   They provide a well defined method for decomposition of complex systems to enhance analysis and understanding,

ii)   They are a natural way of describing the behaviour of entities in a system,

iii)   They explicitly and rigorously depict the interactions of entities needed for the performance of activities,

iv)   They permit any desired degree of decomposition, allowing one to begin a macro-level model and to increase the depth of detail only when needed,

v)   They share with PERT graphs the ability to display interrelationships pictorially for ease of compre-hension and communication,

vi)   They explicitly indicate those points in a system where control can be exercised and system performance affected by management action,

vii)   They provide a rigorous structure for the development of either a system simulation or the manufacturing control software.

These features of ACDs provide the basis for inputting information to CAPS (Computer Aided Programming of Simulation), the code generator for ECSL.

ECSL programs which represent the simulation model of systems have a fixed structure, consisting of five main sections: definition statements, initialisation routine,

activities section, finalisation section, data section. An
ECSL statement is made up of KEYWORDS or COMMANDS and
VARIABLES. The values assigned to the variables describe the
instantaneous state of the system being simulated. The
activities section of the program describes the way in which
these variables are changed over time to represent the
dynamic behaviour of the system. A brief description of each
of these sections is presented later to emphasize language
features and capabilities, and their suitability for
describing and simulating manufacturing systems.

A. ECSL Program flow and control

The ECSL program flow is based on the use of "time cells"
and time attributes. Each definition of entity classes
automatically creates time cells associated with the entity
class. The time cells are storage locations associated with
entities, which hold the relative time at which the entity
will or has become available to participate in the activities. Time cells are thus used to keep track of various
events in the system. When a time cell is positive, it holds
the relative time of the next event associated with the
entity. When it is negative, it holds how long ago the last
event associated with the entity occurred.

The ECSL program of a system's model comprises one or more
activities. Each activity is headed by a concatenation of
tests, which generally include one or more tests involving
time. The satisfaction of the tests, and the availability of
the various entities to take part in an activity are the
necessary conditions for an activity to start.

At the start of program execution, the ECSL processor
translates and executes the definition statements, loads the
first data block and then executes the initialisation

section. The processor scans all time cells of the entities to locate the minimum positive value (the time of next event, DTNOW). The values of all time cells are then reduced by DTNOW, and the clock variable representing the simulation time is increased to the time of the event associated with the entity whose time cell has been reduced to zero. If the ADD switch is on, any recording statements are executed before the queues and sets in the system are updated. The processor then transfers control to the first activity in the program.

The program cycles though all the activities in a sequential order, executing only those activities whose conditions are satisfied, until no more activities can take place at the simulation time. Control is then returned to the processor, which repeats the time cells scan, increases the simulation time to the time associated with the next activity to end and then starts the execution of activities at the new simulation time. This sequence continues until the simulation time reaches the run length specified in the activities statement, or satisfies any of the control exception statements for stopping the run.

If more than one data block exists, then the program is run once for each data block during one program execution.

B. Manufacturing system description

The FMS to be modeled and simulated is used for producing Hydracyclic Gearbox casings. Prior to its installation, the casing were machined by 3 NC-Machining Centers and 17 non-NC machines. The Company expected a threefold increase in demand for the gearboxes and this called for a substantial increase in production capacity. Also the shorter life cycle of the new products, the smaller production volumes, and regular engineering design changes necessitated installing a flexible facility, rather than extending existing facilities. The requirements for the new FMS included operating unmanned for short periods. The simulation study on which this case study is based was carried out at the feasibility and planning stage. The FMS has since been implemented in a slightly different form, and is now being successfully operated [25].

The system proposed consisted of 7 machining centers (5 horizontal and 2 vertical). The machining centers were to be laid out on both sides of a single, bidirectional rail track. Two wire-guided Robotrucks under computer control were to be used for transporting palletised and fixtured components between the machining centers. Each machining center was to have a twin indexing worktable to serve as an input-output buffer. Standard non-NC machines were to be used for preliminary and finishing operations but these non-NC machines are not part of the FMS.

Components from the preliminary area were to be manually loaded onto standard palletised fixtures. These were then to be loaded automatically onto Robotrucks, for transportation to the appropriate machining center. The Robotrucks also picked up machined components for manual re-fixturing if further operations were required or for de-fixturing when all operations were completed.

The FMS was initially used for producing six different component types. A repeating shift pattern was proposed in which the first shift of 8 hours was to be manned, the second shift of 4 hours was to be unmanned and designated for machining long cycle-time components for as long as there were palletised components available. Over 24 hours, this shift pattern gave a total of 8 hours of unmanned operation.

C. The simulation model

As ECSL is an activity-oriented language, the operation of the FMS was first modeled in terms of an activity cycle diagram. The stages in the construction of an activity cycle diagram can be summarised as follows:

i)   Identification of the main entities in the system, determination of the relevant attributes of the entities and system characteristics required to model the system,

ii)   Determination of the activities which the entities in the system engage in, and the queues in which they wait for participation in activities,

iii) Listing and drawing of the life cycle diagram for each entity class,

iv)   Combination of the life cycle diagrams for all the entity classes to form the ACD of the system,

v)   Hand simulation may then be used to check entity flow and model logic.

The entity classes which it was determined to model were:

*   MACHIN: machining centers with indexing worktable,
*   LABOR: pallet loading/unloading personnel,
*   RTRUCK: robotruck for component transport,
*   COMPTS: gearbox components,
*   APALLT/BPALLT: pallets and fixtures.

The activity cycle diagrams for each of the entity classes were drawn out separately as a closed loop of activities and queues and then combined. The combined ACD developed for the FMS is shown in Fig 2. It consists of 13 activities and many queues, entities classes always being in separate queues. A description of two of the entity cycles is now given as a help to understanding the complete ACD.

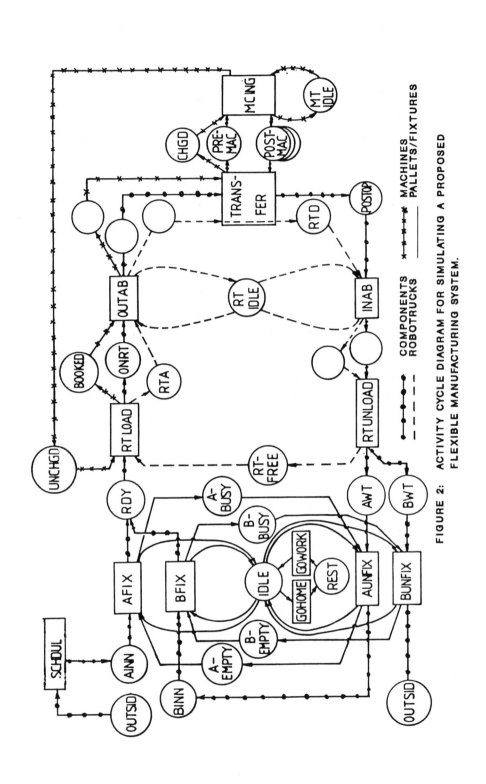

FIGURE 2: ACTIVITY CYCLE DIAGRAM FOR SIMULATING A PROPOSED FLEXIBLE MANUFACTURING SYSTEM.

COMPONENTS •—•—•
ROBOTRUCKS – – – –

MACHINES ×—×—×
PALLETS/FIXTURES ————

1.THE ACD FOR COMPONENTS

The entity class COMPTS models the behaviour of the components in the FMS. The cycle which shows the activities which the COMPTS engage in, and their associated sets/queues is given by the full line with small circles on it.

During the initialisation stage of the running of the model, all COMPTS are placed in the queue OUTSID. The component type mix and loading the system for a particular schedule is specified by the activity SCHDUL. Batches of COMPTS are loaded into the shop by the activity at pre-specified intervals (provided the maximum work-in-process is not exceeded). The COMPTS are then manually fixtured onto pallets (represented by the activity AFIX). The fixtured/palletised COMPTS are kept in the queue RDY.

The loading of components onto robotrucks (activity RTLOAD), and their transportation to the machining centers (activity RTOUT/TRANSFER) is initiated only when a machining center and robotruck are available. The components awaiting machining are held in the sets PREMAC which are associated with each MACHIN. These sets represent the input buffers (shuttles). Components whose operation sequences are complete are held in the output buffer, modeled by the sets POSTMAC. The activity TRANSFER models both the transfer of unmachined components from the robotruck to the input buffer and the transfer of completed components from the output buffer to the robotruck. The robotruck can also travel empty to the machining centers output buffer to collect completed components.

The activity MCING represents the machining of components on the machining centers. The logic for the sequencing of components from the input buffers to the machine and their despatch to the output buffers after machining is incorporated in MCING. The activity also models machine breakdown and repair characteristics.

Machined components in the queue POSTOP may either be transported to the defixturing/depalletizing area (queue BWT) if all operation sequences are completed or to the defixturing area (queue AWT) if more operations are required. Activity BUNFIX models the defixturing/depalletizing of components before their despatch to the queue OUTSID. Activity AUNFIX models the defixturing of components which require additional machining operations, and the routing of the components to the queue BINN where they await refixturing.

The interaction of the COMPTS entity with other entity types in the model can be seen in the combined ACD.

2. ACD FOR PALLETS

Specialised fixtures associated with each different component type are used to hold components onto the pallets and also to aid certain machining operations. The full line in Fig 2 indicates their model sequence. In most FMS, the fixtures stay permanently attached to their pallets so that the fixtures and pallets can be modeled together. This is done in this model for parts of the cycle where they are joined. During this period, pallets are held in the sets ABUSY and BBUSY. For coding convenience, two entity classes model the flow of pallets through the FMS. The APALLT entity class represents the pallets used for the first operation sequence, while BPALLT entity represents pallets used for the second and subsequent operation sequences.

The queues AEMPTY and BEMPTY hold empty APALLT and BPALLET respectively awaiting fixturing. The activities AFIX and BFIX (representing fixturing operations) and the corresponding activities AUNFIX and BUNFIX (representing defixturing and defixturing/depalletizing respectively) were described earlier for the COMPTS entity.

3. ACD FOR OPERATORS

Human operators are required in the FMS for the manual activities of fixturing and defixturing. The operators are represented by the entity class LABOR and are represented by the curved full lines to the left of the ACD. The LABOR entities are held in the set IDLE when they are waiting to load/unload COMPTS. They are held in set REST when not on shift. The activities GOWORK and GOHOME model the scheduling of the duration of shifts and the number of tea/lunch breaks. During the manned shifts, labor is available for the manual activities AFIX, BFIX, AUNFIX, and BUNFIX. The function of these activities was described earlier for the COMPTS entity.

The logic and sequence of the cycles for the robotrucks and the machining centers should be apparent from the descriptions already given.

D. The ECSL coding

This section gives details of how the ACD is represented in the ECSL language. The objective is to highlight the facilities provided by ECSL for modeling, to illustrate how they were used, to give a measure of the effort involved in the modeling and the benefits achieved in using the model.

1. DEFINITION STATEMENTS

The definition statements form the first section of any ECSL program. They are used to declare the classes of entities, the class size, the entity attributes and sets to which each entity may belong. They are also used for declaring the variable names and special ECSL library functions that are used in the model. In the following examples of the definition statement types, ECSL Keywords are underlined:

<u>CLASS</u> 7 MACHIN <u>SET</u> BOOKED CHGD UNCHGD MTIDLE <u>WITH</u> MACNO OP
<u>REAL</u> <u>ARRAY</u> MCTIME(20)
<u>FUNCTION</u> PICTURE NORMAL RANDOM

The first statement declares an entity class MACHIN of size 7 for the seven machines. Machines can be of any type and could include washing and inspection machines. The statement also specifies four sets (or queues), namely BOOKED, CHGD, UNCHGD and MTIDLE to which members of the entity class may belong. The sets may be used for descriptive purposes or for representing queues prior to activities on the entity class life cycle. For example, the set BOOKED holds the list of MACHIN entities that have been assigned definite operations, while MTIDLE represents a queue of MACHIN entities temporarily idle and so available to start another machining (MCING) activity. Finally, the statement specifies two attributes of the entities in the class, MACNO and OP. The attributes are used to give greater detail about individual entities in the class. For instance, the attribute MACNO is used to hold the MACHIN numbers of each machine.

The various entity class definition statements in the program declare the maximum number of entities that are represented by the model. For the FMS model, in addition to the 7 machining centers (MACHIN), we have:

2   robotrucks (RTRUCK)
200   components (COMPTS)

20   pallets (APALLT and BPALLT)

2   operators (LABOR)

ECSL uses integer time increments, consequently, the default variable type is integer. As a result all real numbers must be declared. The second definition statement specifies a real array MCTIME, with the array dimension 20. The array entries hold the machining cycle times for the component types.

The third definition statement defines the ECSL library functions that are used in the model. The function PICTURE is used for specifying a histogram output; NORMAL is used for drawing random variables from a theoretical normal distribution, and RANDOM is used to draw variables at random (within a specified range of numbers). ECSL provides most of the commonly used theoretical probability distributions.

2. INITIALISATION ROUTINE

These statements (along with the data statements), establish the initial conditions for the simulation run. The statements specify the initial values of attributes and which entities are in which sets. At start up, all model variables are given default values of zero or equivalent (ie. blank for strings, true for booleans, and empty for sets). After the first data section is loaded, the initialisation routine is executed to give specific values to some variables. At the end of program execution, if there are further data blocks, the next one is loaded, and the program execution restarted at the initialisation phase. The system variables specified in the data section will be discussed later.

For large and complex models, a large number of parameters are needed to define the initial model state, and it is usually difficult to change all these values using data statements. The data can be stored in a data file which is then read during initialisation. A front-end, interactive keyboard data entry facility can be developed for accepting keyboard data and them storing these in the data file.

3. ACTIVITIES SECTION

The activities section forms the main body of the ECSL program. It consists of three main subsections, namely:
*   dynamics   sector,   *   recording   sector,   *   activity subprograms.

The DYNAMICS sector is only used if it is necessary to represent some continuous processes. No continuous processes were modeled in the FMS, and because most metal cutting type manufacturing systems are discrete in nature and are usually operated in batch mode, the dynamics sector is rarely used.

The RECORDING SECTOR statements are largely used for collecting statistics on the state of the system. They are usually ADD statements and the intermediate calculations necessary to support statistics collection. The statements are executed after each time advance. The simulation run-in period and the ADD switch can be used to control when statistics collection can start. To illustrate the use of RECORDING statements, an example of two statements used in the FMS model are:

ADD MTIDLE TO HIST MTIDLE DURATION

ADD TIME IN IDLE TO HIST MENIDL(20,2,4)

The first statement determines the number of MACHIN entities in the set MTIDLE at the time the statement is executing and adds this to the histogram MCDOWN. The second automatically records the time spent in the queue IDLE by all LABOR entities that pass through the queue in the histogram MENIDL.

Each activity in an ECSL program is concerned with one type of state change of the system. An ACTIVITY BLOCK is written for each of the activities in the ACD. The start of an activity block is denoted by the statement BEGIN, which identifies the activity by name. The structure of a typical ECSL activity consists of the following parts:

a) statements giving the conditions under which the activity can start. These include statements testing state variables, attributes of entities and set membership,

b) statements which represents the change of state at the start of the activity. These normally include those for calculating the duration of the activity,

(c) statements which represent the changes which are necessary at the end of the activity. These include changes in set membership, attributes of entities and state variables.

To highlight specific ECSL features, the coding of two of the activities in the FMS simulation model are now discussed.

a. The AFIX and BFIX activities: These model the manual fixturing of components onto pallets associated with the component types. Component type numbers are assigned to the M and P attributes of the APALLT and BPALLT entities in the DATA section and these are used for matching components to the pallet/fixture types required for the operation sequence.

The compound statement used for matching a BPALLT entity to a corresponding COMPTS has the form:

FIND FIRST BPALLT B IN BEMPTY
  FIND FIRST COMPTS A IN BINN
  P OF BPALLT B EQ N OF COMPTS A

The availability of LABOR entities for the manual AFIX/BFIX activities is checked by the statements:

TIME OF DEPARTURE GT 10

EXISTS (1) LABOR IN IDLE

The first statement tests a time variable which is counted down by the ECSL processor as the CLOCK advances to check that enough time is available to complete the activity prior to the LABOR entities departing from the shift, while the second statement checks that a LABOUR entity is available.

b. RTLOAD activity: This models the loading of palletised components onto robotrucks for transportation to the machining centers. An essential requirement for the FMS was partially unmanned operation. The RTLOAD activity models the loading during the manned and unmanned shifts. For unmanned shifts, components with long cycle times have higher priority. A sample of the program coding for selecting COMPTS for loading during the manned and unmanned shifts is given below:

 * Manned shift:

FIND MACHIN A IN UNCHGD WITH MIN TIME OF MACHIN

  FIND COMPTS B IN RDY WITH MIN PRIORITY OF COMPTS

  COMPTS B IN OPBLST OF MACHIN A

 * Unmanned shift:

FIND MACHIN A IN UNCHGD WITH MIN TIME OF MACHIN

  FIND COMPTS B IN RDY WITH MAX PRIORITY OF COMPTS

  COMPTS B IN OPBLST OF MACHIN A

These two compound statement select the machining center with the minimum remaining cycle time (represented by TIME OF MACHIN). It then selects the component with either minimum priority (manned shift) or maximum priority (unmanned shift), and finally checks that the component can be machined on the selected machining center.

These two activities illustrate how entities are selected

and matched. Once this has been done, the duration of the activity is determined. In this model, the machining times (ie the DURATIONs of the MCING activity) are found by using data in the ARRAY MCTIME. Other durations would be obtained by sampling from a frequency distribution. Statements such as MACHIN A FROM PREMAC INTO POSTMAC AFTER DURATION are used to manipulate the entities between sets.

4. FINALISATION SECTION

The statements in this section are used for producing the output of the simulation run. Consequently, the statements are only executed at the end of a simulation run. A simulation run ends either when the simulation time reaches the run length specified by the ACTIVITIES statement, or a FINISH statement is executed anywhere in the program.

The statements for the output of reports contain keywords which are standard FORTRAN output commands, such as PRINT, PUNCH, and TYPE. These keywords implicitly specify the output device, which may be an internal data file (ie ECSL.MOU), or the output screen on the run terminal. Limited FORTRAN formatting capabilities are provided for producing project specific reports. The output from a run of the FMS model includes work-in-progress levels, the number of gearbox components produced, the utilisation of the machines, the robotrucks and the personnel, failure rates and repair times for machines, and histograms showing the queue sizes and queuing time. A sample output from a run of the FMS model is given under experimental results.

5. DATA SECTION

The data section of an ECSL model may consist of one or more data blocks. Each block starts with a statement, consisting of the single keyword DATA, and contains the initial values of the system variables, attributes and set memberships that constitute the initial state of the model. The data block is executed before model initialisation. For a data section consisting of more than one block, the simulation is run once for each block within one program execution. The final data block is followed by the END statement, which is the last statement of the program.

The ECSL model of the FMS contains only one data block. The block consists of statements for:

a) declaring set memberships for the entities MACHIN, RTRUCK, LABOR, APALLT, BPALLT, and COMPTS,

b) assigning APALLT and BPALLT to COMPTS,

c) specifying the machining time for COMPTS,

d) specifying the priority levels for COMPTS,

e) specifying the durations for some activities,

f) specifying the starting seeds for the random number generator for sampling from statistical distributions used in the model. A different seed is specified for each sampling statement to minimise covariance.

E. Experimental procedure and results

A model once coded has to be checked by verifying it and then validating it. Then it can be run and experiments carried out. The experiments on the FMS model comprised four main parts, each of which related to the design and sizing of the system. These investigated the effect of varying the number of the following resources: machining centers, pallets and fixtures, robotrucks, and labor.

In each experiment, the model was simulated for a period of 7,200 simulation time units, (1 unit = 1 minute), which is equivalent to a 24 hour, 5-day week. A run-in period of 1,440 simulation time units or 24 hours operation was used.

The performance reports collected included:
the total number of gearboxes components produced,
the utilisation of the machining centers,
the utilisation of the robotrucks,
the utilisation of labor.

The performance reports were obtained with the FINALISATION section. A typical result is shown graphically in Fig 3 which shows the effect of the number of machining centers used on the number of gearboxes produced and on the utilisation of other resources in the system.

UTILISATION   %
AND NUMBER OF
GEARBOXES

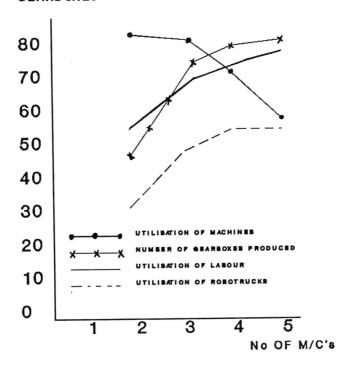

FIGURE 3 EFFECT OF VARIATION ON
NUMBER OF MACHINING CENTRES

The results show that increasing the number of machines used increased the number of gearboxes produced over the preset simulation period but this, and the utilisation of other resources leveled off at over 3 machines.   The number of LABOR, their shift pattern or the number of pallets thus need to be increased to meet the target production level of 90 gearboxes per week.   Further experiments were performed to find the appropriate balance of resources needed to meet this production requirement.   The results showed that the best balance was obtained by the use of the following resources:
*   4 machining centers,   *   3 robotrucks,
* 27 pallets/fixtures,   *   3 labor units.

This alternative is preferred because it meets the production requirement using the lowest number of resource units, and therefore the least investment.   In addition, it gives very good utilisation levels for all the resource types in the system.

This brief review of some of the results obtained illustrates the form of the experimentation and analysis involved in simulation.   In the next section a case study on a simulation package is described.   This package is capable of being used to carry out very similar experiments so these are not presented again.

VIII A SIMULATION PACKAGE CASE STUDY

The second case study is based on SLAM, the Simulation Language for Alternative Modeling.   SLAM permits the features of a simulation package to be demonstrated together with the discrete event and process world views which were not covered in the ECSL case study.   (A comparison of ECSL and SLAM and their world views can be found in Ekere and Hannam [26]).   A graphical network is used to represent the process model.   An example will be given later. An overview of SLAM is presented first, followed by a description of the manufacturing system modeled and details of the model developed.

SLAM is an advanced FORTRAN based software package. It permits the modeler to use discrete event, network/process interaction, continuous modeling perspectives and any combination of the three. In SLAM, a discrete change system can be modeled from the point of view of events or processes, or both at once. Continuous change systems can be modeled using either differential or difference equations. Combined discrete-continuous change systems can be modeled by combining the event or process approach with the continuous approach. As most manufacturing systems of the metal-cutting, batch manufacturing type are discrete and stochastic in nature, only the network/process interaction and discrete event features of SLAM are covered in this case study.

A. Network/Process modeling

The process oriented part of SLAM employs a network structure comprised of specialised node and branch symbols for processes such as queues, server activities and decision points. The modeling task consists of incorporating these symbols into a network of possible paths (including delays) that an entity encounters from its arrival to its departure. The possible paths form the network model (a pictorial representation of a process). The entities flow through the network model waiting for servers at QUEUE nodes, waiting for resources at AWAIT nodes, and incurring delays associated with service operations and transport times. To simulate a network model, the pictorial representation of the system must be transformed into the equivalent SLAM input statements for the SLAM processor.

Some examples of SLAM nodes and their related parameters are:

(i)   CREATE node: CREATE, TBC, TF, MA, MC, M;
      This creates up to MC entities, one every TBC time
      units. The first entity is created at time TF. The
      time of creation is stored as attribute MA of the
      entity, and the entity takes M of the emanating
      branches.

(ii) QUEUE node: QUEUE(IFL), IQ, QC, BALK(NIBL) or
      BLOCK,SLBL;
      Entities wait for service at file number IFL. There
      are initially IQ entities in the queue, whose server is
      at node SLBL. The capacity of the queue is QC and if
      this is exceeded, entities are sent to node NLBL or the
      system is BLOCKed.

(iii) AWAIT node: AWAIT(IFL), RLBL/UR or GLBL, M;
      Entities that require the resource RLBL wait for
      service in an AWAIT node which specifies the number of
      units of resource needed. They wait in file IFL, for
      UR units of resource RLBL or for the gate GLBL to
      open.   The entity is then routed to M emanating
      branches.

(iv) FREE node: FREE, RLBL/UF, M;
      The FREE node releases UF units of resource RLBL. The
      units are made available to entities in AWAIT and
      PREMPT nodes.

The SLAM II Language Manual [2] gives a detailed
description of other network statements.

B. Discrete event and combined models

In the discrete event part of SLAM, the modeler defines
the events that occur at discrete times by writing FORTRAN

routines for the mathematical and logical relationships that describe the changes that each type of event produces. The modeler also uses standard subprograms that SLAM provides to perform common functions such as scheduling events (SCHDL), manipulating files (FILEM), collecting statistics (COLCT), and generating random samples (DRAND). SLAM also provides an executive program which automatically controls the advancing of simulation time, and the order in which the event routines are processed.

The combining of network/process and event models is made possible by the EVENT and ENTER nodes which provide the interface points between the network and event parts of a combined model. SLAM also provides subprograms which allow the modeler to change the status of the network elements from the event subprograms. For example, by FREEing resources, ALTERing resource capacity, STOPping activities, and specifying activity durations through the user function USERF(IFN). Other functions are provided for testing or changing the value of variables associated with the network part of the model from the event part.

C. The SLAM processor

The SLAM processor plays a major role in the interfacing of the network and event features. The network model statements are read in as data by the processor, interpreted and then translated into appropriate variables and storage array entries. The event part FORTRAN routines are complied separately and then linked to the processor. The processor can then call up the user-written event routines automatically at run time. The SLAM processor is written in FORTRAN and consists of an EXECUTIVE and supporting subprograms for network, event and continuous system models, and for any combinations of these. It performs two forms of control. In one, it directs the program into its various modes (for example, model initialisation and model status monitoring), and in the other, it operates within the model and sequences the execution of network, events and continuous system state changes.

The processor keeps track of the simulation time by maintaining a time ordered sequence of entity arrival and joint events in an event calendar. The calendar consists of a list of events, with an event time and an end node of event code as appropriate. The event time specifies the time of entity arrival to the node or time of the occurrence of a joint event. The event code gives the joint event number while the end node gives the node to which the entity is next to arrive. The entries in the calendar are ranked on the Low-Value-first rule, based on the event time.

D. Description of the case system

The system used for the case study is the FMS facility of a major US manufacturer. The FMS is a random manufacturing system which consists of the following:-
* 10 machining centers
* 3 load/unload stations
* 10 machine shuttle units
*   pallets and workpiece fixtures
*   tow-line powered carts for work handling.

Control of the system is handled by two mini-computers. One serves as the DNC computer, and stores and distributes all the NC data to the machining centers. The second computer performs supervisory control functions for the entire FMS. It controls the movement of the materials handling carts and their associated pallets/parts, and

handles   communications   with   the   manual   load/unload
stations.Part processing within the system is accomplished by
routing palletised parts via the tow-line carts to the
various machining centers. Machine shuttles provide the link
between the machining centers and the workhandling system.
Each machining center has a shuttle unit which provides
storage space for two pallets (one on the input side, and the
other on the output). This buffering action helps to isolate
the machining centers from the operation of the workhandling
system.

E. Specification of the simulation model

The purpose of the simulation study was to explore
different sequencing rules and to determine which rules give
the best throughput, for a given combination of part types.
The first stage in the model specification is to identify the
major system components whose characteristics and behaviour
influence the performance of the system.   These were
identified as:
* parts (incorporating pallets/fixtures)
* transportation carts
* workstations (including machines and load/unload stations).
The control computers were not included in the model because
the model is used for investigating the decision processes in
which they are involved.   The operators were also not
included in the model as the system was to be continuously
manned.

The initial conceptual development of the model was
achieved by the use of the SLAM network features using
symbols for characterising entity flows through processes in
a system. The system elements which could not be modeled by
the network approach were subsequently modeled via the event

approach to give a combined network/event model.

In the network modeling approach, the primary system component which flows through the system and interacts with most of the other components is modeled as the entity. Since the parts/pallet are transported by carts, loaded/unloaded by operators and are processed by machining centers, they are the primary component type and are modeled as entities. The other system components are then viewed as forms of resources. The workstations remain stationary, and are only associated with the entity while it is flowing through using them. Machining is therefore modeled as service activities. The carts are acquired by entities through a set of activities, flowing from one activity to another. They are modeled as resource blocks.

The network part of the model is written in SLAM network input language, and primarily models the flow of parts (scheduling and part input, loading/unloading, machining at workstations, and despatch of machined parts). The discrete event part is written in FORTRAN, and comprises a set of routines which model complex decision processes such as selection of alternative stations, model initialisation, front-end model input/output, and the interface between the network and event parts of the model.

F. Network part of the model

The network part consists of five major sections, each of which model specific system features:

(a)   Create blocks: models part arrival into the system
(b)   Input schedule block: models parts input to load stations
(c)   Workstation blocks: models processing at workstations

(d)   Despatch block: models data collection and part
      despatch

(e)   Input control statements: specifically for program
      setup.

Part of the network diagram developed is shown in Fig 4 and
it contains nodes and activities. To illustrate how the
network symbols are translated directly into network
statement, the program for the Create and Workstation blocks
are now given and briefly discussed.

1. CREATE BLOCK

The entities (parts) are created at seven different CREATE
blocks (a block for each part family). The create block for
part family 7 is given below:

```
PT7 CREATE,12.347,0,1,100;
    ASSIGN,ATRIB(2)=7,ATRIB(5)=USERF(3),1;
    ACT,ATRIB(5).EQ.0,OUT;
    ACT,ATRIB(5).NE.0,IBUF;
```

The first creation (or arrival) of part type 7, indicated by
the label PT7, takes place at simulation time 0.0. The time
interval to the next creation is 12.347 time units. The time
of arrival of each entity is stored as the first attribute
ATRIB(1) of the part, and this is used subsequently for
calculating throughput times and for making sequencing and
routing decisions. In Fig 4(i), the CREATE node is the first
symbol shown above the label PT7 which is boxed. The ASSIGN
statement, (the next statement and the next rectangular type
symbol) stores the part type as ATRIB(2) and utilises the
function USERF(3) to check if the order quantity have been
exceeded. A value is assigned to ATRIB(5), which acts as a
flag showing where the entity is to be routed. Two branching
activities then occur, represented in the network

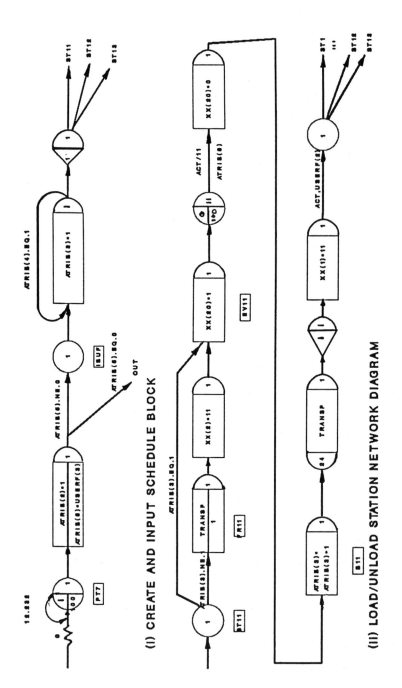

(i) CREATE AND INPUT SCHEDULE BLOCK

(ii) LOAD/UNLOAD STATION NETWORK DIAGRAM

FIGURE 4 SAMPLE NETWORK DIAGRAM FOR FMS SIMULATION

by arrows with the text. These test in turn if ATRIB(5)=0, in which case the part is sent to the output buffer OUT; if ATRIB(5) =/ 0, the part is sent to the input buffer IBUF. The next nodes and arrows give the input schedule block.

An alternative scheme to the create block is the ENTRY block. In this approach, ENTRY statements are used to insert entities and their associated attributes into a specified file. The file must be associated with a QUEUE or an AWAIT node, and the entity is released at the start of simulation or at predetermined times.

2. WORKSTATIONS BLOCKS

There are two categories of workstations in the model: load/unload stations and the machining workstations. At the load/unload stations, the parts are also clamped and refixtured between intermediate operations. The network diagram for a load/unload station is shown in Fig 4,(ii). An corresponding program extract for one of the three load/unload stations is now given:

```
     ;Load/Unload station number 1.
      ST11 GOON,1;
           ACT,ATRIB(3).NE.1,FR11;
           ACT,ATRIB(3).EQ.1,EV11;
           FR11 FREE,TRANSP/1,1;
           ASSIGN,XX(2)=11;
      EV11 ASSIGN,XX(20)=1,1;
           QUEUE(11),0,100;
           ACT/11,ATRIB(6);
           ASSIGN,XX(20)=0,1;
           ASSIGN,ATRIB(3)=ATRIB(3)+1;
      B11  AWAIT(24),TRANSP/1,1;
           ACT;
```

```
EVENT,1,1;
ASSIGN,XX(1)=11;
ACT,USERF(1);
GOON,1;
ACT,USERF(2);
GOON,1;
ACT,ATRIB(4).EQ.1,ST1;
ACT,ATRIB(4).EQ.2,ST2;
      ..
ACT,,ATRIB(4).EQ.13,ST13;
ACT,,ATRIB(4).EQ.0,DONE;
ACT,,,B11 ;
```

Again, each statement has a corresponding node symbol in the network diagram, the node actually displaying the code, the labels are in the boxes and the activities are the arrows with text.

The FREE block represents the release of one unit of resource TRANSP CART. Global variables XX(1) and XX(2) hold the destination and source stations for the cart. XX(20) models the status of the station number 11, and indicates whether it is idle or busy (XX(20) is set to 0 or 1). The station input buffer is represented by QUEUE(11), while the numbered service activity, ACT/11, whose duration is given by ATRIB(6) models the actual processing at the workstation. At the end of processing, the operation counter ATRIB(3) is incremented by 1. The part then awaits transportation in the AWAIT node AWAIT(24). The EVENT,1,1 statement links to the discrete event routine to select the next workstation in the operation routing. The Function USERF(1) and USERF(2) are used to read up the duration for the transportation activity. The conditional activity branchings then routes the part to the appropriate station, represented by ATRIB(4).

The network block for the machining station models the part sequencing on the machine shuttle (input/output buffer), and the machining operations (milling, drilling, etc). A network block is written for each of the 10 machining centers in the FMS. The program for machining workstations are very similar to those for load/unload stations, but differs in the sense that parts must arrive on transport carts (the resources TRANSP).

G. Discrete event part of the model

The primary role of the event part of the combined model developed for the FMS is to provide an interface between the network model, the SLAM run time executive/processor and the model user. The event routine also provides additional modeling facilities for programming complex decision processes such as workstation selection and event type selection.

The main discrete event routines used are:
(a)   Subroutine INTLC: sets up initial conditions,
(b)   Function USERF: used to make programming inserts in the network model,
(c)   Subroutine EVENT: used for event selection, for example workstation selection event, and data collection/analysis events,
(d)   Subroutine STSELECT: used to specify the workstation and the processing time for specific part types.

Other FORTRAN routines and subprograms were developed for the model entry and exit, model status monitoring and display, but these are not necessarily part of the SLAM discrete event facilities. Because these routines and subprograms are standard FORTRAN, no time will be spent describing them further.

H. Review of SLAM

Using a high level language as the foundation of a simulation package gives packages a power to model detail which may not be matched by simulation languages. However, because they are not fully tailored to simulation applications, programming with a package can be more demanding and the resulting code can be longer than with a simulation language.

SLAM's network structure provides five node types for modeling resources, three node types for gates associated with resources and other node types for conventional entity flow and entity flow decisions. The small number of node types and their generalised nature gives a concise network language that is easy to learn and use. In fact the modeler can build up any conceivable system by combining network symbols to represent processes in the system.

Although the use of generalised symbols brings with it ease of modeling, the representation of more complex logic is difficult because of the limited number of node types. In providing features for combining network and discrete event routines, thus increasing modeling flexibility, SLAM provides more features than most simulation packages.

These case studies have each used one language and one package to illustrate features of simulation software and their use. Examples and discussion of a larger range of simulation software and its use can be found in Talavage and Hannam [27]. Other means of modeling manufacturing systems are now reviewed.

IX MATHEMATICAL MODELS

Mathematical models try to capture the essence of a manufacturing system in mathematical terms. One of the more common ways of doing this is through the use of queuing theory. This is particularly applicable to manufacturing systems in which there is competition for resources such has just been discussed. The mathematical model has to represent the resources, the queues and the items or batches and their interactions in the form of the routes that the items follow between the resources. This is done by a network-of-queues analysis. Talavage and Hannam [27] describe network-of-queues approaches and classify them into 3 types;- the classical, the mean value analysis (MVAQ) and the operational analysis.

A useful computerised implementation of the classical method is CAN-Q (Computer Analysis of Network of Queues), developed by Solberg [28]. The user/modeler specifies the form of a system to CAN-Q in terms of the number of workstations (machines and processes visited), the number of parts in the system, processing times and operational sequences and average transport times. Some of the system is specified in terms of probabilities. Running the system enables performance measures to be made about the operation of the system, the utilization of resources, the length of the queues at the workstations and the throughput time of parts at stages in the system. CAN-Q has been particularly applied to the design of flexible manufacturing systems. It has been found to be able to model the behaviour of FMS more closely than the assumptions on which CAN-Q is based would suggest likely.

The MVAQ method is so called because it uses the mean

values of statistical distributions associated with the problem, rather than samples from the full distribution. An implementation of mean-value analysis method is described by Suri and Hildebrant [29]. The paper has an Appendix giving theory and algorithms for MVAQ. Mathematical models have to make simplifying assumptions and generalizations about the behaviour of a system so their results should reflect this. For example, MVAQ assumes that a part is removed from a workstation as soon as it has been processed, whereas competition for workhandling devices means this is not always true. Like CAN-Q, experience with MVAQ has shown that its use often gives more accurate results than would be expected. Even so, such modelers are generally more used in the early stages of the design of a system, when alternative configurations are being investigated and when initial evaluations of the number of machines needed, the frequency of handling and likely machine utilizations are being explored.

Mathematical models can potentially include many techniques used by operational researchers, such as linear programming, integer programming, dynamic programming, goal programming, shortest path technique, maximal-paths technique, etc. As a model of a system is only ever claimed to be representative, the degree of success achieved by modeling using these techniques or a combination of these techniques does vary. Researchers are still investigating the most appropriate methods to use and the type of problem it is best to use them on.

Mathematical models have the advantage of being quicker and easier to use than simulation and not requiring as much input data. Once ideas have been more firmed up and the operation of a particular configuration with a particular range of parts needs to be investigated, then it is appropriate to move to simulation to investigate a configuration in more detail.

Two particular operational aspects of the behaviour of manufacturing systems which have received particular attention and been mathematically modeled are those of scheduling and balancing. The schedule to be used in smaller manufacturing systems is often investigated by using simulation and the results of the simulation for a period ahead are then fed into the scheduler. In larger manufacturing systems, some form of mathematical algorithm may be used and these have been the topic of a significant amount of research. One scheduling algorithm which has been well publicised was devised by Goldratt [30] and subsequently built into the OPT (Optimised Production Technology) scheduling package. This package particularly addresses the machines or processes that give rise to bottlenecks (and hence queues) and how these should most effectively be scheduled. The topic of scheduling is covered elsewhere in these volumes so will not be considered further here.

The balancing problem relates to the design of flow-line systems such as transfer lines and link lines. The essence of the transfer-line problem is to design the line in terms of the number of workstations necessary and to distribute the machining and related washing, gauging and indexing operations across those workstations at a cycle time per station to give the necessary production rate. The calculation is complicated by the fact that cutting tools wear and need to be changed, thus stopping production at regular periods. Cutting tools can be made to last longer by cutting more slowly. However, cutting more slowly necessitates the addition of more workstations to complete

the total operations required. Thus an economic and
technological balance has to be found so that each
workstation and the system operates economically while
ensuing the first off cost of the total line is as low as
practicable. These systems do not involve variability or
competition for resources as every part usually visits each
workstation in turn and each workstation is dedicated to a
particular task. It is not thus a problem which is typically
investigated by simulation. The assembly-line balancing
problem is similar except it involves human resources and
these can be varied to a degree at every assembly station. A
recent review of the literature by Ghosh and Gagnon [31]
included over 180 references many of which were published in
the 1980s. This illustrates the interest in the topic and
the difficulty posed to the current authors in starting even
to summarise the field. Other aspects of manufacturing
systems which have similarly been investigated include the
control of inventory, batch sizes and plant layout including
transportation paths around a plant.

X NETWORKS and GRAPHICAL MODELS

The benefits of graphical models for conceptualising the
operation of manufacturing systems have already been shown
with the discussion of activity cycle diagrams. There are
other graphical models that have been used effectively and
which, like activity cycle diagrams, can form a front end to
a computerised procedure for analysing the behaviour of a
system. These include: PERT (Program Evaluation and Review
Technique), Precedence networks; GERT (Graphical Evaluation
and Review Technique) and CAM-1's IDEF. Chapters could be
written on these techniques so they will only be listed here.

XI PETRI-NETS

The final technique to be reviewed that has been used for the modeling of manufacturing systems is Petri-nets. The inventor of Petri nets was Carl Petri who developed them to help with theorising about asynchronous communications in computer systems. The concepts developed by Petri have been extended by others working on Information Systems Theory [32]. Their use has subsequently spread from modeling computer systems to modeling other types of system. (The capability of modeling techniques to model systems from many disciplines has been previously stressed.) The features of Petri nets will first be described in terms which are already familiar from earlier discussions of activity and event based models.

Petri nets can be represented diagrammatically and they exploit elements which are common to both activity and event based simulation languages, though in different ways. The terminology of Petri nets is summarised below in terms of activity and event based modeling.

Activity based modeling	Event based modeling	Petri net terminology
Activity cycle diagram	Flow diagram	Petri net structure graph
Activity	(Start event	A transaction
	(Duration	A place
	(End event	A transaction
Queue	Queue	A place

Transitions are represented by bars and places by circles. They are linked by arcs (arrows) as indicated in Fig.5. Thus an activity rectangle is represented in a Petri net by 2 transactions (bars) and a place (a circle). As with activity and event based models, the logical sequence of A-Q-A-Q or start event, end event, start event sequence is followed except with Petri-nets, this becomes a t-p-t-p sequence of transition (arrival say), place (input queue), transition (start event), place (in-activity), transaction (end event), place (output queue for last sequence of t-p-t and input queue for next sequence).

input   transaction   place   transaction   output
queue   queue

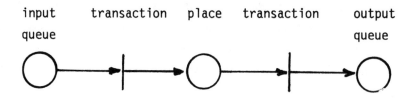

Fig. 5   Elements of a Petri net graph

With circles representing both in-activity 'places' and queuing 'places', the graph needs to be studied to distinguish between the two different types of place.

Fig.6 shows a simple Petri net for batches being processed through a set-up operation using a load station and a setter

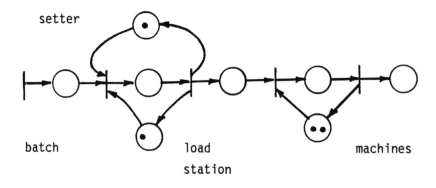

Fig.6   A simple Petri net graph

and then passing to a queue prior to being machined by one of two machines. The dots in the circles (effectively queues) are called tokens and they represent the setters (one), the load station (one) and the machines (two). The batches which continuously move through the net are not explicitly represented. Two forms of arc representation are shown and either can be used.

Activities and start events always have conditions which need to be satisfied for them to take place. At the simplest, it is the presence of a token in an input queue. When all the necessary input tokens are present, the related transition is 'enabled' or it 'fires' (ie the activity or the start event starts). The tokens are then removed from the input places and the subsequent conditions will be represented by tokens in output places. The net is executed by continuous firing.

The mathematics of a net is expressed in terms of set theory[32] where a net structure C, is a four-truple, C=(P,T,I,O) where

P is a set of places $P = \{p_1, p_2, p_3 \ldots p_n\}$
T is a set of transactions $T = \{t_1, t_2, t_3 \ldots t_n\}$
I is an input function which defines the input places
   for each transaction, eg $I(t_1) = \{p_1, p_2, p_3\}$
O is an output function which defines the output places
   following a transaction eg $O(t_3) = \{p_4, p_5\}$

A Petri net structure is then formulated. It consists of the set of places (P), the set transactions (T), the input function comprising all the inputs $I(t) = \{p\}$ and the output function comprising all the outputs. $O(t) = \{p\}$. The size of this structure will vary with the number of transactions and places in the net. Once the Petri net graph and structure have been formed, the tokens can be allocated to the net. A particular assignment of tokens is called a 'marking'.μ. The number assigned to places in the net will depend on the number of entities characterised by different parts of the net.

A Petri net graph can be substantially more complicated than an activity cycle because entities do not have attributes as such for matching entities which come together in particular transactions. Thus an operator who operates particular machines would have a separate sub-graph from other operators who operated other machines.

A Petri net is executed by firing those transactions which can fire which is determined by the positioning of the tokens. Firing leads to a re-distribution of the tokens. However, firings are 'instantaneous' transactions, this deriving from their computer science base. Here, the sequence was important to check, time was not of the essence. In modeling manufacturing systems, the time-base and its variabilities are fundamental and this has to be addressed in modeling manufacturing systems with Petri nets.

This description of Petri nets has to end here because the topic is a large one and its discussion has to be limited at some point. The length of this description compared with those on simulation probably corresponds to the extent of the use of Petri nets for modeling manufacturing systems compared with the use of simulation.

REFERENCES

1. J.R. Ernshoff and R.L. Sisson, "Design and Use of Computer Simulation Models", Macmillan Publishing Co. Inc. (New York), (1970).

2. A.A.B. Pritsker, "Introduction to Simulation and SLAM II", 3rd Edition, Wiley and Systems Publishing, New York and West Lafayette, IN, (1986).

3. J.E. Lenz, "MAST: A Simulation Tool for Designing Computerised Metalworking Factories", Simulation, February, 51-58, (1983).

4. C.R. Standridge, "Performing Simulation Projects with The Extended Simulation System (TESS)", Simulation, Vol.45, n.6, pp.283-291, (1985).

5. A.A.B. Pritsker, "The GASP IV Simulation Language", John Wiley and Sons, New York, (1975).

6. P.J. Kiviat, A. Villanuava and H.M. Markowtz, "The SIMSCRIPT II Programming Language", Prentice-Hall Pub. Co., New Jersey, (1968).

7. Anon, See-Why Programmers Guide, Istel Ltd., Redditch, (1987).

8. G.K. Hutchinson and A.T. Clementson, "Manufacturing Control Systems: An Approach to Reducing Software Costs", Proc.Int.Conf. on Manufacturing Science and Technology of the Future, MIT, Cambridge, MA, USA, (1984).

9. P.R. Hills, "SIMON - A Computer Simulation Language in ALGOL", in Digital Simulation in Operational Research, Hollingdale S.H. (Editor), American Elsevier Pub. Co., New York, (1967).

10. S.C. Mathewson, "DRAFT II/SIMON Manual", Dept. of Management Science, Imperial College, London, (1982).

11. Insight International Ltd., "Genetik User Manual and GENETIK Simulation Extension User Manual", Woodstock, Oxfordshire, England, (1987).

12. S.C. Mathewson, "Simulation Program Generators", Simulation, Vol.23, no.6, pp.181-189, (1975).

13. T.G. Poole and J.Z. Szymankiewicz, "Using Simulation to Solve Problems", McGraw-Hill, UK Ltd., London, (1977).

14. D.J. Dahl, "SIMULA - An ALGOL Based Simulation Language", Com. of the ACM, Sept., pp.671-678, (1966).

15. G. Gordon, "The Application of GPSS V to Discrete System Simulation", Prentice-Hall, New Jersey, (1975).

16. A.A.B. Pritsker, "Modelling and Analysis using Q-GERT Networks", Halsted Press and Pritsker and Associates, New York, (1977).

17. C.D. Pegden, "Introduction to Siman, Version 3.0", Systems Modelling Corp., State College, PA, (1985).

18.   Systems Modelling Corporation, "CINEMA User's Manual",
      State College, Pennsylvania, USA.

19.   CACI Inc., "SIMFACTORY User Manual", Los Angeles,
      (1986).

20.   P.J. Nolan, J.M. Fegan and T.S. Moloney, "Application
      of AI in Simulation of Manufacturing Systems", Proc.
      3rd Int. Conf. Irish Manufacturing Committee, Galway,
      Sept., 782-804, (1986).

21.   B.J. Shroer and T. Tseng, "An Intelligent Assistant for
      Manufacturing System Simulation", Int.Jn.Prod.Res.,
      27,10,1665-1683, (1989).

22.   R.E. Shannon, "Simulation: The Art and the Science",
      Prentice-Hall, New Jersey, (1985).

23.   D.P. Christy and H.J. Watson, "The Application of
      Simulation: A Survey of Industrial Practice",
      Interfaces, 13,15,47-52, (1980).

24.   A.T. Clementson, "ECSL Users Manual", Clecom Ltd,
      Birmingham(UK), (1982).

25.   A. Khochar, "Leyland Bus puts Heller FMS into Use", FMS
      Magazine, 5(1), 11-13, (1987).

26.   N.N. Ekere and R.G. Hannam, "An Evaluation of
      Approaches to Modelling and Simulating Manufacturing
      Systems", Int.Jn.Prod.Res., 27,3,599-612, (1989).

27.   J. Talavage and R.G. Hannam, "Flexible Manufacturing
      Systems in Practice, Applications, Design and
      Simulation", Marcel Dekker Inc., (New York), (1988).

28.   J.J. Solberg, "Can-Q Users' Guide, Report No.9, NSF
      Grant, No. APR 74, 15256, Purdue University, (1980).

29.   R. Suri and R.R. Hildebrant, "Modelling Flexible
      Manufacturing Systems using Mean-Value Analysis", Jn.
      of Mfg.Sys., Vol.3, No.1, 27-38, (1984).

30.   M.E. Goldratt and J. Cox, "The Goal, Excellence in
      Manufacturing", North River Press, Inc., Croton-on-
      Hudson, NY, (1984).

31.   S. Ghosh and R.J. Gagnon, "A Comprehensive Literature
      Review and Analysis of the Design, Balancing and
      Scheduling of Assembly Systems", Int.Jn. of Prod.Res.,
      Vol.27, No.4, 637-670, (1989).

32.   J.L. Peterson, "Petri Net Theory and the Modeling of
      Systems", Prentice-Hall Inc., Englewood Cliffs, NJ,
      (1981).

Knowledge-Based Simulation Environment Techniques
A Manufacturing System Example

Tae H. Cho
Jerzy W. Rozenblit
Bernard P. Zeigler

AI Simulation Group
Dept. of Electrical and Computer Engineering
The University of Arizona
Tucson, Arizona 85721

I. INTRODUCTION

The need for interdisciplinary research in AI and Simulation has been recently recognized by a number of researchers. In the last several years there has been an increasing volume of research that attempts to apply AI principles to simulation [1, 2, 3, 4].

This work was supported in part by Siemens Corporation, Princeton, New Jersey.

Expert systems are operational in several application areas where expertise is essential in tackling practical problems. Simulation of systems also requires a lot of expert's knowledge and skill. Ideas about how expert systems and simulation relate to each other and how they can be combined to the advantage of the simulationist are discussed in [5, 6].

This chapter describes a methodology for building rule based expert systems to aid discrete event simulation. It also shows how expert systems can be used in design and simulation of manufacturing systems.

The use of rule based systems to aid simulation can be categorized by the following three cases:

Case 1: The rule based expert systems are used for simulation model building according to a given goal (objective) and for evaluation of simulation runs that generate alternative simulation scenarios. The Knowledge Based Simulation System (KBS) and Simulation Craft are simulation systems of this kind [7, 8].

The KBS is one of the earliest AI-based simulation systems. KBS uses expert systems to assist the simulationist during the entire simulation process. After a user defines a model, KBS interprets it and sets up an event calendar of notices ordered by execution time in order to carry out simulation. After a simulation run, a rule base is executed to determine model modifications for the next run. The most important aspect of KBS is its focus on the automated analysis expert system that can conduct experiments and rate different scenarios to make a recommendation for the selection of a model [7].

Simulation Craft is a simulation expert system under development at Carnegie Group Inc. [8] which augments the simulation expertise by infusing AI techniques into the simulation life cycle. There are three basic experts embedded in the system: Model Building Expert, Model Execution Expert, and Model Analysis Expert. The Model Building Expert performs a consistency and completeness check needed in model building. The Model Execution Expert is primarily responsible for issues related to design of a simulation experiment. Depending upon the simulation goal, this module decides the number of runs and the associated alternatives to be evaluated. It helps the user in deciding the initial conditions and the simulation run length. The Model Analysis Expert generates and evaluates alternative experiments for a given simulation goal. Using statistical concepts, it provides an explanation of a scenario and does a situational rule-based analysis that incorporates the merged knowledge of experts in various domains of interest concerning the simulation problem.

Case 2: The behavior of the objects created for simulation is expressed by production rules. ART-ROSS is a good example of this case [9]. ART-ROSS is an integrated rule and object based implementation and extension of the ROSS language. ROSS combines SIMSCRIPT/SIMULATION-like techniques with object oriented programming. Each ROSS entity function is based on the Actor-Message-Actor paradigm: an actor (entity) sends messages to another actor which consults its behavior list to determine a response to the message. The response is in the form of another message sent to the original sender or some other actor [10]. ROSS was transformed into ART-ROSS by integrating its object oriented facilities with Inference Corporation's ART knowledge-base creation tool [4].

Several types of objects exist within the ROSS language including the clock, generic class objects, and individual objects representing instances of the generic objects. The clock object has attributes for simulation time. The individual objects have one default attribute, a list of things to do, each of which contains the time at which an action is to take place in the simulation and a pattern describing what the action is.

The procedural implementation of ROSS was converted to the rule-based implementation in ART. Rules gave a very natural way to express the behavior of objects, particularly in relation to, and in combination with, other objects. A complex series of queries in the form of messages between objects can be easily modeled by a series of patterns matching against attributes of any number of objects on the left hand side of a simple IF-THEN rule [9].

Case 3: A simulation model needs expert system-like processing within the model to solve the problem. O'Keefe [6] has suggested four ways for combining simulation modeling and expert systems: embedded, parallel, cooperative and intelligent front/back ends. The most obvious combinations involve embedding an expert system within a simulation model, and vice versa. Two examples show reasons for embedding an expert system within a simulation model in design of advanced automation and robotic systems. First example concerns job routing in an automated production and assembly facility. The routing is done by utilizing all the data available in a computerized manufacturing cell [11]. In the second example, visual identification of objects in an automated chemical lab by mobile seeing robots requires an expert system since the visual data (physical attributes such as the size and shape of the object to be identified) gathered by image processing mechanisms have to be analyzed for the identification of the objects [12].

Embedding simulation within an expert system is useful because the expert system may use one or more time-dependent variables, and thus it needs simulation results to update their values. This situation has been applied in real-time military applications where the system needs to know the position of ships, aircraft, etc. [6]. In real-time manufacturing applications, the control system may need to know the probable future states and/or positions of automated guided vehicles, transporters, etc. [13].

Several issues must be solved to efficiently develop simulation models with embedded expert systems in terms of time, cost, and validity. The issues are: (a) The interruptibility of the expert system components of simulation models is an issue in the model construction. The two examples mentioned above (the third case of expert system use) show that the inference process of the expert system must stop and wait until the simulation model returns the requested result. Often, the expert system has to handle requests from more than one source due to cost and resource limitations, e.g., only one expert system can be used for handling several requests due to budget restrictions. In this case, the inferencing has to be interrupted to process more urgent requests first. (b) Expert systems have to interact with simulation models efficiently, i.e., there should be a systematic way of communicating between the expert system and simulation models for rapid and valid model development. (c) Expert systems should be distributed among simulation models if needed (distributed expert systems), which suggests multiple existence of expert systems within the same simulation environment. (d) Expert systems should be treated no differently from other simulation models in building complex models from the existing components (hierarchical modular construction of models [14, 15, 16]).

This chapter presents an approach to embedding expert systems within an object oriented simulation environment. The basic idea is to create classes of expert system models that can be interfaced with other model classes. An *expert system shell for the simulation environment* (ESSSE) is developed and implemented in DEVS-Scheme *knowledge-based design and simulation environment* (KBDSE) which combines artificial intelligence, system theory, and modeling formalism concepts [17].

The KBDSE and its underlying formalisms are explained in the background section. Sections III and IV describe how the expert systems are implemented in the OOP (object oriented programming) paradigm and how they are interfaced to KBDSE. In Section V, the interruptibility of ES models (expert system models, i.e., the models created under the ESSSE) is explained. Section VI presents the application of ES models to flexible manufacturing system (FMS) modeling.

II. KNOWLEDGE BASED DESIGN AND SIMULATION ENVIRO-NEMNT (KBDSE)

KBDSE allows a modeler to keep models in an organized library in a modular form, enabling hierarchical assembly and disassembly as required in investigating design alternatives. The KBDSE environment is based on two formalisms: *discrete event-system specification* (DEVS) and *system entity structure* (SES) formalism. The KBDSE allows the modeler to specify explicitly the structure of a simulation model using SES formalism and its behavior using DEVS formalism. Structural and behavioral specifications of a model can be saved in a structural knowledge base called *entity structure base* (ENBASE) and behavioral knowledge base called *model base* (MBASE), respectively (Figure 1).

DEVS formalism

The DEVS formalism developed by Zeigler [15] is a theoretical, well grounded means of expressing hierarchical, modular discrete-event models. In DEVS, a system has a time base, inputs, states, outputs, and functions. The system functions determine next states and outputs based on the current states and inputs [14, 15, 18]. In the formalism, an atomic model is defined by a structure:

$$M = <X, S, Y, \delta_{int}, \delta_{ext}, \lambda, t_a >$$

where X is an external input set, S is a state variable set, Y is an external output set, δ_{int} is an internal transition function, δ_{ext} is an external transition function, λ is an output function, and t_a is a time advance function.

$$DN = <D, \{M_i\}, \{I_{i,j}\}, \{Z_{i,j}\}, select >$$

where D *is a set of component name*, M_i is a component basic model, I_i is a set of influencees of i, $Z_{i,j}$ is an output translation function, and *select* is a tie-breaking function. Such a coupled model can itself be employed in a larger coupled model.

Several atomic models can be coupled to build a more complex model, called a coupled-model. A coupled model tells how to couple several models together to form a new model. The DEVS formalism also defines a coupled model in modular form as the following structure:

SES formalism

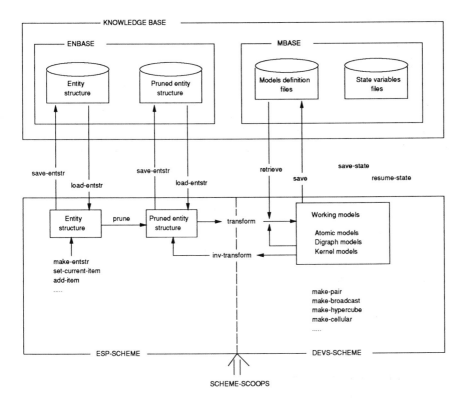

FIGURE 1.   KMDSE

A SES is a structured knowledge representation scheme that contains knowledge of decomposition, taxonomy, and coupling relationships of a system necessary to direct model synthesis [15]. Formally, the SES is a labeled tree with attached variable types that satisfy six axioms - *alternating mode, uniformity, strict hierarchy, valid brothers, attached variables and inheritance.*

There are three types of nodes in the SES - *entity, aspect,* and *specialization* - which represent three types of knowledge about the structure of systems. The *entity* node, having several *aspects* and/or *specializations,* corresponds to a model component that represents a real world object. The *aspect* node (a single vertical line in the labeled tree of Figure 2) represents one possible *decomposition* of an entity. Thus the children of an aspect node are entities, distinct components of the decomposition. The *specialization* node (a double vertical line in Figure 2) represents a way in which a *general* entity can be categorized into *specialized* entities. A *multiple entity* represents the set of all members of an entity class and it is a special entity that consists of a collection of homogeneous components. The aspect of such a multiple entity is called *multiple decomposition* (a triple vertical line in Figure 2) [19, 20].

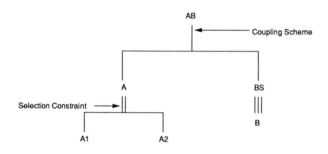

Figure 2. System Entity Structure (SES)

KBDSE automatically synthesizes hierarchical, modular models from the model base resident components under the direction of structural knowledge in the entity structure base. An entity structure can be synthesized into a simulation model by the operation called transform.

As shown in Figure 1, model objects expressed in DEVS-SCHEME must reside in working memory in order to be simulated.

The class specialization hierarchy in DEVS-SCHEME is shown in Figure 3. All classes are subclasses of the universal class *entities*, which provides tools for manipulating objects. *Models* and *processors,* the main subclasses of entities, provide the basic constructs needed for modeling and simulation. Models are further specialized into the major classes *atomic-models* and *coupled-models,* which realize atomic DEVS and coupled DEVS, respectively. The *coupled-models*, in turn, are specialized into more specific cases, a process which may be continued indefinitely as the user builds up a specific model base. Class *processors,* on the other hand, have three specializations: *simulators, co-ordinators* and *root-co-ordinators.* These carry out the simulation of a model in a manner which follows the hierarchical simulation concepts [15, 16].

Due to the object-oriented realization, subclasses of existing classes and new classes can be readily added to DEVS-SCHEME as required. As a result the DEVS-SCHEME environment:

(i) supports modular, hierarchical model operation,

(ii) allows independent testing of components models,

(iii) separates models from experiments, and

(iv) supports distributed simulation.

Details of all classes in DEVS-SCHEME along with their instant/class variables and methods are available [16, 19].

In the next section, we describe an additional layer of our environment, which facilitates the amalgamation of expert system and simulation techniques.

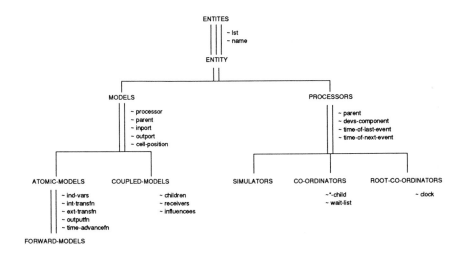

Figure 3.   Class hierarchy in DEVS-Scheme

III. OBJECT ORIENTED PROGRAMMING (OOP) FOR DISTRIB-UTED EXPERT SYSTEM ENVIRONMENT (DESE)

The DESE is designed and implemented in SCOOPS (Scheme's object oriented programming system) [21]. The term *distributed expert system environment* reflects the fact that expert systems created under the environment have distributed control (inference engine) and data base (rules and facts). The DESE consists of OOP classes, methods and other utility functions for creating distributed expert systems (DESs). The DESs created under the DESE are instance objects of the general classes in the DESE. Many researchers have pointed out OOP's benefits such as the ease of reuse, modularity, and extensibility [22, 23]. The artificial intelligence community has also shown interest in OOP, as evidenced by many object-oriented extensions to conventional AI programming languages such as Lisp and Prolog [24].

The OOP classes of DESE are shown in Figure 4. The most general class is *entities* which provides utilities for manipulating instances (objects) for its subclasses. The class *entities* can be shared among all OOP subclasses which belong to their own problem domain. The highest level classes (most general classes) in DES are *knowledge-bases* and *inference-engines*. The class *knowledge-bases* is further specialized into *synthesis-kb* and *classification-kb* to distinguish between synthesis expert systems and classification expert systems.

The distinction between synthesis expert systems and classification expert systems is a conventional one; these are two main types of expert systems and inferencing methods. Examples of classification expert systems are MYCIN [25] and DENDRAL [26]; that of a synthesis expert system is XCON [27]. The corresponding two inference engine classes are *synthesis-ie* and *classification-ie*. The inference engines class for the *synthesis-kb* is not always *synthesis-ie*. The same characteristic holds true for the *classification-kb*. As new expert systems for new applications are required, a new class(es) may be

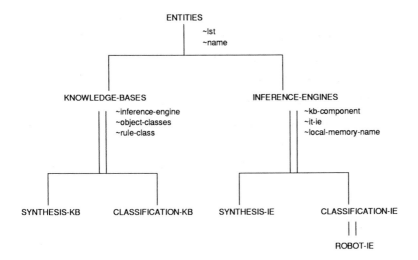

Figure 4.   Class hierarchy of DESE

added to the existing classes shown in Figure 4. For example, a new class *robot-ie* may be required in a new application. The expert system created for this application consists of instances from *classification-kb* and *robot-ie* class.

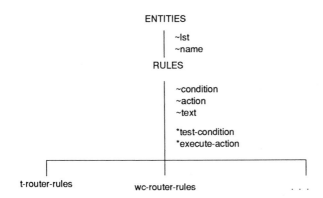

~ : class or instance variable
* : method

Figure 5.   Class hierarchy of rules

The rules and facts are also instance objects in OOP but they are not direct subclasses of *knowledge-bases*. Both of them have their own class specialization hierarchy as shown in Figures 5 and 6 where the leaf nodes represent some example classes created as required in creation of new DESs. These leaf classes, except for *parameters*, shown in Figures 5 and 6 are part of the classes required for modeling an automated testing facility in which PCBs (printed circuit boards) are routed according to decisions made by the expert systems involved in routing process. We describe this example in detail in Section VI.

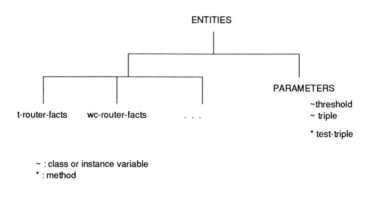

Figure 6.   Class hierarchy of rules and facts

The class *parameters* allows the creation of fact objects for the uncertainty management mechanism which employs a modified version of Dempster-Shafer operators to provide well-defined evidence accumulation properties [28].

The instance objects involved in the composition of a DES are a knowledge-base instance, an inference-engine instance, and rules and facts instances. The term knowledge-base instance means the instance object of the class *inference-engines* or its subclasses. The inference-engine instance does not directly access the rules and facts instances. Instead, the inference engine instance refers to the knowledge-base instance which has links to rules and facts instances so that the inferencing methods defined for the class *inference-engines* or its subclasses can be used for different rules and facts. The creation of a new DES involves generation of rules and facts for a given problem domain unless new classes are added to the class hierarchy of the DESE shown in Figure 4. The name of a DES is the same as the name of a knowledge-base instance which organizes all the instances within the expert system being created.

Figure 4 also shows the attached variables (instance variable or class variable) of each class. The class variable *lst* of class *entities* is a class variable which stores all the instances created in each class. The roles of instance variables are explained in Table I. The instance variable *local-memory-name* is used when a copy of a DES is made. This copy can access its own set of facts which is different from the original DES's set of facts.

The instance variable *it-ie* points to the inference table which stores the current inferencing information of DES. The DES returns *it-ie* when the inferencing is done. The type of data structure for the inference table is the *structure* of the Scheme language which is similar to the record in Pascal or the property list in Lisp. The inferencing information gathered in the inference table are fired rules, goal state, attribute list, etc. The inference table is also used as input to the DES at the beginning of inferencing. In this case the attribute list contains the initial value of the facts.

An example SCOOPS code for the creation of a DES, including rules and facts is shown in Appendix. This DES is one of expert system models used in the Automated Test Facility (ATF) model (Section VI).

Table I. Instance variables of classes in DESE.

Class	Instance Variable	Usage
entities	name	name of the instance
knowledge-bases	*inference-engines*	name of the inference engine instance
"	*object-classes*	list of classes of the fact instances
"	*rule-class*	name of the class of the rule instances
inference-engines	*kb-comp.*	name of the instance of the knowledge base component which is the same as the DES's name
"	*it-ie*	points to an inference table which stores the inferencing state
"	*local-memory-name*	used to access each DES's set of fact instances

IV. INTERFACING DESE AND KBDSE: ATOMIC-EXPERT-MODELS AND FORWARD-EXPERT-MODELS

The interface of DESE to KBDSE is accomplished by creating OOP classes called *atomic-expert-models* and *forward-expert-models*. These classes inherit properties (methods and instance variables) from the class *knowledge-bases,* and *atomic-models* or *forward-models*. The last two classes are shown in the *class hierarchy* of DEVS-Scheme (Figure 3). Thus, the DEVS models created in the class *atomic-expert-models* or *forward-expert-models* have a DES as its subcomponent in addition to the DEVS components that *atomic-models or forward-models* have.

Figure 7 shows the class *atomic-expert-models* which inherits all the variables and methods from both classes *knowledge-bases* and *atomic-models*. The DEVS models can be replicated by the *make-copy* method which produces isomorphic copies of the DEVS model [29]. The *make-copy* method of class *atomic-expert-models* makes a copy of a DES component as explained in the previous section and an isomorphic copy of a DEVS component.

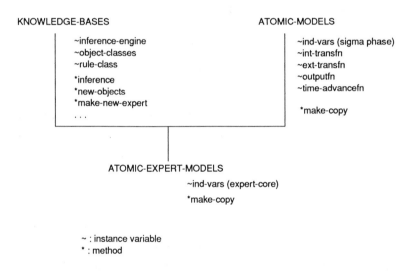

Figure 7.   Interface of DES to KBDSE

The components of atomic-expert-models are shown in Figure 8. The DEVS-COMP is the collection of elements for the atomic model of a DEVS system specification. The EXPERT-CORE is the DES created for an atomic-expert-model.

The utilities for creating a DES under the DESE and embedding it within DEVS models are called *expert system shell for the simulation environment* (ESSSE). Since *atomic-expert-models* have all the components that DEVS

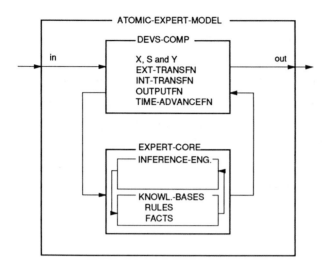

Figure 8.   Composition of atomic-expert-class model.

models have (other than the DES component) these models can be handled by KBDSE just like any other DEVS models once they are created. Thus, all the methods and utilities of the KBDSE are also applicable to the models for these two classes including modular, hierarchical construction of complex models called *coupled models*.

One important feature in creating *coupled-models* is multiple decomposition which needs isomorphic copies of a model. A copy of an ES (expert system) model (ES models are the models which have an embedded DES, i.e., *atomic-expert-models* and *forward-expert-models* class models) has a copy of an inference engine and facts. However, rules are shared among the original ES model and its copies. Having a separate copy of an inference engine and facts is necessary in order to keep track of separate inferencing states in a distributed simulation environment.

Figure 9 shows initialization of the instance variables and model state variables when a copy of a model is made. The newly created copies of a set of fact instances have names which are created by extending the name of the original model. These names are shown as A-M', B-M', etc. of the model M'. The name of the state variable EXPERT-CORE is initialized to ES-M' which is the copy of ES.

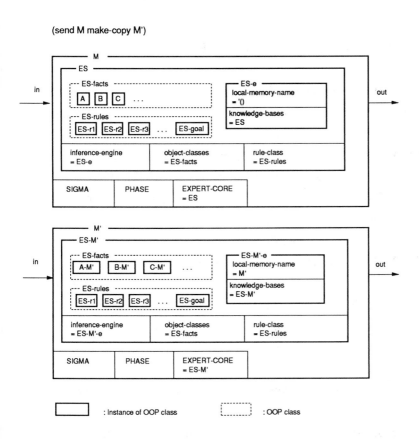

Figure 9.   Copy of an atomic-expert-model and its variable
             initialization at creation time

The name of rule instances within the copied model M' is the same as that of model M, which reflects the fact that instances of rules are not created when a copy of ES model is made. The rules are shared among the original model and its copies. Other changes are made for instance variables *inference-engine* and *knowledge-bases,* and state variable of the ES model, *expert-core.*

V. INTERRUPTIBILITY OF ES MODELS

An important property of expert systems embedded in a simulation environment is to provide interruptibility of the ES models. The interrupt capability is needed when there is an urgent task or a high priority task arrives at the model while the ES model is processing a less urgent task. Then, the current task is interrupted and its inferencing state is saved. The urgent task is processed first. When the model finishes processing of the urgent task, it resumes the interrupted process.

There is transfer of control between the DES component (expert-core) and the DEVS component within an ES model. The transfer of control from the *expert-core* to a DEVS component happens after the end of each basic inference cycle. We check for the arrival of a high priority external input event. This is similar to the cyclic polling of an interrupt. The basic inference cycle can be defined by firing a single rule, firing certain number of rules, elapse of a certain predefined simulation time, or reaching a subgoal.

An ES model with the interrupt capability can be easily modelled by using several copies of ES models. Figure 10 shows a model with interruptibility. The interruptible model, lco-ESMODELSl, represents a single expert system capable of being interrupted externally. Several virtual copies of ES models are used in constructing the model. The controller distributes an external input event, a task which needs the ES inferencing process, to one of ESMODELs, say ESMODEL0. If a new, high priority external input event arrives at the lco-ESMODELsl before processing at ESMODEL0 is finished, the controller sends an interrupt message to ESMODEL0 and then delivers the high priority task to one of idle virtual copies of the ES model, say ESMODEL1. The inference processing in ESMODEL0 is interrupted at the current processing state and processing in ESMODEL1 begins. When the processing at ESMO-DEL1 is finished, the output is sent to the controller for output. Then, the controller sends a resume message to ESMODEL0. The ESMODEL0 can resume the interrupted process since all states at which the model was interrupted were saved.

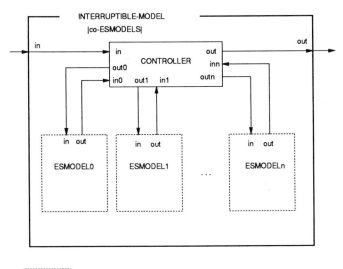

Figure 10.   Composition of an interrupible model constructed using virtual copies of ES models

VI. AUTOMATED TEST FACILITY (ATF) DESIGN MODEL

In this section, we illustrate the application of the conceptual framework presented so far to simulation modeling of manufacturing systems. We focus on a printed circuit board test architecture developed by Anderson [30] and enhanced by Rozenblit et al. [31].

The design model construction process begins with developing a representation of design components and their variants. To appropriately represent the family of design configurations, we have used the *system entity structure* [14, 32].

Beyond this, procedural knowledge is available in the form of production rules [33, 34]. They can be used to manipulate the elements in the design domain by appropriately selecting and synthesizing the domain's components. This selection and synthesis process is called *pruning* [14, 32, 35]. Pruning results in a recommendation for a *model composition tree,* i.e. the set of hierarchically arranged entities corresponding to model components. A composition tree is generated from the system entity structure by selecting a unique entity for specializations and a unique aspect for an entity with several decompositions.

The final step in the framework is the evaluation of alternative designs. This is accomplished by simulation of models derived from the composition trees. Performance of design models is evaluated through computer simulation in the DEVS-SCHEME environment [16]. In what follows, we show the application of out simulation environment to design modeling of a flexible test architecture.

Assume that an Automated Test Facility (ATF) for testing printed circuit boards is being designed. The facility should have devices that are configured for testing a specific type of board. Different configurations of the ATF may be generated, depending on the type of boards being tested.

As illustrated in Figure 11, the major subcomponents (workstations) of the ATF include: test cells, transport devices, production stores, and auxiliary facilities [36]. A test cell can be an in-circuit tester or a functional tester. A transport device can be: a conveyer, a crane, or an automatic guided vehicle (agv). A production store can be: a post assembly dock, a scrap store, or a stock store. A burn-in, an inspection cell, and a repair workstation are auxiliary facilities.

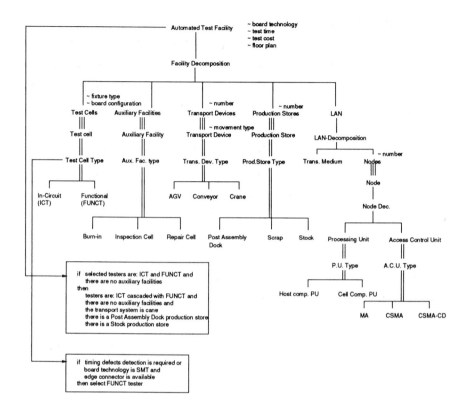

Figure 11. SES of an Automated Test Facility with example selection rules

Flexible testing would consist in generating a configuration of components (from the set of the ATF's components) for testing a particular type of PCB. For example, in order to test a bare-board (board with circuit connections etched on it but without any attached devices), only an inspection facility that checks for shorts and opens may be sufficient. On the other hand, another type of board may be tested using all of the ATF's components.

An explanation of the components of the ATF and their functions follows: An in-circuit tester tests each separate device on a board by applying test signals to a device and sensing the results from its output. A functional tester tests a board as a complete, functional entity by applying inputs and sensing outputs through the board's edge connector.

Transport devices are used to move boards from one workstation to another. A production store is where the boards are held. For example, if a board cannot be repaired, then it is sent to a scrap store. A burn-in facility is used to test the board's dynamical operation at an elevated temperature. An inspection facility is used to examine a board after it has been tested or burned-in. A repair facility is used to rebuild the board on the site of the ATF.

After the board is processed by a workcell, the processing information is sent to a host computer through a local area network (LAN). The host computer stores all the processing information in the data base and downloads routing information for the next workcell.

LAN also provides an ability to communicate among test cells, auxiliary facilities, production stores and the host computer. The transmission medium carries the information from node to node. The types of transmission media can be *twisted pair cable, coaxial cable, optical fiber* and so on. The node in LAN is either a *host computer* or *cell computers* that are linked to each workcell. The model is decomposed into *processing unit* and *access control unit*. The processing unit is the actual information processing module whereas access control unit is the module that takes care of I/O interface to a transmission medium. The host computer processing unit is different from the cell computer processing unit in that the former monitors and collects all the processing information of the each workcell and provides necessary facilities such as providing software for a specific type of PCB and making routing decision. The latter just monitors its workcell, sends processing information to the host computer and receives routing information from it.

The types of access control units depend on a contention-based access method. The simplest access method is medium access (MA), where the status of a network is not checked. Therefore, if a node sends a data packet when the network is used by another node, both packets are destroyed (this is called *collision*). In an improved version of this technique, a node checks the status of the network before transmission. This is called *carrier sense* and the method is called carrier sense medium access (CSMA). A further improvement can be achieved by enabling the node to continue to listen to the network while it is transmitting to prevent a collision due to transmitting packets exactly at same time. This technique is called carrier sense medium access/collision detect (CSMA/CD) [37].

Given a system entity structure such as depicted in Figure 11, we can generate alternative configurations of the ATF. The problem here is: given a set of parameters, test design attributes, select a set of workstations from which ATF will be composed. Different design parameters generate different arrangements of the ATF. Rozenblit and Zeigler [36] illustrate several possible architectures of the ATF. Here, we shall focus on one example in which we apply the expert system simulation shell (ESSSE).

A. APPLICATION OF EXPERT SYSTEM MODELS TO FLEXIBLE MANUFACTURING SYSTEM (FMS) MODELING

Figure 12 represents a candidate architecture for the ATF design. This architecture has been recommended for the following set of design parameters [36]: The board assembly technology is conventional, devices under test can be isolated, the bed-of-nails fixture and an edge connector are available, the timing detection is required, and the dynamical operation at elevated temperature as well as on site repair facility are desired.

The architecture of Figure 12 was augmented by the Host Computer (HC) and LAN as illustrated in Figure 14.

The testing process of printed circuit boards (PCBs) of various types is done by routing the PCBs through the workstations. The routing decision is made at the HC based on various PCB-related data (pcb type, sequence scheduling, previous workstation or cell type, etc.) and dynamic processing information (queue length, operational level, previously tested PCB type, utilization, software availability, etc.) gathered by the HC from workcells. The transport of PCBs is done by the conveyor, AGV, and crane.

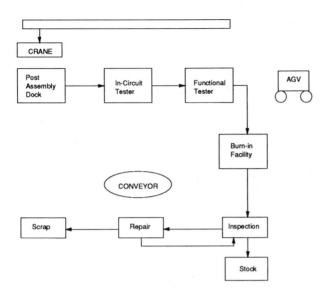

Figure 12. Architecture of Automated Test Facility

The SES of ATF for DEVS modeling is shown in Figure 13. The ATF is a DEVS model built with WCUS (WorkCell Units), BUS and HC (Host Computer) as sub-components as shown in Figure 14. WCUS consists of several WCU. The number of WCU in WCUS is specified by pruning of ATF. Each WCU is composed of CARRIER and WCELLS (WorkCells), where CARRIER is specialized into either CRANE, AGV and CONVEYOR as shown in Figure 13. The WCELL (WorkCell) is decomposed into CP (Cell Processor), CC (Cell Computer) and RB (Routing Buffer). CP is the actual processing element which is one of tester, auxiliary facility or production store type. The type of CP decides functionality of each WCELL, i.e., the type of CP decides whether the WCELL is a test cell, an auxiliary cell or a production store cell. Figure 16 shows the WCELL with ICT (In-circuit Tester) as its cell processor, where the WCELL can be distinguished as a test cell. The number of WCELLs in each WCU is also specified during the pruning process. Figure 15 shows the ATF composition tree.

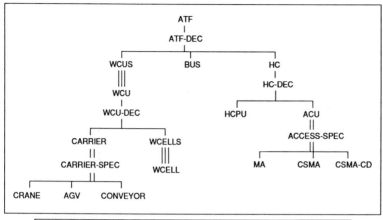

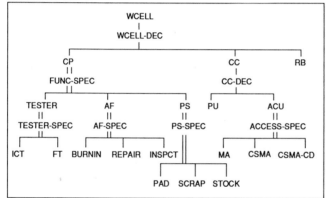

ATF: Automated Test Facility
WCU: Work Cell Unit
HC: Host Computer
ACU: Access Contol Unit
HCPU: Host Computer Processing Unit
PU: Processing Unit
MA: Medium Access
CSMA: Carrier Sense Medium Access
CSMA-CD: Carrier Sense Medium Access/Collision Detect

WCELL: Work Cell
CP: Cell Processor
CC: Cell Computer
RB: Routing Buffer
AF: Auxiliary Facility
PS: Production Store
ICT: In-Circuit Tester
FT: Functional Tester
PAD: Post Assembly Dock

Figure 13.   System Entity Structure of ATF for DEVS Modelling

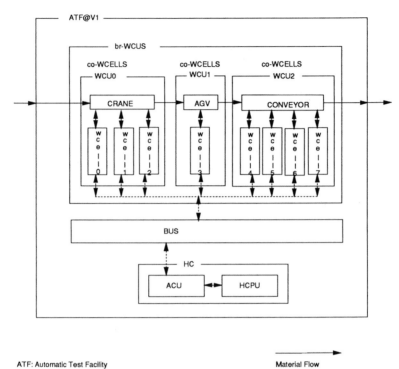

ATF: Automatic Test Facility
WCU: Work Cell Unit
WCELL: Work Cell
HC: Host Computer
HCPU: Host Computer Processing Unit
ACU: Access Control Unit

Material Flow

Information Flow

Figure 14. ATF model

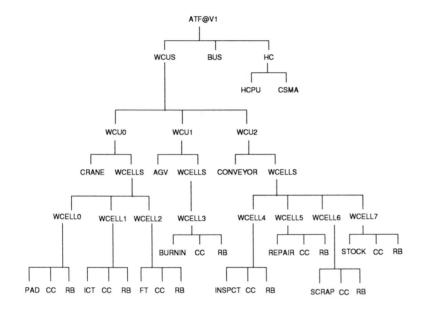

Figure 15. Pruned SES of ATF Architecture

The above description shows how ATF is decomposed and how some of its components can be specialized out of several choices. The model can also be explained in relation to material flow and information flow that flow through the model [38]. Below, we describe the functionality of the model components.

Models related to material flow: The material is a PCB (PCBs) which is carried by either CRANE, AGV or CONVEYOR models. The CRANE and AGV are basically the same models with two priority queues for each model. One for newly entering PCBs to the WCU and the other one for transporting PCBs within the WCU. The queue for newly incoming PCBs has a lower priority than the other. The CONVEYOR model is different from other carrier type models in that it has an unlimited capacity in carrying the PCBs. This reflects a pure delay in which the delay due to waiting for processing is not involved. Thus, the CONVEYOR does not need a FIFO queue for waiting

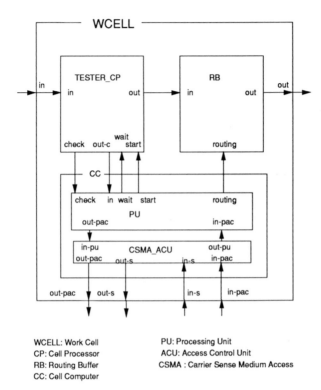

WCELL: Work Cell   PU: Processing Unit
CP: Cell Processor   ACU: Access Control Unit
RB: Routing Buffer   CSMA : Carrier Sense Medium Access
CC: Cell Computer

Figure 16.   Testcell type of workcell model

PCBs. Once a PCB has arrived at WCELL, the CP sends it to RB (routing buffer) after the processing is completed. Meanwhile CC monitors and records necessary information.

The processing time of CP consists of the setup time and testing time if the CP is either a tester or an auxiliary facility. For a production store, the processing time is the only parameter for the process delay. The setup time is needed if the type of PCB that has just arrived is different from that of the previous PCB. The testing time is the time required for actual processing of a PCB regardless of its type.

The first WCELL that a PCB enters is the WCELL with PAD (Post Assembly Dock) model as a cell processor. The PCB in PAD is carried to a testing cell after the processing time of PAD according to the routing information. The WCELL with INSPCT (inspector) decides whether the board is faulty, not faulty or not repairable at all. If it is faulty, then the PCB is sent to WCELL with REPAIR. If it is not-faulty, it is sent to the STOCK workcell. The unrepairable PCBs are sent to the SCRAP workcell. After going through all the required WCELLS, the PCB is placed at WCELL with STOCK model. It leaves the ATF model after the processing time of STOCK.

Models related to information flow: The LAN (Local Area Network) in the ATF model handles the information flow. The nodes in this LAN are host computer (HC) model and the cell computer (CC) model in each WCELL. The messages are carried in packets between CCs and HC.

Upon the arrival of a new PCB to the CP, CC checks whether specific software for processing a certain type of board is available. If it is not available at the current WCELL, CC sends a *sw-request* message to HC for downloading of the required software. The HC sends the requested software in the *sw-download* message. When CP receives the software, it tells the CP to start the processing. If the specific software for a PCB is already available then the *sw-request* message is not needed. After the PCB is processed by the CP, it is sent to RB where it waits for routing information. Meanwhile CC sends PCB processing data to the HC through *pcb-info* message. When HC receives the *pcb-info* message, it stores the data in the data base, decides which WCELL should be executed next (i.e., makes a routing decision) and sends the *routing-info* message to the CC of WCELL which has sent the *pcb-info* message. Upon receipt of the message, the CC sends the routing information to the RB.

The I/O access is handled by access control unit (ACU) of CC. Three types of contention based network access control methods are implemented. These are: MA, CSMA and CSMA-CD [37]. In order to recover from loss of message due to collision, the CP stays in the waiting state for only a given number of time units and the PCB is placed at the back of the queue after the maximum allowable wait time has expired.

The summary of packet messages and their formats are shown below.

pcb-info   from cell computer to host computer

processing information recorded from cell processor by cell computer and sent to host computer to be used for calculation of % utilization and routing information

Format: (message-type   source   destination   PCB-name PCB-type arrival-time start-proceessed-time departure-time <fault-status>)
* <> only from INSPCT workcell to HC

sw-request   from cell computer to host computer

cell computer checks whether the software for processing incoming PCB is available; if it is not, the cell computer sends this message to the host computer for a downloading request

Format: (message-type   source   destination   PCB-name PCB-type)

sw-download   from host computer to cell computer

host computer sends the requested software to cell computer for processing of PCB at CP

Format: (message-type   source   destination   PCB-name software-type)

routing-info   from host computer to cell computer

host computer decides which workcell should be the next one to process PCBs and sends this information to the CC for routing of PCBs through workcells in ATF

Format: (message-type source destination PCB-name PCB-type next-work-cell)

To evaluate the ATF by simulation, we define circumstances under which a model is to be observed and experimented with. Such conditions are called *an experimental frame*. Zeigler [14] has shown that an experimental frame can be realized as a coupling of three components: a generator (supplying a model with an input segment reflecting the effects of the external environment upon a model), an acceptor (a device monitoring a simulation run), and a transducer (collecting and processing model output data). The specification of experimental frames in the DEVS-Scheme environment is equivalent to that of specifying basic models and their corresponding couplings.

Experimental frames reflect I/O performance design requirements. The tester performance criteria include the following measures: average arrival rate, average flow time, average work in process, production rate, individual workcells' utilization and number of collisions on BUS.

The experimental frame for ATF model consists of GENR (generator) and TRANSD (transducer). There are three types of generators, these are A-GENR, B-GENR and AB-GENR as shown in SES of EF-ATF (Figure 17). Figure 18 shows how AB-GENR is constructed. A-GENR generates PCB1 type of printed circuit boards and B-GENR generates that of PCB2. AB-GENR generates both PCB1 and PCB2 type of printed circuit boards. The actual output format generated from these generators is '(PCB-id, PCB-type), e.g.,

'(G1 PCB1) or '(G5 PCB2) and so on. The arrival rate of PCBs (jobs) on ATF is decided by the parameter called *inter-arrival-time* of each generator. Currently fixed *inter-arrival-time* is implemented.

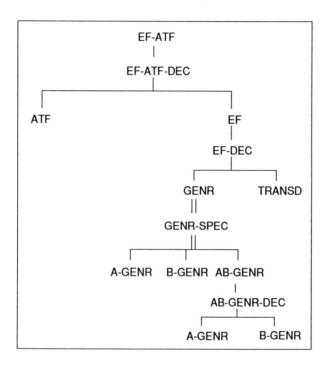

Figure 17.   SES of EF-ATF

The data items collected by TRANSD are lists of all the PCBs that arrived at ATF model and all the processed (tested) PCBs associated with time stamps within a given simulation run time. The simulation run time is determined by the parameter *observation-interval* at TRANSD. Based on these collected data items and the *observation-interval*, performance measures like *arrival rate* (jobs/time unit), *average flow time* (time units), *production rate* (jobs/time unit), *% utilization* of each WCELL (workcell) and *average work-in-process* (jobs) are calculated [39]. The *% utilization* is calculated by the HC (host computer) of the ATF model and sent to TRANSD.

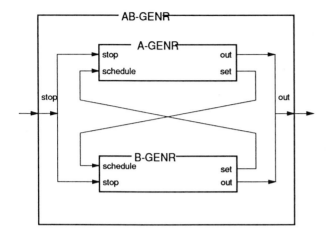

Related parameters for a-genr and b-genr:
inter-arrival-time, pcb-num, and pcb-alone

Figure 18. Composition of AB-GENR

Figure 17 shows the SES for ATF model coupled with an experimental frame (EF), so the name of the root entity is now EF-ATF. EF consists of transducer (TRANSD) and generator (GENR), where TRANSD calculates and summarizes all important simulation results. The observed simulation results are: the arrival rate of PCB (jobs/time unit), output rate (jobs/time unit), system utilization (%), mean service time per job (time unit) and WIP (Work In Process).

The important parameters in the ATF model that affect ATF performance are carrier (crane, agv and conveyor) processing time, cell processor (tester and auxiliary facility) processing time (or setup time and test time), PCB inter-arrival time at the generator, yield ratio at the inspector, the host and cell computers processing time, bus transmission time etc.

Table III shows actual simulation output of the ATF model of Figure 14 with values of parameters assigned as given in Table II.

Table II. Important parameters with their value

CARRIER processing time	5
CP (Tester and Auxiliary Facility Type) processing time	4 (setup time) 10 (testing time)
CP (Production Store Type) processing time	1
HC and CC processing time	1
BUS transmit time	0.2
yield ratio observed at INSPCT	70 %

* Values are in time units otherwise specified

Table III. Simulation results of ATF (MA access control method)

The arrival list: ((((G142 PCB2) 335.) ((G141 PCB1) 330.) ((G140 PCB2) 305.) ((G139 PCB1) 300.) ((G138 PCB2) 275.) ((G137 PCB1) 270.) ((G136 PCB2) 245.) ((G135 PCB1) 240.) ((G134 PCB2) 215.) ((G133 PCB1) 210.) ((G132 PCB2) 185.) ((G131 PCB1) 180.) ((G130 PCB2) 155.) ((G129 PCB1) 150.) ((G128 PCB1) 120.) ((G127 PCB1) 90.) ((G126 PCB1) 60.) ((G125 PCB1) 30.) ((G124 PCB1) 0))

The tested list: ((((G132 PCB2) 326.4) ((G128 PCB1) 278.6) ((G127 PCB1) 204.8) ((G126 PCB1) 173.4))

Avg. arrival rate : 0.0544 (jobs/time unit)

Avg. Flow Time: 132.05 (time units)

Avg. WIP: 7.1848 (jobs)

Production Rate : 0.0114 (jobs/time unit)

% utilization:

WCELL0<WCELLS<WCU0   4.8822

WCELL1<WCELLS<WCU0   50.5456

WCELL2<WCELLS<WCU0   38.4836

WCELL3<WCELLS<WCU1   31.5910

WCELL4<WCELLS<WCU2   35.6117

WCELL5<WCELLS<WCU2   14.9339

WCELL6<WCELLS<WCU2   0.

WCELL7<WCELLS<WCU2   1.14876

number of collisions on BUS: 4

obervation period: 0 - 350 (time units)

* arrival list - list of inputs to ATF with time stamps.

　　 where, for example ((G142 PCB2) 335.) denotes that

　　 G142 is the pcb name, PCB2 is the pcb type and

　　 335. is the arrival time

* tested list - list of outputs from ATF with time stamps.

* The result shows 4 collisions on BUS.

Total of 8 routing or pcb-info messages are lost due to these collisions.

* WorkCell distribution of ATF

wcell0 - post assembly dock cell

wcell1 - in-circuit tester cell

wcell2 - functional tester cell

wcell3 - burnin cell

wcell4 - inspector cell

wcell5 - repair cell

wcell6 - scrap cell

wcell7 - stock cell

The decision making capability of routing at HC of ATF model is given by the ES models as shown in Figure 19. The routing process consists of two steps: First, the *type routing* process decides the functional type (selects a type from the various functional types of IN-CIRCUIT TESTER, FUNCTIONAL TESTER, BURNIN TESTER, INSPECTION, REPAIR, POST ASSEMBLY DOCK, STOCK, or SCRAP) of next workcell. Second, the *cell routing* process selects a specific workcell (station) given the selected type of workcells based on the status of workcells, and types of the PCB being routed. The type router model performs the type routing process and the cell router model performs the cell routing process. The Type Router shown in Figure 19 is an *atomic-expert-model*.

The class definition file for rules and facts, and instances of rules and facts for type router model are shown in Appendix. The code is written in SCOOPS. Also shown in Appendix are class definitions of *knowledge-base* and *inference-engines,* methods of inference engines, and the definition file for *t-router* which is a DES (expert-core) of type router model. Figure 20 shows all the classes involved in the creation of the ES model *tr* (name of type router model). The classes for the cell router are the same as the type router except for rules and fact classes.

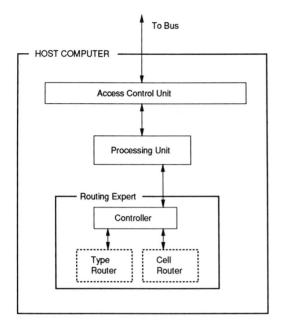

Figure 19.   Distributed Architecture of KB-Routing

The other expert systems involved in the routing task of ATF are sequence scheduling, monitoring, and error-handling processes [40]. A brief explanation of these processes follows: The initial sequence scheduling is done based on the current workcells' (workstations) status, type of PCB, and test level of the PCB. If the due date has to be checked, sequence simulation process must be carried out on a PCB to get a rough estimate time for the end of the processing. The sequence scheduling expert can perform this task. The routing expert system routes the PCBs according to the sequence scheduling. The routing process is divided into two steps as explained before. The monitoring expert system monitors the model for machine breakdowns or improper routing. The error-handling expert system either re-schedules the PCBs in the wait queue or makes those PCBs wait until the machine is repaired, depending on the average repair time. In building simulation models for these tasks, the ES

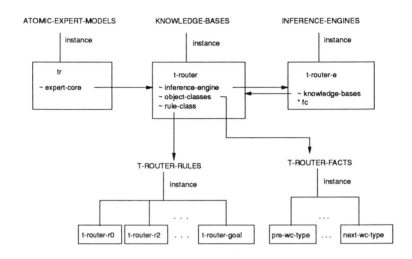

Figure 20.   The SCOOPS (scheme object oriented programming system) classes
involved in the creation of ES model tr (type router)

models can be used as shown in Figure 21 and Figure 22. In making these
expert systems work together to solve the problem, two approaches can be
taken. The distributed architecture approach, Figure 21, and the black board
architecture approach, Figure 22.

Currently, the *type router* atomic-expert-model has been implemented
and tested independently. Other atomic-expert-models shown in the Figures
6.11 and 6.12 are under development.

The phrase *distributed processing* is applied to systems with widely
different structures, software environments, interconnect characteristics,
degrees of cooperativeness, and dispersions of data [41]. The common characteristics of distributed processing systems are distribution of the control and

data, less communication among processing modules (expert systems) compared to the centralized processing systems. The distributed architecture shown in Figure 21 is designed to incorporate these common characteristics. The concepts of distributed systems can be found in many articles like [41, 42, 43].

Historically, the blackboard model arose form abstracting features of the HEARCH-II speech-understanding system developed between 1971 and 1976. The blackboard model of problem solving is a highly structured special case of opportunistic problem solving. In addition to opportunistic reasoning as a knowledge application strategy, the blackboard model prescribes the organization of the domain knowledge and all the input and intermediate and partial solutions needed to solve the problem. The blackboard model is usually described as consisting of three major component:

The knowledge sources (KSs). The knowledge needed to solve the problem is partitioned into knowledge sources, which are kept separate and independent.

The blackboard data structure. The problem solving state data are kept in a global database, the blackboard. The KSs produce changes to the blackboard that lead incrementally to a solution to the problem. Communication and interaction among the knowledge sources take place solely through the blackboard.

Control. The KSs respond opportunistically to changes in the blackboard. The characteristics of a black board system are common, shared knowledge base among knowledge sources and global control within the system [44]. The detail descriptions of the black board systems can be found in [45, 46, 47, 48].

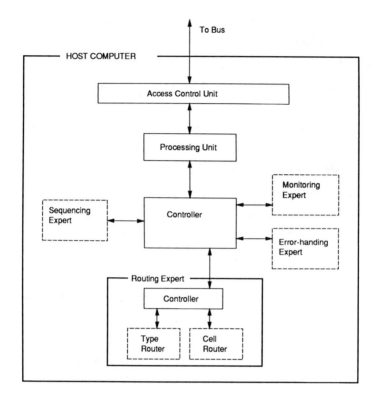

Figure 21.   Distributed Architecture of KB-Routing

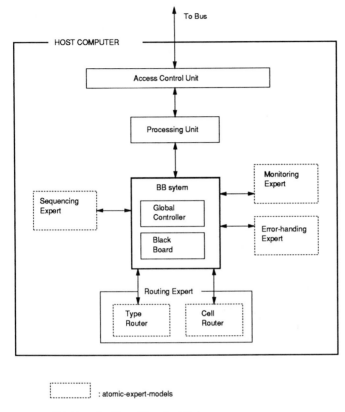

: atomic-expert-models

Global Controller - Focus of Control and Monitor
Shared, Common Data among KSs

Figure 22.   Black Board Architecture of KB-Routing

VII. SUMMARY

This chapter has reviewed recent efforts in combining artificial intelligence and simulation modeling techniques. A knowledge-based design and simulation environment (KBDSE) has been augmented with an expert system layer which facilitates intelligent decision making in model processing.

An example of a flexible printed board test facility has been described. This example is representative of a manufacturing machining process and can be easily extended to assembly operations. A model of the test architecture was developed and simulated using the KBDSE. Current work focuses on applying the expert system layer to find an optimal routing of printed circuit boards through the proposed model.

VIII. REFERENCES

1.　Vaucher, J.G. 1985., "Views of Modelling: Comparing the Simulation and AI Approaches", *Proceedings SCS Multiconference in Artificial Intelligence, Graphics, and Simulation,* pp. 3-7 (1985).

2.　Oren, T.I. and B.P, Zeigler, "Aritificial Intelligence in Modeling and Simulation: Directions to Explor", *Simulation* 48:4, pp. 131-134 (1987).

3.　Tzafestas, S.G., "Knowledge Engineering Approach to System Modelling, Diagnosis, Supervision and Control", *Simulation of Control Systems. Selected Papers from the IFAC Symposium,* pp. 15-22 (1987).

4.　Tsatsoulis, C., "A Review of Artificial Intelligence In Simulation", *SIGART Bulletin,* Vol. 2, No. 1 (1991).

5.　Arons, H. de Swann, "Expert Systems in the Simulation Domain", *Mathematics and Computers in Simulation,* Vol. XXV (1983).

6.　O'Keefe, R., "Simulation and Expert Systems - A taxonomy and Some Examples", *Simulation,* Vol. 46, pp. 10-16 (1986).

7.　Reddy, Y.V.R., and Fox, M.S., "The Knowledge-Based Simulation System", *IEEE Software,* 3:2, pp. 26-37, March (1986).

8.　Sathi, N., Fox, V.B. and Baskaran, J.B., "Simulation Craft: An Artificial Intelligence Appraoch to the Simulation Life Cycle", *Proceedings of the Summer SCS Conference,* pp. 773-778 (1986).

9.　McFall, M.E. and Klahr P., "Simulation with Rules and Objects", *Proceedings of the 1986 Winter Simulation Conference,* pp. 470-473 (1986).

10.　Klahr, P., "Expressibility in ROSS, an Object-Oriented Simulation System", *Proceedings of AI Applied to Simulation Conference,* pp. 136-139 (1985).

11. Matsuo, H., Shang, J.S. and Sullivan, R.S., "A Knowledge-Based System for Stacker Crane Control in a Manufacturing Environment", *IEEE Transactions on Systems, Man, and Cybernetics,* Vol. 19, No. 5, Sep./Oct. (1989).

12. Luh, C. J., and Zeigler, B. P., "Hierarchical Modelling of Mobile, Seeing Robots", *Proceedings SPIE conference on Intelligent Robots and Computer Vision VIII: Systems and Applications,* Vol. 1193, pp. 141-150 (1990).

13. Shannon, R.E., "Knowledge-base simulation techniques for manufacturing", *Knowledge-based Systems in Manufacturing (ed. Andrew Kusiak),* Taylor & Francis (1989).

14. Zeigler, B.P., *Multifacetted modelling and discrete event simulation,* Academic Press, Orlando, FL, USA (1984).

15. Zeigler, B.P., *Theory of modelling and simulation,* John Wiley, NY, USA, 1976, reissued by Krieger, Malabar, Fl, USA (1985).

16. Zeigler, B.P., *Object-Oriented Simulation with Hierarchical, Modular Models,* Academic Press, San Diego, CA, USA (1990).

17. Rozenblit, J.W., Hu, J., Kim, T.G. and Zeigler B.P., "Knowledge-based Design and Simulation Environment (KBDSE): Foundation Concepts adn Implementation", *Journal of The Operatioanl Research Society,* Vol. 41, No. 6 (1990).

18. Concepcion, A.I. and Zeigler, B.P., "The DEVS formalism: hierarchical model development", *IEEE Trans. Soft. Eng.,* Vol. 14, No. 2, pp. 228-241, Feb. (1988).

19. Kim, T.G. And Zeigler, B.P., "Knowledge-based environment for investigating multicomputer architectures", *Information and Software Tchnology,* Vol. 31, No 10., Dec. (1989).

20. Lee, C.G., "Simulation-based investigation of hierarchical multilevel architecture", *PhD dissertation,* University of Arizona, Tucson, AZ, USA, Aug. (1990).

21. Texas Instruments, *PC-Scheme Manual,* TI Inc., Austin, Texas (1987).

22. Yonezawa, A. and Tokoro M., *Object-Oriented Concurrent Programming,* MIT Press, Cambridge, MA (1987).

23. Keene, S.E., *Programming in Common Lisp Object-Oriented Systems,* Addison-Weslsy, Ma (1988).

24. Alpert, S. R., Woyak, S. W., Shrobe, H. J. and Arrowood, L. F., "Object-Oriented Programming in AI", *IEEE Expert*, pp. 6-7, Dec. (1990).

25. Buchanan, B.G. and and Shortliffe, E.H., *Rule-Based Expert-Based Programs: The MYCIN Experiments of the Standford Heuristic Programming Project,* Addison-Wesley, Reading, MA (1984).

26. Lindsay, R., Buchanan, B.G. and Feigenbaum, E.A. and Lederberg, J., *Applications of Artificial Intelligence for Chemical Inference:* The DENDRAL Project, McGraw-Hill Book Company, New York (1980).

27. McDermott, D., "R1: A rule-Based Configurer of Computer Systems", *Artificial Intelligence,* Vol. 19, No. 1 (1982).

28. Zeigler, B.P., "Some Properties of Modified Dempster-Shafer Operators in Rule Based Inference System", *Int. J. General Systems,* Vol 14., pp. 345-356 (1988).

29. Kim, T.G., "A Knowledge-based environment for hierarchical modelling and simulation", *Techical report AIS-7 (PhD Thesis),* University of Arizona, Tucson, AZ (1988).

30. Anderson, K.R., and Diehl, G.W., "Rapid Modelling: In the Design of A New PCB Manufacturing System", *Proceedings of the 1989 Winter Simulation Conference,* Vol. 48, pp. 818-826 (1989).

31. Rozenblit, J.W., Zeigler, B.P. and Cho, T.H., "Design and Simulation of a Flexible Test Architecture, Technical Report summitted to Siemens Corporate Research and Support", *AI and Simulation Group, University of Arizona, Tucson* (1990).

32. Rozenblit, J.W., "A Conceptual Basis for Integrated, Model-Based System Design", *Ph.D. Thesis,* Department of Computer Science, Wayne State University, Detroit, Michigan (1985).

33. Nilsson, N.J., *Principles of Artificial Intelligence,* Tioga, Palo Alto, CA (1980).

34. Winston, P.H., *Artificial Intelligence, 2nd Ed,* Addison-Wesley, Massachusetts (1977).

35. Rozenblit, J.W. and Y Huang, "Constraint-Driven Generation of Model Structures", *Proceedings of 1987 Winter Simulation Conf.,* Atlanta, GA, pp. 604-611 (1987).

36. Rozenblit, J.W. and Zeigler, B.P., "Knowledge-Based Simulation Design Methodology: A Flexible Test Architecture Application", *Transactions of The Society for Computer Simulation,* Vol. 7, No. 3, pp. 195-228, (1990).

37. Tangney, B. and O'Mahony, D., *Local Area Networks and Their Applications,* Prentice Hall (1987).

38. Anderson, K.R., and G.W. Diehl, "Rapid Prototyping: Implications for Business Planning", *Proc. of the 1988 Winter Simulation Conference,* San Diego, pp. 691-696 (1988).

39. Anderson, K.R., "A Method for Planning Analysis and Design Simulation of CIM Systems", *Proc. of the 1987 Winter Simulation Conference,* Atlanta, GA, pp. 715-720 (1987).

40. O'Grady, P.J. and Lee, K.H., "An Intelligent Cell Control System for automated manufacturing", *Knowledge-base Systems in Manufacturing (ed. Andrew Kusiak),* Taylor & Francis (1989).

41. Lorin, H., *Aspects of Distributed Computer Systems,* Wiley Inter-Science, second edition (1988).

42. Coulouris G.F. and Dollimore J., *Distributed Systems, Concepts and Design,* Addison-Wesley (1988).

43. Sharp, J.A., *An Introdution to Distributed and Parallel Processing,* Blackwell (1987).

44. Nii, H.P., "Blackboard systems: the blackboard model of problem solving and the evolution of blackboard architecture", *The AI Magazine,* Part I, pp. 38-53, Summer (1986).

45. Saul, G., "Flexible manufacturing system is CIM implemented at the shop floor level", *Industrial Engineering 35,* June (1985).

46. Erman, L.D., Hayes-Roth, F., Lesser, V.R., and Reddy, D.R., "The HEARSEY-II speech understanding system: Integrating knowledge to resolve uncertainty", *Computing Surveys,* Vol. 12, No. 2, pp. 213-253, June (1980).

47. Pang, G.K.H., "A Blackboard System for the Off-line Programming of Robots", *Journal of Intelligent and Robotics Systems,* pp. 425-444 (1989).

48. Dodhiawla, R., Jagannathan, V., Baum, L. and Skillman, T., "The First Workshop on Blackboard Systems", *AI Magazine,* pp. 77-80 (1989).

IX. APPENDIX

```
;;; t-router.s

;;; router for functional types of wcells in ATF system
;;; create an instance of type router expert system
(mk-exp knowledge-bases t-router (t-router-facts) t-router-rules inference-
engines )
;;; loading knowledge bases
(load "\\scheme\\dis-exp\\kb\\tr-cl.s")
(load "\\scheme\\dis-exp\\kb\\tr-rl.s")
(send t-router-e set-it-ie (make-inference-table))


;;; tr-cl.s

;;; fact classes of type router
(define-class t-router-facts
  (classvars)
  (instvars (value 'u))
  (options
    gettable-variables
    settable-variables
    inittable-variables
  )
  (mixins entities)
)
(compile-class t-router-facts)

;;; create instance objects for parameters of t-router rules
(mk-ent t-router-facts wc-types)       ;; workcell types
(mk-ent t-router-facts pre-wc-type)    ;; previous workcell type
(mk-ent t-router-facts next-wc-type)   ;; next workcell type (goal parameter)
(mk-ent t-router-facts c-next-wc-type) ;; candidate next workcell type
(mk-ent t-router-facts pcb-name)       ;; name of being routed
(mk-ent t-router-facts fault-status)   ;; fault-status reported by inspector
(show-class t-router-facts)


;;; t-rl.s
;;; t-router-rules

;;; class for the type-router rules
(define-class t-router-rules
  (classvars)
  (instvars)
  (options
    settable-variables
```

```
    gettable-variables
    inittable-variables
  )
  (mixins rules) )
(compile-class t-router-rules)

;;----------------------------------------------------------------
;; make rules of class t-router-rules
;;----------------------------------------------------------------
;; The rules are written for rule ordering conflict resolution of
;; forward chaining.
;; More specific rules have lower rule numbers.

;; t-router-r1
(mk-ent t-router-rules t-router-r1)  ;; create r1 for t-router
;; if previous workcell type is repair
(send t-router-r1 set-condition
  '(equal? (send pre-wc-type get-value) 'repair)
)
;; then next workcell type is inspector workcell
(send t-router-r1 set-action
  '(send next-wc-type set-value 'inspct)
)
(send t-router-r1 set-text " t-router-r1 ")

;; t-router-r2 (mk-ent t-router-rules t-router-r2)
(send t-router-r2 set-condition
  '(and (equal? (send pre-wc-type get-value) 'inspct)
      (equal? (send fault-status get-value) 'faulty)
  )
)
(send t-router-r2 set-action
  '(send next-wc-type set-value 'repair) )
(send t-router-r2 set-text " t-router-r2 ")

;; t-router-r3
(mk-ent t-router-rules t-router-r3)
(send t-router-r3 set-condition
  '(and (equal? (send pre-wc-type get-value) 'inspct)
      (equal? (send fault-status get-value) 'not-faulty)
  )
)
(send t-router-r3 set-action
  '(send c-next-wc-type set-value (get-next-wc pre-wc-type wc-types))
)
(send t-router-r3 set-text " t-router-r3 ")
```

```
;; t-router-r4
;; to prevent a pcb from routed to the repair when pcb is not faulty ;; this happens
when repair is place right next to the inspct.
(mk-ent t-router-rules t-router-r4)
(send t-router-r4 set-condition
  '(and (equal? (send pre-wc-type get-value) 'inspct)
      (equal? (send c-next-wc-type get-value) 'repair)
      (equal? (send fault-status get-value) 'not-faulty)
  )
)
(send t-router-r4 set-action
  '(send next-wc-type set-value (get-next-wc c-next-wc-type wc-types))
)
(send t-router-r4 set-text " t-router-r4 ")

;; t-router-r5
(mk-ent t-router-rules t-router-r5)
(send t-router-r5 set-condition
  '(not (equal? (send pre-wc-type get-value) 'u))
)
(send t-router-r5 set-action
  '(send c-next-wc-type set-value (get-next-wc pre-wc-type wc-types))
)
(send t-router-r5 set-text " t-router-r5 ")

;; t-router-r6
;; if c-next-wc-type is null then pre-wc-type was the last type in ATF
(mk-ent t-router-rules t-router-r6)
(send t-router-r6 set-condition
  '(null? (send c-next-wc-type get-value))
)
(send t-router-r6 set-action   '(send next-wc-type set-value 'out) )
(send t-router-r6 set-text " t-router-r6 ")

;; t-router-r7
;; if candidate wc type is not null, i.e., if pre-wc-type is not
;; the end of the ATF workcell.
(mk-ent t-router-rules t-router-r7)
(send t-router-r7 set-condition
  '(not (null? (send c-next-wc-type get-value)))
)
(send t-router-r7 set-action
  '(send next-wc-type set-value (send c-next-wc-type get-value))
)
(send t-router-r7 set-text " t-router-r7 ")
```

```
;; t-router-goal
(mk-ent t-router-rules t-router-goal)
(send t-router-goal set-condition
  '(not (equal? (send next-wc-type get-value) 'u))
)
(send t-router-goal set-action
  '(list (send next-wc-type get-value) (send pcb-name get-value))
)
(send t-router-goal set-text " t-router-goal ")

;; define helper methods for t-router-rules classes
;; (utility fuction used by rules)
(define-method (t-router-rules get-next-wc) (pre-wc-type wc-types)
  (cadr (member (send pre-wc-type get-value)
          (send wc-types get-value)
      )
  )
)
```

FAULT DETECTION AND ISOLATION IN
AUTOMATIC PROCESSES

Paul M. Frank
Ralf Seliger

University of Duisburg
Department of
Measurement and Control
Bismarckstr. 81, BB
W-4100 Duisburg 1, FRG

I. INTRODUCTION

The spectacular progress in computer technology over the last decades allows to control technical processes and systems of continuously increasing complexity. Naturally, the issues of reliability, operating safety and, last but not least, environmental protection are of major importance, especially if potentially dangerous processes like chemical reactors, nuclear power plants or aircrafts are concerned.

In order to improve the safety of automatic processes, they must be supervised such that occuring failures or faults can be accomodated as quickly as possible. Failures or faults are malfunctions hampering or disturbing the normal operation of an automatic process, thus causing an unacceptable deterioration of the performance of the system or even leading to dangerous situations. They can be classified as component-faults (CF), instrument-faults (IF) and actuator-faults (AF). The first two steps towards a failure accomodation are the *detection* and the *isolation* of the fault in the system under supervision. The term *detection* denotes in this context the knowledge of the time at which a fault has occurred, while *isolation* means the determination of the fault location in the supervised system, i.e. the answer to the question which instrument, actuator or component failed.

There are two distinct approaches to Fault-Detection and -Isolation (FDI):
The first one is the hardware- or physical- redundancy approach. In the case of Instrument-FDI for instance, information about the time of occurrence and the location of faults is gained by means of redundant sensors. Each instrument is installed threefold and a two-out-of-three decision logic is employed in order to decide whether a certain instrument works properly or not.

The second approach is characterized in the literature by the terms analytical- or

CONTROL AND DYNAMIC SYSTEMS, VOL. 49

software- redundancy. Here, the capabilities of modern computer technology are not only utilized to control a process but also for fault-diagnosis.

The general method can basically be outlined as follows:

The knowledge of a static or dynamic mathematical model of the process under supervision is used to gain insight into the actual system. To this end, the nominal system behaviour, derived from the mathematical model is compared to the actual system behaviour, known from the measurements. The result of the comparison allows conclusions on when and where a fault has occurred or whether the system operates fault free. In the simplest approach this is achieved by checking the limit values of certain charactaristic signals or by evaluating static relations in order to check for plausibility. More sophisticated techniques make use of dynamic state observers or parameter estimation methods.

Generally it can be stated:

The aim of analytical FDI techniques is the generation and evaluation of fault accentuated signals on the basis of the available measurements and a mathematical model of the system. The generated signal is to reflect a fault in the sense that information on both the fault-time and the fault-location in the system under supervision can be extracted. A signal of this type will be called residual in the sequel. The problems of generating and evaluating residuals as well as various solution techniques are the subjects of this contribution.

The major advantage of the analytical redundancy approach as compared to hardware redundancy is that, on principle, no additional hardware components are needed in order to realize a FDI algorithm. Any analytical FDI algorithm can basically be implemented additionally on the computer which is used in the first place to control the process to be supervised. Moreover, the measurements necessary to control the process are in many cases also sufficient for the FDI algorithm so that no additional sensors have to be installed. Under these circumstances, only additional storage capacity and computer time is needed for the implementation of an analytical FDI algorithm.

Over the last two decades the problem of FDI using analytical redundancy has been studied extensively in the literature. Especially mentionable are the fault detection filter [4], [38], the innovation test on the basis of Kalman filters [65] or Luenberger observers [26], [42], the dedicated observer approach [9], the unknown input observer approach for linear and nonlinear systems [67], [54], [55] the parity space approach [7], the frequency domain approach [15] and the parameter identification technique [37].

Since the FDI algorithms using analytical redundancy operate on the basis of mathematical models of the supervised systems, there is the danger of false alarms caused by model-plant mismatches due to modelling errors or simplifying and idealizing assumptions. Thus, one of the most essential requirements imposed on analytical FDI algorithms is the need for robustness to disturbances as well as to model-plant mismatches. This problem has been recognized for example in [27] and [65]. Effective solutions have been suggested in [17], [19], [41], [49] using linear models and in [67], [54], [55] on the basis of certain classes of nonlinear models.

For major survey studies and books on analytical FDI techniques we refer to [2], [3], [23], [26], [29], [32], [35], [49], [57].

Even though FDI using analytical redundancy is a relatively young branch in the area of control and computer technology, there is a significant number of application studies in the literature. Applications of the unknown input observer approach to a robot and a three-tank system are reported in [67]. A comparative study of various state estimation techniques applied to a steam generator has been performed in [53]. Applications to aircrafts and turbofan engines are investigated in [50] and [12], respectively. Various applications of Instrument-FDI for jet engines are reported and compared in [45]. In [35] and [36] parameter estimation techniques are applied to a system consisting of a D.C. motor and a centrifugal pump.

II. MODELLING

A. GENERAL REMARKS

The very first step in the analytical redundancy approach is to set up a mathematical model of the automatic process to be supervised. In order to determine the type of model that is needed, one must distinguish between fault detection (FD) and fault isolation (FI). The FD problem can usually be tackled succesfully with a so-called *representative* model.

A *representative* model describes the functioning of a process in terms of a mathematical model with a preassigned, usually simple structure, so that the actual behaviour of the process is approximated with satisfactory accuracy. Clearly, one may use different models for the same process. Due to its simplicity as well as to its ability to describe the functional bahaviour of the process adequately, this kind of model is often used in control.

However, it is important to note that for FI a functional model of the process must be provided that is differently organized than the functional model used for control. This type of model is called *diagnostic* model.

A *diagnostic* model provides the knowledge of the physical properties of a process in terms of its structural funcionality. It describes analytically the dynamic behaviour under different working conditions with all interactions between different parts of of the system. To acquire this kind of model, one takes into consideration the actual architecture of the process and uses known physical laws that govern the behaviour of the individual devices such as energy and mass balances. Such a model provides the deepest insight into the process.

It should be noted at this point that it is always possible to transform a diagnostic model into a representative model by structural simplifications or by combining individual devices mathematically. While these techniques are generally allowed in control they must be handled with care when FI problems are to be solved.

Despite the superiority of diagnostic models to representative models with respect to FI, model-based FDI often restricts itself to the latter type of models because of their simplicity.

If a representative model is considered as a basis for a FI algorithm, the question will be: How well can faults be modelled in the representative model? In other words, how well is a fault then structureable? Obviously, in general there are some faults that cannot be structured by a representative model. In this case, a diagnostic model is inevitable if all the faults are to be isolated. This means that providing as much insight in the process as possible must be the ultimate goal of any analytical modelling for FDI.

In Fig. 1 the difference between the terms of representative and diagnostic models is illustrated using the example of an electrical motor which is connected to a mechanical load by a gear.

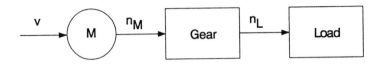

Fig. 1. Motor with mechanical load

A sufficient representative model of this system is, for instance, a 1st-order differential equation, describing the dynamic relation between the motor input voltage v and the mechanical speed n_L of the load. Such a model provides an appropriate basis for the design of a FD algorithm that is able to indicate the time of occurrence of a fault, because the model describes the dynamic of the system sufficiently accurate. On the other hand, based on this model almost no information can be gained about the exact location of a fault within the system, i.e. information whether the fault occurred in a motor winding, the gear or the load. The reason for this drawback is: The various physical parameters affected by faults occuring in the different components of the system are combined in form of just two mathematical parameters defining the 1st-order differential equation, i.e. gain and time constant. In order to isolate the faults listed above, the respective units of the system where faults might occur must be modelled in a strict seperate manner, even though there are parts, like the gear, which do not include any dynamics. This will result in a diagnostic model.

A third type of model, the so-called *knowledge-based* models, may often be a useful complementary tool for the purpose of providing insight into the process.

A knowledge-based model is a heuristic representation of a process in terms of a qualitative model. All available information and knowledge are given in form of facts and rules of general form. Based on such heuristic models, an expert system can be used for fault diagnosis. For more details we refer to [49].

B. PROCESS MODEL DESCRIPTION

The mathematical system representation which is used throughout this contribution is the dynamic state space description given by a system of 1st-order ordinary differential equations.

$$\dot{x} = f(x, u, f_a, f_c, d_1) \tag{1}$$
$$y = h(x, u, f_s, d_2) \tag{2}$$

where x, u and y denote the $n \times 1$ state vector, the $m \times 1$ input vector and the $p \times 1$ output vector, respectively. This description can, if necessary, be transformed into an equivalent set of nth-order ordinary differential equations describing the input-output behaviour of the system. The vector signals f_a, f_c and f_s are the AF, CF and IF, respectively. The expressions d_1 and d_2 denote vectors of unknown or unmeasure-able external signals acting on the process dynamics and on the sensors. They may represent noise or other disturbances. Moreover, modelling errors or model-plant mis-matches can conveniently be described in this form. We will elaborate on the issues of fault- and disturbance-modelling in the next section. However, it should be noted at this point that in general the time histories of both the faults and the unknown inputs are assumed to be completely unknown.

The unknown inputs and disturbances which are always present in one form or another, introduce the issue of robustness to the problem of FDI using analytical redundancy. In order to avoid false alarms caused by unknown inputs and not by faults, it is necessary to design FDI algorithms which are robust to the unknown inputs and sensitive to the faults. This problem and various approaches to a solution will be discussed in some detail in the following sections.

On some occasions we will refer to a discrete time model instead of the continuous time description in Eq. (1). In those cases an appropriate model will be a system of 1st-order difference equations, i.e. the time derivative operator on the left-hand side in Eq. (1) must be replaced by a shift operator.

The model discussed above can be structured according to the block diagram rep-resentation displayed in Fig. 2.

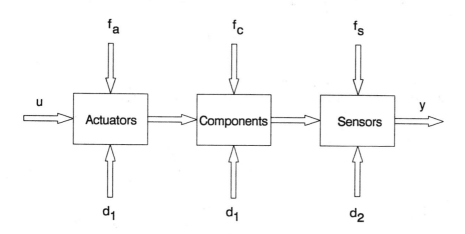

Fig. 2. System representation

Besides a partition of the complete system in actuators, components and sensors, the figure shows the three possible classes of faults AF, CF and IF as well as the known and unknown input signals u and $d_{1,2}$ and the output signal y.

C. FAULT- AND DISTURBANCE-MODELLING

In this section we motivate how unknown inputs as well as faults can be modelled in terms of parameter uncertainties of a system [54]. Assume that faults and model-plant mismatches (modelling errors, model uncertainties) can be linked to variations of the model parameters contained in the parameter vectors θ_f and θ_d, respectively. These vectors can be introduced explicitly as arguments of the vector function $g(x, u)$ which describes either the system dynamics or the output relations.

$$g(x, u) = g(x, u, \theta_f, \theta_d) \tag{3}$$

Suppose that the parameters θ_d as well as, in the fault free case, the parameters θ_f assume known nominal values θ_{d0} and θ_{f0} respectively. A Taylor series approximation of Eq. (3) which is terminated after the 1st-order term yields

$$
\begin{aligned}
g(x, u, \theta_f, \theta_d) \approx \quad & g(x, u, \theta_{f0}, \theta_{d0}) \\
& + \frac{\partial g(x, u, \theta_{f0}, \theta_{d0})}{\partial \theta_f} \Delta \theta_f \\
& + \frac{\partial g(x, u, \theta_{f0}, \theta_{d0})}{\partial \theta_d} \Delta \theta_d
\end{aligned}
\tag{4}
$$

Now the faults as well as the unknown inputs can be identified by the corresponding parameter variations $\Delta \theta_{f,d}$. These definitions may be extended to time varying systems, where even the nominal parameters are functions of time.

If the discussion is restricted to linear time-invariant systems, a faulty and uncertain model can be described by [67]

$$\dot{x} = Ax + Bu + Ed + Kf \tag{5}$$
$$y = Cx + Du \tag{6}$$

where, for the sake of simplicity, the measurements are free of faults and disturbances. All matrices are of appropriate dimensions and assumed to be known. The matrix E contains the deviations of the elements of the matrices A and B from their nominal values caused by model uncertainties. Now the uncertainties, represented in the model by unknown input signals, are given by [67]

$$Ed = (\Delta A \quad \Delta B) \begin{pmatrix} x \\ u \end{pmatrix} \tag{7}$$

The term Kf can be modelled in a similar manner. Uncertainties which can be modelled according to Eq. (5) are called structured uncertainties [22]. In other words, the distribution matrix E is assumed to be known.

In the frequency domain one can additionally define unstructured uncertainties in terms of transfer functions [22]. To this end the Laplace transformed input-output relation of a linear system is considered.

$$
\begin{aligned}
y(s) = \quad & (G_u(s) + \Delta G_u(s))u(s) + (G_d(s) + \Delta G_d(s))d(s) \\
& + (G_f(s) + \Delta G_f(s))f(s)
\end{aligned}
\tag{8}
$$

The unstructured uncertainties are represented by the unknown transfer matrices $\Delta G_{u,d,f}(s)$, while the transfer matrices $G_{u,d,f}(s)$ are known.

III. CONCEPT OF MODEL-BASED FDI

As already outlined in the introduction, the ultimate goal of all FDI algorithms is to generate so-called residuals carrying information on the time and the location of occuring faults. To this end the known inputs u and the measured outputs y are processed in order to verify certain relations which must hold if no fault occurs. The evaluation of these so-called redundancy relations results in the residual which must be different from zero in case a fault occurs and zero otherwise. The system that performs these operations is called residual generator. In order to isolate faults, the residual must be processed further by some kind of decision function and an appropriate decision logic [29]. The block diagram displayed in Fig. 3 illustrates the general structure of the outlined FDI concept.

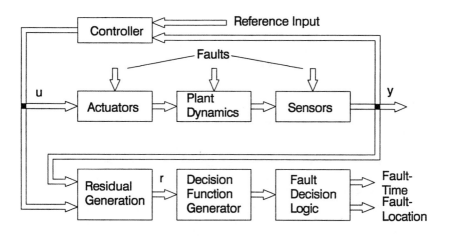

Fig. 3. Concept of FDI using analytical redundancy

In the last section faults have been defined as external unknown signals acting on the dynamics or on the sensors of the process to be supervised. For a more precise mathematical definition of the Fault-Detection (FD) and the Fault-Isolation (FI) concept suppose that the fault signal f which is to be detected consists of l different faults f_i, $i = 1 \ldots l$ which are to be isolated.

$$f = (f_1 \ldots f_l)^T \tag{9}$$

A residual generator is then defined by a generally vector valued relation

$$r = G(u, \dot{u}, \ldots, y, \dot{y}, \ldots) \tag{10}$$

where

$$r \neq 0 \quad \text{if and only if} \quad f \neq 0 \tag{11}$$

After a successful detection ($r \neq 0$) the FI problem can be solved by suitable decision functions and a suitable decision logic defined by relations $R_i(r)$, $i = 1 \ldots l$ where

$$R_i(r) \neq 0 \quad \text{if and only if} \quad f_i \neq 0, \quad i = 1 \ldots l \tag{12}$$

Note that these definitions are based on the idealizing assumption that the residual will vanish completely if no faults occur. Naturally, in reality this cannot be achieved. In these cases more sophisticated techniques like constant or adaptive thresholds, decorrelation methods etc. must be incorporated in the decision functions. We will discuss these issues which are closely related to the robustness problem in the following sections.

IV. RESIDUAL GENERATION

A. OBSERVER-BASED METHODS

1. OBSERVER-BASED FDI IN LINEAR SYSTEMS

This section deals with various approaches to observer-based FDI in systems which can be represented sufficiently accurate by linear time-invariant models.

a. Dedicated Observer Scheme: The utilization of observers is a somewhat self-suggesting approach to model-based FDI, because an observer basically consists of a model of the underlying process. The principle of FDI using observers is based on comparing the estimated output of the supervised process (observer- or Kalman filter-output) with the actual process output available by measurements. The difference between the two signals which is called estimation error in observer theory and innovation in conjunction with Kalman filters, is used as a residual. Provided the nominal model of the process on which the observer is based is sufficiently accurate, the residual will be close to zero in case no fault occurs. Only if faults occur, there will be a discrepancy between the process and the model causing the residual to differ significantly from zero. Considering the faulty linear model of a dynamic process

$$\dot{x} = Ax + Bu + K_1 f_1 \tag{13}$$
$$y = Cx + K_2 f_2 \tag{14}$$

this becomes clear. The identity observer for this system has the structure [47]

$$\dot{\hat{x}} = (A - HC)\hat{x} + Bu + Hy \tag{15}$$
$$\hat{y} = C\hat{x} \tag{16}$$

The state estimation error $e = \hat{x} - x$ and the output estimation error $\epsilon = \hat{y} - y$ are goverened by

$$\dot{e} = (A - HC)e + HK_2 f_2 - K_1 f_1 \tag{17}$$
$$\epsilon = Ce - K_2 f_2 \tag{18}$$

Besides asymptotically decaying transients caused by initial deviations between system state and observer state, the output estimation error is only affected by the faults $f_{1,2}$. Therefore this signal can be used as a residual.

$$r = \epsilon \tag{19}$$

Note that the system must be known perfectly for Eq. (17) and (18) to hold. If this can not be guaranteed, the residual will differ from zero even in case no fault occurs.

If only IF are considered, i.e. $f_1 = 0$, then the FI problem can be solved by the so-called *dedicated observer scheme* [10]. The idea is that not only one, but a bank of observers is devised to supervise a given process. Each observer is driven by the input signal u and one sensor only. Thus, each observer is dedicated to a single instrument. It is desired to reconstruct the complete output y with each observer. If this is not feasible, as many components of y as possible are estimated. Since there are as many residuals r_i, $i = 1 \ldots p$ as there are sensors, it is possible to isolate the IF $f_2 = (f_{21} \ldots f_{2,p})^T$ uniquely by processing the residuals with an appropriate decision logic. For each of the p instruments a so-called error function [69]

$$F_i = \prod_{j=1}^{p} r_{i,j} \prod_{j=1}^{p} r_{j,i}, \quad i = 1 \ldots p \tag{20}$$

can be defined. Here $r_{i,j}$ denotes the jth component of the ith residual. It can be easily verified that F_i will be different from zero if and only if the ith sensor fails.

Fig. 4 displays the so-called *generalized dedicated observer scheme* which also consists of a bank of observers [24].

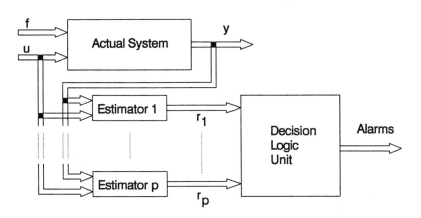

Fig. 4. Generalized dedicated observer scheme

In contrast to the *dedicated observer scheme* the ith observer is driven by all but the ith sensor, resulting in residuals

$$r_i = R(f_{21} \ldots f_{2,i-1} \quad f_{2,i+1} \ldots f_{2,p}), \quad i = 1 \ldots p \tag{21}$$

Similar to the *dedicated observer scheme*, a decision logic can be derived on the basis of the p residuals [26], allowing for a unique isolation of the IF. As compared to the *dedicated observer scheme*, the *generalized dedicated observer scheme* provides more robustness to unknown disturbances and modelling errors [29].

   b. Fault Detection Filter: The fault detection filter [4], [38] is of the same structure as the identity observer in Eq. (15). Considering the residual defined by Eq. (17) and (18), different faults f_{ij} will result in residuals pointing in different directions in the space spanned by the components of the residual. This property can be enhanced by a suitable choice of the observer gain matrix H [41], [47]. Under certain conditions it might even be possible to decouple the faults completely, i.e. residuals corresponding to different faults appear mutually orthogonal. The complete decoupling provides a unique FI scheme.

   Fig. 5 illustrates the directional properties of two residuals corresponding to two different faults.

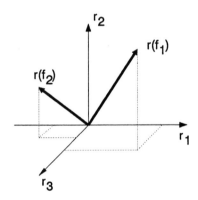

Fig. 5. Geometrical interpretation of the fault detection filter

   The direction of the residual remains unaffected by the size or time history of a fault f_{ij}.

   A drawback of the fault detection filter scheme is the fact that unknown inputs accounting for disturbances or model uncertainties are not considered in its framework [64]. Therefore the fault detection filter approach does not provide robustness.

   c. Robust FDI via Parity Space Approach: The basic idea of the parity space approach is to evaluate and check so-called parity or consistency relations using the actual input and output measurement data [7], [13], [41], [52], [58]. The parity relations are derived from the mathematical model of the automatic process. The residual derived from the parity relations includes all available redundant information inherent in the model [26]. If only static balance equations are used as parity relations, one arrives at the simple plausibility checks which are widely used by the industry.

It was shown for instance in [26] and [68] that, provided certain conditions are fulfilled, the residual can be completely decoupled from unknown inputs. If model-plant mismatches are represented by unknown input signals, it is thus possible to generate residuals which are robust to modelling errors.

The parity relations can be derived from a disturbed linear time-invariant discrete time state space model of the automatic process

$$x_{k+1} = Ax_k + Bu_k + E_d d_k + K_d f_k \qquad (22)$$
$$y_k = Cx_k + Du_k + E_m d_k + K_m f_k \qquad (23)$$

where the unknown inputs are denoted by d_k and the faults by f_k. The distribution matrices $E_{d,m}$ and $K_{d,m}$ of the unknown inputs and the faults must be known.

Since the distribution matrices $E_{d,m}$ are assumed to be known one refers to the unknown inputs d_k as structured uncertainties. As already mentioned in Section II,B, there are also unstructured uncertainties which will be discussed in Section IV,A,1,f.

Equations (22) and (23) can be rewritten as

$$\begin{pmatrix} y_{k-s} \\ y_{k-s+1} \\ \vdots \\ y_k \end{pmatrix} = H_0 x_{k-s} + H_1 \begin{pmatrix} u_{k-s} \\ u_{k-s+1} \\ \vdots \\ u_k \end{pmatrix} + H_2 \begin{pmatrix} d_{k-s} \\ d_{k-s+1} \\ \vdots \\ d_k \end{pmatrix} + H_3 \begin{pmatrix} f_{k-s} \\ f_{k-s+1} \\ \vdots \\ f_k \end{pmatrix} \qquad (24)$$

where s is called time horizon and

$$H_0 = \begin{pmatrix} C \\ CA \\ CA^2 \\ \vdots \\ CA^s \end{pmatrix} \qquad (25)$$

$$H_1 = \begin{pmatrix} D & 0 & \cdots & \cdots & 0 \\ CB & D & 0 & \cdots & 0 \\ CAB & CB & D & \cdots & 0 \\ \vdots & \ddots & \ddots & \ddots & \vdots \\ CA^{s-1}B & \cdots & \cdots & CB & D \end{pmatrix} \qquad (26)$$

$$H_2 = \begin{pmatrix} E_m & 0 & \cdots & \cdots & 0 \\ CE_d & E_m & 0 & \cdots & 0 \\ CAE_d & CE_d & E_m & \cdots & 0 \\ \vdots & \ddots & \ddots & \ddots & \vdots \\ CA^{s-1}E_d & \cdots & \cdots & CE_d & E_m \end{pmatrix} \qquad (27)$$

$$H_3 = \begin{pmatrix} K_m & 0 & \cdots & \cdots & 0 \\ CK_d & K_m & 0 & \cdots & 0 \\ CAK_d & CK_d & K_m & \cdots & 0 \\ \vdots & \ddots & \ddots & \ddots & \vdots \\ CA^{s-1}K_d & \cdots & \cdots & CK_d & K_m \end{pmatrix} \qquad (28)$$

Based on this description a residual can be generated. To this end one recalls that Eq. (24) must hold if evaluated on-line using actual measurement data u_k and y_k.

Thus, an appropriate residual is defined by [68]

$$
r_k = v^T \left(\left(\begin{array}{c} y_{k-s} \\ y_{k-s+1} \\ \vdots \\ y_k \end{array} \right) - H_1 \left(\begin{array}{c} u_{k-s} \\ u_{k-s+1} \\ \vdots \\ u_k \end{array} \right) \right) \tag{29}
$$

The vector v^T must be selected such that the residual is neither affected by any initial conditions x_{k-s} nor by the unknown inputs d_k. On the other hand the residual must be sensitive to the faults f_k. The robustness to the initial conditions and the unknown inputs is guaranteed if and only if v^T satisfies the conditions

$$
v^T H_0 = 0 \tag{30}
$$
$$
v^T H_2 = 0 \tag{31}
$$

This yields

$$
r_k = v^T H_3 \left(\begin{array}{c} f_{k-s} \\ f_{k-s+1} \\ \vdots \\ f_k \end{array} \right) \tag{32}
$$

Clearly, the residual reflects the faults if

$$
v^T H_3 \neq 0 \tag{33}
$$

However, it is possible that even if this condition is fulfilled different independent faults compensate for each other, thus preventing the residual from reflecting them. In order to avoid this, the following more restrictive sensitivity condition must be imposed on the vector v^T [68].

$$
v^T H_{3,i} \neq 0 \tag{34}
$$

where

$$
H_{3,i} = \left(\begin{array}{ccccc} K_{m,i} & 0 & \cdots & \cdots & 0 \\ 0 & K_{m,i} & 0 & \cdots & 0 \\ 0 & 0 & K_{m,i} & \cdots & 0 \\ \vdots & \ddots & \ddots & \ddots & \vdots \\ 0 & \cdots & \cdots & 0 & K_{m,i} \end{array} \right), \quad i = 1 \ldots l_1 \tag{35}
$$

with $K_{m,i}$ denoting the ith column of the matrix K_m and

$$
H_{3,i} = \left(\begin{array}{ccccc} 0 & 0 & \cdots & \cdots & 0 \\ CK_{d,i} & 0 & 0 & \cdots & 0 \\ CAK_{d,i} & CK_{d,i} & 0 & \cdots & 0 \\ \vdots & \ddots & \ddots & \ddots & \vdots \\ CA^{s-1}K_{d,i} & \cdots & \cdots & CK_{d,i} & 0 \end{array} \right), \quad i = 1 \ldots l_2 \tag{36}
$$

with $K_{d,i}$ denoting the ith column of the matrix K_d. l_1 is the number of IF. l_2 is the combined number of CF and AF. Thus, $l = l_1 + l_2$ is the dimension of the fault vector f_k.

If Eq. (34) holds, the residual is sensitive to almost all but the so-called detection concealed faults [67]. Independently of the initial conditions and the unknown inputs, the residual will be different from zero if and only if a fault occurs.

In order to solve the FI problem, the set of all faults to be detected is partitioned into subsets of faults which are to be isolated. The number of different residual generators to be designed is then equal to the number of fault subsets. Each residual generator is designed sensitive to just one subset of faults, while the remaining faults are treated in the same manner as the intrinsic unknown inputs d_k.

It is worth to point out that the parity space approach results in a so-called dead-beat observer [68], i.e. an observer with poles located at the origin of the complex z-plane.

Geometrically, the parity approach can be interpreted according to Fig. 6. The distance between the observed value and the nominal trajectory corresponds to the value of the residual.

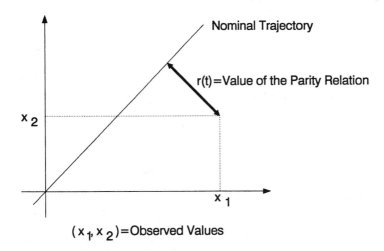

Fig. 6. Geometrical interpretation of the parity space approach

Eq. (30) and (31) constitute requirements for the complete decoupling of the residual from the unknown inputs. Since these conditions are rather restrictive, there will be many cases where no solution exists. Wünnenberg [67] proposed an optimal solution when no ideal solution can be found. To this end a performance index is defined which reflects a compromise between the effect of faults and unknown inputs on the residual. The performance index reads

$$p = \frac{w^T V_0 H_2 C_d H_2^T V_0^T w}{w^T V_0 H_3 C_f H_3^T V_0^T w} \tag{37}$$

The nominator and the denominator of p reflect the influence of the unknown inputs and the faults, respectively. C_d and C_f are weighting matrices. V_0 is the base of all solutions v^T of Eq. (30). Thus, the vector w^T singles out the optimal solution

$$v^T = w^T V_0 \tag{38}$$

of all possible solutions V_0 [68], thus minimizing p. Differentiating Eq. (37) with respect to w^T shows that the minimization problem can be solved via the generalized eigenvalue-eigenvector problem

$$w^T(V_0 H_2 C_d H_2^T V_0^T - p V_0 H_3 C_f H_3^T V_0^T) = 0 \qquad (39)$$

The minimal value of the performance index is equal to the minimal eigenvalue p. The corresponding eigenvector is w^T [67].

d. Robust FDI via Unknown Input Observers: The unknown input observer is a well-known tool in control theory used in conjunction with linear time-invariant systems subject to unknown inputs [20]

$$\dot{x} \;=\; Ax + Bu + E_1 d_1 + K_1 f_1 \qquad (40)$$
$$y \;=\; Cx + E_2 d_2 + K_2 f_2 \qquad (41)$$

Slightly modified, the unknown input observer approach can conveniently be employed for the purpose of robust FDI [31], [51], [57], [67].

A generalized observer for the given system is described by

$$\dot{\hat{z}} \;=\; F\hat{z} + Ju + Gy \qquad (42)$$
$$r \;=\; L_1 \hat{z} + L_2 y \qquad (43)$$

Suppose for the moment the system is operating fault free, i.e. $f_1 = f_2 = 0$. Then the observer state $\hat{z}$ is subject to the condition [67]

$$\hat{z}(t = t_0) = Tx(t = t_0) \quad \Rightarrow \quad \hat{z} = Tx \; \forall \; t \qquad (44)$$

where $T \in R^{l \times n}$, $l < n$ is a constant matrix. In other words, the observer is supposed to estimate a linear transformation $z = Tx$ of the original state x.

In order to detect the faults $f_{1,2}$ the residual r must satisfy the requirement [67]

$$f_1 \;\neq\; 0 \text{ or } f_2 \neq 0 \quad \Rightarrow \quad r \neq 0 \qquad (45)$$
$$f_1 \;=\; 0 \text{ and } f_2 = 0 \quad \Rightarrow \quad r(t \to \infty) = 0 \qquad (46)$$

These conditions must hold for all x, u and $d_{1,2}$. The estimation error

$$e = \hat{z} - Tx \qquad (47)$$

and the residual are governed by the equations

$$\dot{e} = \quad Fe + Ju + GCx + GE_2 d_2 + GK_2 f_2 \qquad (48)$$
$$\quad - \;\; TAx - TBu - TE_1 d_1 - TK_1 f_1$$
$$r = \quad L_1(e + Tx) + L_2 Cx + L_2 E_2 d_2 + L_2 K_2 f_2 \qquad (49)$$

Now the requirements imposed on the estimation error and the residual can be translated into conditions to be satisfied by the matrices defining the observer and the residual.

$$TA - FT = GC, \ F \text{ stable} \tag{50}$$
$$J = TB \tag{51}$$
$$TE_1 = 0 \tag{52}$$
$$GE_2 = 0 \tag{53}$$
$$L_2E_2 = 0 \tag{54}$$
$$L_1T + L_2C = 0 \tag{55}$$

On these conditions e and r evolve according to

$$\dot{e} = Fe + GK_2f_2 - TK_1f_1 \tag{56}$$
$$r = L_1e + L_2K_2f_2 \tag{57}$$

It can be seen that the residual remains unaffected by the unknown inputs $d_{1,2}$, the state x and the input u.

In order to avoid cancellations of different components of the fault signals, it must be required that [67]

$$\text{rank}(TK_1) = \text{rank}(K_1) \tag{58}$$
$$\text{rank}\left(\left(\begin{array}{c} G \\ L_2 \end{array}\right)K_2\right) = \text{rank}(K_2) \tag{59}$$

If these requirements are fulfilled, the faults will be reflected by the residual independently of their time evolution. Necessary and sufficient conditions for the existence of solutions of Eq. (50)-(55), (58) and (59) have been derived in [67].

The block diagram displayed in Fig. 7 illustrates the discussed concept.

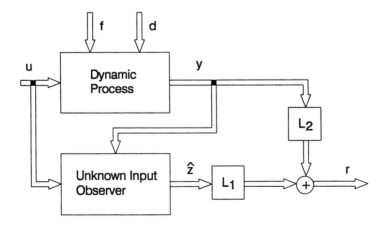

Fig. 7. FDI concept using an unknown input observer

For the purpose of FI, one proceeds according to Section IV,A,1,c, i.e. the observer is sensitized only to those faults which are to be detected at a time, while the other possible faults are treated like the intrinsic unknown inputs $d_{1,2}$. It is also conceivable to utilize the dedicated observer scheme or the generalized dedicated observer scheme outlined in Section IV,A,1,a, in order to isolate faults.

e. Robust FDI via Observer Eigenstructure Assignment: This method utilizes the technique of eigenstructure assignment which is well-known from control theory for the design of FDI observers [51]. The underlying process model is assumed to be linear and time-invariant. It is subject to AF and IF. Moreover, the model accounts for uncertainties and disturbances by the unknown inputs d.

$$\dot{x} = Ax + Bu + Ed + Bf_a \qquad (60)$$
$$y = Cx + f_s \qquad (61)$$

Based on this model an identity observer is set up

$$\dot{\hat{x}} = A\hat{x} + Bu + G(y - \hat{y}) \qquad (62)$$
$$\hat{y} = C\hat{x} \qquad (63)$$

The estimation error $e = x - \hat{x}$ evolves according to

$$\dot{e} = Ae - G(y - \hat{y}) + Ed + Gf_s + Bf_a \qquad (64)$$

The residual will be defined by means of the estimation error

$$r = He \qquad (65)$$

where H is a constant weighting matrix to be specified in the sequel. Suppose for the moment $f_a = 0$. In the absence of IF the estimation error will be zero, if H satisfies the following conditions [51]

$$H = \{X \mid x^T(\nu I - A + GC) = 0\} \qquad (66)$$
$$\text{for some complex number } \nu$$
$$H = WC \qquad (67)$$
$$\text{for some suitably designed matrix } W$$
$$\text{rank}(H) = \text{rank}(C) \qquad (68)$$

A matrix H, satifying these conditions defines a so-called invariant submanifold [51]. The residual is then given by
$$r = W(y - \hat{y}) \qquad (69)$$

Note that if the estimation error is zero regardless of d, the observer is robust to the unknown inputs. Complete insensitivity to f_s, f_a and d will be achieved on the conditions

$$H(A - GC)^i = 0, \quad i = 1 \ldots n - 1 \qquad (70)$$

$$HG \;=\; 0 \qquad\qquad (71)$$
$$HB \;=\; 0 \qquad\qquad (72)$$
$$HE \;=\; 0 \qquad\qquad (73)$$

In order to solve the FI problem with respect to certain sensor or actuator faults, the observer must be sensitized to those faults. This means that the direct transmission terms HG or HB must be designed to provide windows for the transmission of the faults to be detected. The aim of sensitizing the rows of HG or HB to certain instruments or actuators can be achieved by considering the dual control problem corresponding to the estimation error described by Eq. (64). The output estimation error $y - \hat{y}$ is then understood as the forcing signal of the system. This problem can be treated by the well-known methods of eigenvalue and eigenvector (eigenstructure) assignment. For details on the solution we refer to [51].

Notice the difference between the approach outlined in this section and the un-known input observer method discussed previously: In the previous section, based on a generalized reduced order observer, the state estimation error has been decoupled from the unknown inputs by a linear state transformation $z = Tx$. Subsequently, based on the estimated transformed state and the available measurements, a robust residual has been generated. In contrast to that approach, here not the state esti-mation error but the output estimation error is decoupled from the unknown inputs by multiplication with a suitable matrix W, thus generating a robust residual. The state estimation error is still subject to the unknown inputs, because the underlying structure is a full order identity observer.

f. Robust FDI via Frequency Domain Methods: In the previous sections we have discussed various time domain methods for the development of residual gen-erators. However, in analogy to linear control theory where both time and frequency domain methods are widely used for system design, frequency domain methods can be employed for the development of robust observer-based FDI algorithms as well [16], [59]. Especially mentionable are the recently developed techniques based on H_∞-theory [15], [22], [29] which will be the subject of this section.

Consider the linear time-invariant model [15]

$$\dot{x} \;=\; Ax + Bu + E_1 d + K_1 f \qquad\qquad (74)$$
$$y \;=\; Cx + Du + E_2 d + K_2 f \qquad\qquad (75)$$

The signal f denotes a q-dimensional fault vector. d incorporates structured uncer-tainties and disturbances represented by unknown inputs.

Applying the Laplace transform and writing the result in form of transfer functions yields on the assumption $d = 0$

$$y(s) = G_u(s)u(s) + G_f(s)f(s) \qquad\qquad (76)$$

where

$$G_u(s) \;=\; C(sI - A)^{-1}B + D \qquad\qquad (77)$$
$$G_f(s) \;=\; C(sI - A)^{-1}K_1 + K_2 \qquad\qquad (78)$$

A general residual generator is given by

$$r(s) = F(s)u(s) + H(s)y(s) \tag{79}$$

It is subject to the properties

$$f(s) = 0 \quad \Rightarrow \quad r(s) = 0 \tag{80}$$
$$f(s) \neq 0 \quad \Rightarrow \quad r(s) \neq 0 \tag{81}$$

where $F(s)$, $H(s) \in RH_\infty$. RH_∞ denotes the set of all stable and proper transfer matrices. For the purpose of FI it must be required that [15]

$$f_i(s) = 0 \quad \Rightarrow \quad r_i(s) = 0, \quad i = 1 \ldots q \tag{82}$$
$$f_i(s) \neq 0 \quad \Rightarrow \quad r_i(s) \neq 0, \quad i = 1 \ldots q \tag{83}$$

It has been shown in [15] that all residual generators can be parametrized according to

$$r(s) = Q(s)(\hat{M}_u(s)y(s) - \hat{N}_u(s)u(s)) \tag{84}$$

where $Q(s)$ denotes the parametrization. $\hat{M}_u(s)$ and $\hat{N}_u(s)$ build a left-coprime factorization of $G_u(s)$ which can be calculated by the algorithm given in [22]. $Q(s)$ is subject to the condition

$$Q(s)\hat{N}_f(s) = I \tag{85}$$

where $\hat{N}_f(s)$ is determined from $G_f(s)$ by the algorithm provided in [22]. If the faults are to be isolated $Q(s)$ must be selected such that

$$Q(s)\hat{N}_f(s) = \mathrm{diag}(t_1(s) \quad \ldots \quad t_q(s)) \tag{86}$$

The design problem is solved, if a RH_∞ matrix $Q(s)$ can be found satisfying the stated requirements [29]. If certain existence conditions which are derived in [15] are fulfilled, this problem can be solved by the algorithms provided in [22].

Fig. 8 displays a block diagram where $Q(s)$ is interpreted as post-filter of the pre-processed measurement data u and y.

In order to achieve robustness to the unknown inputs d, a performance index according to [15], [29] can be defined. A proper selection is

$$J = \frac{\| \partial r(s)/\partial f(s) \|}{\| \partial r(s)/\partial d(s) \|} \tag{87}$$

A maximization of J will result in a residual which is maximally sensitive to the faults and minimally sensitive to the unknown inputs.

This approach also provides means for an optimal isolation of the faults [14]. To this end a bank of residual generators is designed such that one residual is maximally sensitive to only one fault and minimally sensitive to the others.

The maximization of J yields, similar to the time domain solution in Section IV,A,1,c, a generalized eigenvalue-eigenvector problem in the frequency domain where the maximal eigenvalue is the optimal value of the performance index and the corresponding eigenvector is the optimal residual selector. Eventually, the optimization results in a realizable optimal frequency selecting post-filter $Q(s)$ [14], [29].

Consider now unstructured uncertainties, characterized by the unknown transfer

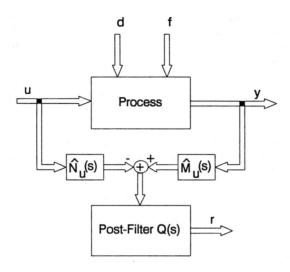

Fig. 8. Block diagram of the H_∞-FDI scheme

matrices $\Delta G_u(s)$, $\Delta G_d(s)$ and $\Delta G_f(s)$ as introduced in Section II,B. Suppose, the uncertainties are bounded and the bounds are known.

$$| \Delta G_u(j\omega) | \leq \delta_u(\omega) \tag{88}$$

$$| \Delta G_d(j\omega) | \leq \delta_d(\omega) \tag{89}$$

$$| \Delta G_f(j\omega) | \leq \delta_f(\omega) \tag{90}$$

If these uncertainties remain unconsidered while deciding either in favor or against an alarm, false alarms will be the consequence. According to [29] it is possible to determine a minimum detectable fault on the basis of the bounds of these uncertainties. The minimum detectable fault is a function of the post-filter $Q(s)$. Thus, it is in turn possible to design a post-filter $Q(s)$ such that the minimum detectable fault becomes minimal. For details on this subject we refer to [29].

2. OBSERVER-BASED FDI IN NONLINEAR SYSTEMS

In this section we discuss several observer-based methods for residual generation which are to be employed, if the system to be supervised cannot be represented by a linear model with sufficient accuracy. If the process under consideration does not operate at a constant operating point, but is subject to transients covering a wide range of working conditions, a linearization about a constant operating point will produce significant model-plant mismatches thus causing false alarms. Therefore nonlinear models must be used in order to devise suitable observer-based FDI algorithms.

a. Nonlinear Fault Detection Filter: This approach to FDI for nonlinear systems has been introduced in [24]. Similarly to the linear fault detection filter outlined in Section IV,A,1,b, it is based on an identity observer for the supervised system. The model of the automatic process is given in form of the nonlinear state space description

$$\dot{x} = f(x, u) + f_1 \tag{91}$$
$$y = c(x, u) + f_2 \tag{92}$$

where $f_{1,2}$ denote the possible faults modelled by external input signals. Provided $f_{1,2} = 0$, a nonlinear identity observer for this system has, according to [70], the following structure

$$\dot{\hat{x}} = f(\hat{x}, u) + H(\hat{x}, u)(y - \hat{y}) \tag{93}$$
$$\hat{y} = c(\hat{x}, u) \tag{94}$$

A 1st-order Taylor series approximation reveals that in case $\| e \|$ is sufficiently small, the estimation error $e = \hat{x} - x$ evolves according to the linear differential equation

$$\dot{e} = F(t)e - H(\hat{x}, u)f_2 - f_1 \tag{95}$$

where

$$F(t) = \frac{\partial f(\hat{x}, u)}{\partial \hat{x}} - H(\hat{x}, u)\frac{\partial c(\hat{x}, u)}{\partial \hat{x}} \tag{96}$$

A suitable residual is defined by

$$r = \hat{y} - y \tag{97}$$

A linearization of Eq. (97) about $e = 0$ yields

$$r = \frac{\partial c(\hat{x}, u)}{\partial \hat{x}}e - f_2 \tag{98}$$

It can be seen that the faults $f_{1,2}$ are reflected in r either directly or indirectly via e. The design problem that remains to be solved consists of two steps:

1. Find a time-variant matrix $H(\hat{x}, u)$ such that $e = 0$ is an asymptotically stable equilibrium point of Eq. (95). In many practical situations even a constant matrix H may be sufficient for this purpose [23].

2. Exploit the remaining degrees of freedom in $H(\hat{x}, u)$ to decouple the effects of the different faults on the residual in order to allow FI.

Alternatively to step 2, a dedicated observer approach similar to the one outlined in Section IV,A,1,a can be pursued for FI purposes.

It should be noted, however, that no general algorithm providing a solution of these problems is known. Since the problem is nonlinear, complex numerical and computational problems may arise. Moreover, the problem of robustness to modelling errors and disturbances remains unsolved by this approach.

In [70] an algorithm for the stabilization of the system

$$\dot{e} = F(t)e \tag{99}$$

is proposed. It is based on a stability theorem that can be found in [5].

In Fig. 9 the structure of the discussed FDI concept is displayed in form of a block diagram.

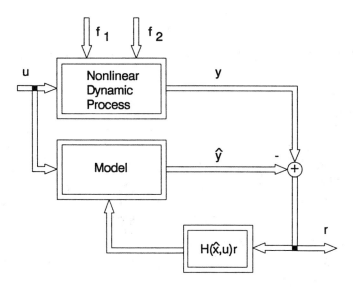

Fig. 9. Residual generation by nonlinear identity observers

b. Robust FDI via Nonlinear Unknown Input Observers: The linear unknown input observer discussed in Section IV,A,1,e can be extended easily to a certain class of nonlinear systems [67]. This class is characterized by the following state space description

$$\dot{x} = Ax + B(y, u) + E_1 d_1 + K_1 f_1 \tag{100}$$
$$y = Cx + E_2 d_2 + K_2 f_2 \tag{101}$$

Note that the nonlinear term $B(y, u)$ depends only on y and u, i.e. signals which are directly available by measurements. It is therefore possible to compensate completely for the nonlinearity by reproducing it exactly by the observer

$$\dot{z} = F\hat{z} + J(y, u) + Gy \tag{102}$$
$$r = L_1 \hat{z} + L_2 y \tag{103}$$

r is again the residual. The conditions which are to be met by the observer matrices are identical to those in Eq. (50). They are repeated here for the convenience of the reader. The only difference is that $J(y, u)$ is now a nonlinear function and not a constant matrix as before.

$$TA - FT = GC, \ F \text{ stable} \tag{104}$$

$$J(y,u) = TB(y,u) \tag{105}$$

$$TE_1 = 0 \tag{106}$$

$$GE_2 = 0 \tag{107}$$

$$L_2E_2 = 0 \tag{108}$$

$$L_1T + L_2C = 0 \tag{109}$$

If these requirements can be fulfilled, the dynamics of the residual are described by

$$\dot{e} = Fe + GK_2f_2 - TK_1f_1 \tag{110}$$

$$r = L_1e + L_2K_2f_2 \tag{111}$$

The drawback of this elegant extension of the linear FDI theory to nonlinear systems is the fact that the class of systems described by models fitting Eq. (100) is rather small. Many technical or physical systems cannot be modelled naturally in this form. If this is the case, the given physical model must be transformed into the required form by a suitable nonlinear state space transformation. The existence conditions for these transformations are very restrictive. Consequently, the class of models which are actually transformable is also small. Even if the existence conditions can be satisfied, finding the transformation will be hampered by the necessity to solve nonlinear partial differential equations [6].

A different approach [54], [55] to FDI in nonlinear systems which extends the class of transformable systems by leading to weaker existence conditions, is based on the following, more general model

$$\dot{x} = A(x) + B(x)u + E(x)d + K(x)f \tag{112}$$

$$y = C(x) \tag{113}$$

where the unknown inputs and faults are modelled according to Section II. A nonlinear transformation $z = T(x)$ is sought, seperating the disturbed from the undisturbed portion of the state. This seperation will be achieved if and only if

$$\frac{\partial T(x)}{\partial x}E(x) = 0 \tag{114}$$

This relation constitutes a system of linear partial differential equations to be solved simultanously by $z = T(x)$. The theorem of Frobenius [37] can be exploited to formulate existence conditions for solutions of Eq. (114) [54], [55].

Suppose now solutions $z = T(x)$ of Eq. (114) exist. On the assumption, that a relation $x = \Psi_0(z, y^*)$ exists, the model can be rewritten as

$$\dot{z} = \frac{\partial T(x)}{\partial x}(A(x) + B(x)u + K(x)f)|_{x=\Psi_0(z,y^*)} \tag{115}$$

where the output transformation $y^* = C^*(y)$ denotes a subset of the set of measurements $y = C(x)$ with $\dim(y^*) < \dim(y)$. Suppose furthermore, a relation $R(T(x), C(x)) = 0$ exists. Then a nonlinear observer similar to the one in [70] can be set up in order to estimate the undisturbed portion z of the state x. The observer is of the structure

$$\dot{\hat{z}} = \frac{\partial T(\hat{x})}{\partial \hat{x}}(A(\hat{x}) + B(\hat{x})u) + H(\hat{z}, y, u)R(\hat{z}, y)|_{\hat{x}=\Psi_0(\hat{z},y^*)} \tag{116}$$

where the feedback matrix $H(\hat{z}, y, u)$ provides design freedom for the stabilization of the differential equation governing the dynamics of the estimation error $e = \hat{z} - z$.

This observer is called nonlinear unknown input observer or disturbance decoupled nonlinear observer [55].

The relation $R(\hat{z}, y)$ can be used conveniently as a residual.

$$r = R(\hat{z}, y) = R(z + e, y) \tag{117}$$

The estimation error e evolves according to

$$\dot{e} = \rho(e, t) - \frac{\partial T(x)}{\partial x} K(x)f \tag{118}$$

where the system of nonlinear differential equations $\dot{e} = \rho(e, t)$ must be designed such that the equilibrium point $e = 0$ is, at least locally, asymptotically stable. The residual will then converge to zero if $f = 0$. On the other hand, all faults f will be reflected by the residual, if

$$\text{rank}\left(\frac{\partial T(x)}{\partial x} K(x)\right) = \text{rank}(K(x)) \tag{119}$$

If in additon to Eq. (114) and Eq. (119), the conditions

$$\frac{\partial T(x)}{\partial x} A(x) = FT(x) + \Phi_0(C(x)) \tag{120}$$

$$\frac{\partial T(x)}{\partial x} B(x) = \Phi_1(C(x)) \tag{121}$$

are also satisfied, a residual generator with stable linear error dynamics can be designed [54]. F is a stable constant matrix. $\Phi_0(C(x))$ and $\Phi_1(C(x))$ are suitable output transformations. The residual generator is then given by the following equations

$$\dot{\hat{z}} = F\hat{z} + \Phi_0(y) + \Phi_1(y)u \tag{122}$$

$$r = R(\hat{z}, y) \tag{123}$$

In this case, the estimation error and the residual evolve according to

$$\dot{e} = Fe - \frac{\partial T(x)}{\partial x} K(x)f \tag{124}$$

$$r = R(T(x) + e, C(x)) \tag{125}$$

It is important to note that, in contrast to the nonlinear observer in Eq. (93) which is only locally stable in a neighborhood of the equilibrium point $e = 0$, the observers in Eq. (102) and Eq. (122) are globally stable. Regarding the observer in Eq. (116) no such statement can be made.

Existence conditions for the outlined nonlinear observers and residual generators based on the Theorem of Frobenius [37], [21] can be found in [54] and [55].

Fig. 10 displays a block diagram of the nonlinear observer-based FDI concept.

For the purpose of FI the set of all considered faults can be devided into subsets of faults which are treated like the intrinsic unknown inputs d if they are not to be detected by a certain residual generator. For a more detailed description of this FI strategy see Section IV,A,1,c. It is also conceivable to employ the dedicated observer

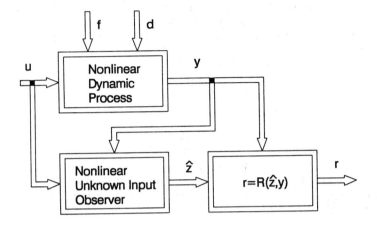

Fig. 10. FDI via nonlinear unknown input observers

scheme or the generalized dedicated observer scheme. For details on this subject we refer to Section IV,A,1,a.

B. IDENTIFICATION-BASED METHODS

The FDI techniques discussed in Section A were based on state observers, i.e. the state of an automatic process is being estimated and the estimates are subsequently compared to actual measurement data. Discrepancies between estimates and measurements provide means for a decision whether or not a fault has occurred. Yet, the mathematical models the observers are based on consist not only of the states and the relations between them but also of mathematical parameters comprising information about the actual physical parameters like friction, mass, resistance, viscosity etc. Thus, a natural alternative approach to model-based FDI is the on-line estimation of the parameters of the process under supervision using some suitable parameter estimation technique such as the least squares algorithm. The estimated parameters are further investigated in order to reach conclusions on the condition of the process.

1. FDI VIA PARAMETER ESTIMATION

The parameter estimation approach to FDI which is to be discussed in this section has been proposed in [35], [36]. The models normally used as a basis for the parameter estimation procedure are linear input-output differential equations with lumped parameters

$$\frac{d^n y}{dt^n} = \sum_{i=0}^{m} b_i \frac{d^i u}{dt^i} - \sum_{i=0}^{n-1} a_i \frac{d^i y}{dt^i} \tag{126}$$

where y and u are deviations from a certain operating point of the output and the input signals, respectively [35]. The vector of mathematical parameters

$$\theta = (a_{n-1} \cdots a_0 \quad b_m \cdots b_0)^T \tag{127}$$

is related to the physical parameters p by a vector function

$$\theta = f(p) \tag{128}$$

that must be determined . Measuring the input and output signal as well as determining the time derivatives at the discrete times $t = kT_0$, $k = 0 \ldots N$, where T_0 is the sampling time, yields $N + 1$ equations of the form

$$y_k^n = \psi_k^T \hat{\theta} + e_k, \quad k = 0 \ldots N \tag{129}$$

where $\hat{\theta}$ is a vector containing the estimated mathematical parameters $\hat{a}_i$ and $\hat{b}_i$. The vector ψ_k consists of the input and the output data and their time derivatives which are to be measured or calculated. The term e_k denotes the so-called equation error [35]. This procedure eventually results in a vector equation

$$y^n = \Psi \hat{\theta} + e \tag{130}$$

The data matrix Ψ consists of $N + 1$ rows of data vectors ψ_k^T. A suitable estimation algorithm is the well-known nonrecursive least squares method given by

$$\hat{\theta} = (\Psi^T \Psi)^{-1} \Psi^T y^n \tag{131}$$

Next, the estimated physical parameters

$$\hat{p} = f^{-1}(\hat{\theta}) \tag{132}$$

are determined by calculating the inverse function of Eq. (128) where it is assumed that the inverse function exists. Eventually, the parameter variation

$$\Delta p = \hat{p} - p_{nom} \tag{133}$$

is calculated, comparing the estimated physical parameters $\hat{p}$ with the known nominal values of the physical parameters p_{nom}. Based on Δp, it can be decided when and where a fault has occurred.

The algorithm outlined in this section is displayed in Fig. 11.

2. ROBUST FDI VIA HYPOTHESES TESTING

This method [8] allows to consider systems with unmodelled dynamics characterized by an unstructured uncertainty $G_\Delta(z)$ according to

$$y(z) = (G(z, \theta) + G_\Delta(z))u(z) \tag{134}$$

where $G(z, \theta)$ is the transfer matrix of the explicitly modelled portion of the system depending on a parameter vector θ and the complex variable z. The uncertainty can be treated as a stochastic disturbance acting on the nominal model, provided the

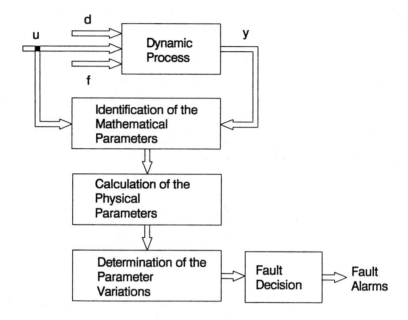

Fig. 11. FDI via parameter estimation

impulse response of the unmodelled part of the system is a realization of a stochastic process with zero mean [8]. As in the previous section, the parameters are estimated using the nonrecursive least squares algorithm. Suppose the parameter vector assumes the nominal value $\theta = \theta_n$ in the fault free case and the value $\theta = \theta_f$ after a fault has occurred. Now two hypotheses are defined

$$H_0 \quad : \quad \theta_n = \theta_f \tag{135}$$

$$H_1 \quad : \quad \theta_n \neq \theta_f \tag{136}$$

where H_0 is the hypothesis assuming fault free system operation, while H_1 assumes the occurrence of a fault. Under the hypothesis H_0 the covariance

$$S = \mathrm{Cov}(\hat{\theta}_n - \hat{\theta}_f) \tag{137}$$

is calculated, where $\hat{\theta}_{n,f}$ are the estimated parameters. The robust FD procedure is now based on the idea of comparing the known value of $(\hat{\theta}_n - \hat{\theta}_f)(\hat{\theta}_n - \hat{\theta}_f)^T$ with the expected value, i.e. with the covariance S [26]. The hypothesis test can be performed using several different residual definitions. One possible test is defined by

$$r = (\hat{\theta}_n - \hat{\theta}_f)^T S^{-1}(\hat{\theta}_n - \hat{\theta}_f) \tag{138}$$

Subsequently to the calculation, r is to be compared to a preassigned fixed threshold λ. If $r < \lambda$, the hypothesis H_0 is accepted.

Another suitable choice is

$$r_i = \frac{|\hat{\theta}_{ni} - \hat{\theta}_{fi}|}{C_{ii}} \tag{139}$$

These definitions must be evaluated for all components of the vectors $\theta_{n,f}$. The results are also to be compared to a threshold.

A third alternative is to form a Cholesky factorization $S = XX^T$ of the covariance matrix and to evaluate

$$r = X^{-1}(\hat{\theta}_n - \hat{\theta}_f) \tag{140}$$

The magnitude of each component of r is again compared to a fixed threshold. If all components remain below the threshold, the hypothesis H_0 will be accepted [8].

Notice that the latter two tests also provide information on the location of the fault within the supervised system, thus providing means for the isolation of different faults, while the first test only solves the FD problem.

V. RESIDUAL EVALUATION

Throughout Section IV,A we have assumed ideal conditions in the sense that the residuum will vanish completely, if no fault occurs. In actual applications however, this will never be the case. It is therefore necessary to develop residual evaluation methods and algorithms which are to be applied subsequently to the residual generation. The algorithms must lead to reliable decisions in the sense that they enable distinctions between situations where a fault actually has occurred and situations where the residual was excited by some disturbance which does not constitute a failure in the supervised process. The aim of these methods is to keep the false alarm rate due to disturbances as small as possible while assuring at the same time that actual faults will not remain undetected. In this section, various approaches to this problem will be discussed.

A. GENERALIZED LIKELIHOOD RATIO TECHNIQUE

The general likelihood ratio technique is based on the idea that the supervised process is subject to stochastic disturbances. In this case Kalman filters are usually employed for the purpose of state estimation as well as for the generation of residuals. The residual is then defined by the innovation of the Kalman filter [47], [65]

$$r_k = y_k - C\hat{x}_{k|k-1} \tag{141}$$

where $\hat{x}_{k|k-1}$ is the Kalman estimate of the state. $y_k = Cx_k$ denotes the system output. Provided, the system- and the sensor-noise signals are white with zero-mean, the innovation sequence, i.e. the residual will also be white noise with zero mean [32]. Based on this fact the generalized likelihood ratio technique can be devised. Considering faults f_k acting on the system, the residual can be formulated as

$$r_k = r_{0,k} + G_k f_k \tag{142}$$

where $r_{0,k}$ is white noise with zero mean. The faults act on the residual through the matrix G_k. To decide whether a fault has occurred, a set of hypotheses is established.

$$H_0 \quad : \quad r_k = r_{0,k} : \qquad \text{no fault has occurred}$$
$$H_i \quad : \quad r_k = r_{0,k} + G_{i,k}f_{i,k} : \quad \text{the } i^{th} \text{ fault has occurred}$$

In order to decide in favor of one hypothesis, the so-called likelihood ratio is to be evaluated. To this end the maximum likelihood estimates $\hat{f}_{i,k}$ of f_k are calculated on the assumption that H_i is true [58]. The results are substituted into the definition of the likelihood ratio

$$L_{i,k} = \frac{p(r_{01}, \ldots, r_{0,k} | H_i, f_k = \hat{f}_{i,k})}{p(r_{01}, \ldots, r_{0,k} | H_0)} \tag{143}$$

where $p(.)$ denotes the probability density function of the underlying stochastic process. The result is compared to a certain predetermined threshold T_r [58]. Based on the rule

$$L_{i,k} > T_r : \quad H_i \text{ is accepted} \tag{144}$$

$$L_{i,k} < T_r : \quad H_0 \text{ is accepted} \tag{145}$$

it is then decided in favor or against the occurrence of the fault $f_{i,k}$.

This process is illustrated in Fig. 12.

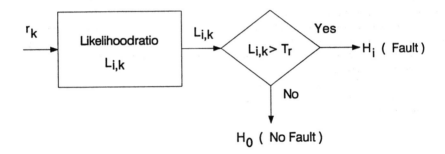

Fig. 12. Generalized likelihood ratio technique

B. MULTIPLE HYPOTHESES TESTING

The multiple hypotheses testing method is related to the general likelihood ratio technique which has been outlined in the previous section in the sense that for both methods the innovations of Kalman filters are used for the residual generation. The major difference between the two methods is that for multiple hypotheses testing various process models, corresponding to each possible fault mode are set up. Each fault mode is subject to the previously defined hypotheses H_0 and H_1. Based on these models a bank of Kalman filters is designed [1], [58]. The innovations of the Kalman filters are in turn used to compute the conditional probabilities for each hypothesis [58]. A hypothesis will be accepted if the corresponding conditional probability is close to one.

C. DECORRELATION METHOD

On certain assumptions it is possible to suppress model uncertainties by a so-called decorrelation filter [40]. Suppose that we have q residuals which are again generated by Kalman filters. They are then described by the equations

$$r_{i,k} = \Delta m_{i,k} + f_{i,k} + n_{i,k}, \quad i = 1 \ldots q \tag{146}$$

where the signals $f_{i,k}$ represent the sensor faults, the $\Delta m_{i,k}$ denote the model uncertainties and the $n_{i,k}$ are the measurement noise signals. Now each $r_{i,k}$ can be predicted or decorrelated from one or more so-called reference signals $r_{j,k}, \quad i \neq j$ [40]. Assuming that the faults do not occur at the same time, a fault f_i affecting r_i will not be contained in the other signals and can therefore not be predicted [40]. On the other hand, the model uncertainties $\Delta m_{i,k}$ will affect all or many components of r_k and may be suppressed by decorrelation.

Consider for instance a system with two sensors, both of them subject to noise and faults as displayed in Fig. 13.

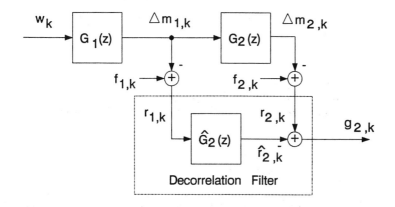

Fig. 13. Model for the error signals r_k

The block diagram shows the structure for a prediction of $r_{2,k}$ from $r_{1,k}$. The uncertainties $\Delta m_{1,2,k}$ are generated by white noise w_k and two adaptive shaping filters $G_{1,2}(z)$. On the condition that the sensor faults and the model uncertainties are uncorrelated, the signal $g_{2,k}$ equals $f_{2,k}$. Thus, $g_{2,k}$ can be used as an estimate of the sensor fault $f_{2,k}$.

D. THRESHOLD SELECTOR

One of the most straight forward ways to deal with residuals excited by disturbances and model-plant mismatches as well as by faults, is to define a fixed threshold T_r to which the residual is compared. If the residual surpasses the threshold, the occurrence of a fault will be assumed. If, on the other hand, the residual remains below the threshold, the supervised process is considered free of faults. The problem is that the sensitivity to faults will be intolerably reduced if the threshold is selected to high, while choosing the threshold to low will increase the rate of false alarms [26]. To overcome this difficulty, the application of Markov theory was proposed for threshold determination [60]. Another simple and logical approach is that of adaptive threshold selection which has been demonstrated to be suitable for IFD under a class of modelling errors [49].

This method has been generalized in [19], where the concept of the threshold selector was introduced. In the frequency domain a linear time-invariant model with unstructured modelling errors can be written as

$$y(s) = (G_u(s) + \Delta G_u(s))u(s) + (G_f(s) + \Delta G_f(s))f(s) \tag{147}$$

where the unknown transfer matrices $\Delta G_{u,f}(s)$ represent the unstructured model uncertainty. For this model the dynamics of the residual generator are given by [18]

$$r(s) = Q(s)(\hat{M}_u(s)(G_f(s) - \Delta G_f(s) - \hat{M}_u(s)\Delta G_u(s)u(s)) \tag{148}$$

From this relation the following threshold selector can be constructed

$$T_r(s) = \mu Q(s)\hat{M}_u(s)u(s) \tag{149}$$

where μ is the bound of $\Delta G_u(s)$, i.e.

$$\| \Delta G_u(s) \| < \mu \tag{150}$$

A fault is declared if $\| r(s) \| > \| T_r(s) \|$.

In Fig. 14 this decision strategy is illustrated by a block diagram.

VI. KNOWLEDGE-BASED FDI METHODS

Knowledge-based methods (expert systems) complement the existing analytical and algorithmic methods of fault-diagnosis [26], [56]. They open a new dimension of possible fault-diagnosis for complex processes with incomplete process knowledge. While the algorithmic methods use quantitative analytical models, the expert system approach makes use of qualitative models based on the available knowledge of the system. The combination of both strategies allows the evaluation of all available information and knowledge of the system for fault-diagnosis. Such additional knowledge may be, for instance, the degree of ageing, the operational environment, used tools, the history of operation, fault statistics etc.

The core of the architecture of a fault-diagnosis system using knowledge-based methods is an on-line expert system which complements the analytical model-based method by the knowledge-based method using heuristic reasoning. The resulting overall fault-diagnosis system consists of the following architectural components:

1. The knowledge base (knowledge of facts and rules)

2. The data base (information about the present state of the process)

3. The inference engine (forward or backward reasoning)

4. The explanation component (to inform the user on why and how the conclusion was drawn)

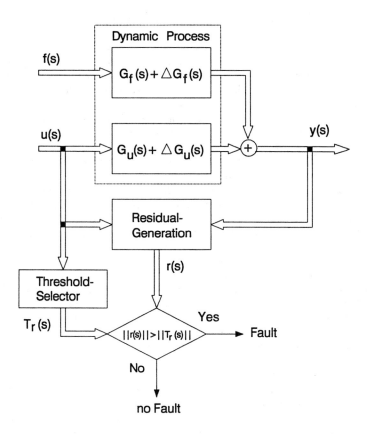

Fig. 14. Threshold selector scheme

The resulting on-line system combines the analytical redundancy FDI methods described in the previous sections with the method of fault-diagnosis by evaluation of heuristic knowledge about the process. This task is performed in the inference engine which has to combine heuristic reasoning with algorithmic operations in terms of the evaluation of analytical redundancy. The inference engine has access to:

1. The analytical knowledge in terms of the mathematical model (structure and parameters)

2. Heuristic knowledge of fault propagation, fault statistics, operation and environmental conditions, process history etc.

3. The actual data (inputs, outputs, operating conditions etc.)

For more details we refer to [56].

VII. ASPECTS OF PRACTICAL APPLICATION

Applications of the previously described fault-diagnosis methods to practical processes will encounter various problems which must be solved before a successful and reliable on-line operation of the FDI algorithms can be achieved. The most important requirement is certainly the avoidance of false alarms on the one hand and the detection security of actual faults on the other hand. If these requirements are not satisfied to a sufficient degree, the FDI algorithms will not be suitable for an installation in safety sensitive processes or, if installed, they will simply be turned off by the operating personnel.

As already discussed in Section II, it is crucial to provide process models which are as accurate and structured as possible to minimize the on principle unavoidable modelling errors. It will therefore often be the case that the available models which have already been utilized successfully for the controller design etc., do not fulfill the stronger requirements imposed on the models needed for fault-diagnosis. If this is the case, the process to be supervised must be modelled more accurately and detailed before FDI algorithms using analytical redundancy can be applied. This includes the task of fault- and disturbance modelling.

Most of the algorithms known up to date are devised on the basis of linear process models. Particularly in case the process is actually nonlinear, there will arise the problem of false alarms due to model-plant mismatches caused by linearization errors if the system is in a state of transition between different operating points. If the process is to operate in more then one operating point, consequently more than one linearization must be performed and the linear FDI algorithm must be adjusted accordingly.

To avoid this drawback, one can alternatively fall back on the recently developed algorithms suitable for nonlinear processes which have the advantage that they are not subject to linearizations and linearization errors in the first place.

It has been shown in various publications that the existence conditions for fault-diagnosis algorithms become weaker as the number of available independent measurements increases [31], [55]. Besides the already implemented sensors which are necessary for the control routines, it might therefore be required to provide additional instruments in order to fulfill the theoretical conditions for a realization of a particular FDI algorithm.

As far as the sensors are concerned, one must moreover consider the problems brought up by measurement noise. Besides the model-plant mismatches, measurement noise also introduces the necessity of proper threshold selection in order to reach reliable decisions, i.e. decisions by which both false alarms as well as missing the detection of actual failures is avoided. Another source of noise to be taken into account in this context is introduced by sampling and discretizing the measurements for the numerical processing by the computer.

For economical or other reasons it might be impossible to provide additional hardware components for supervision algorithms. In those cases it is generally desired to implement the FDI algorithms on the computers which are used in the first place to control the process, thus avoiding the necessity of additional hardware. Since both control and FDI algorithms must work on-line, the additional computer operations to be performed for the fault-diagnosis algorithms must be placed into the remaining computer time slots which are not used by the control algorithms. Particularly for the nonlinear FDI techniques this may present problems, because, for a given system, the number of mathematical operations to be performed on-line for a nonlinear algorithm is usually higher than for a linear technique. Moreover, the nature of the mathematical operations and the design of the algorithm is usually more complicated in the nonlinear than in the linear case. This argument shows that, in practice, there is a trade-off for the advantages of the nonlinear techniques which are in terms of reliability superior to the linear methods due to the overall reduction of model-plant mismatches.

Last but not least, it is necessary to keep the algorithms from being too complex and involved, because the operating personnel, who usually does not consist of specialists in the field of fault-diagnosis, must be able to understand and tune them [37].

VIII. PRACTICAL APPLICATIONS

In this section we present some examples where FDI techniques have been practically applied to automatic processes. It will be seen that the results are encouraging for further applications on a larger scale.

A. THREE-TANK SYSTEM

Fig. 15 shows the experimental set up of a three-tank system. The system consists of three cylindrical tanks with identical cross sections A, being filled with water. The tanks are interconnected by circular pipes. The water inlet is controlled by two pumps which are driven by an electronic power device with an appropriate control strategy implemented on a micro-computer. The system can be modelled conveniently by the mass balances of the three tanks represented by three nonlinear differential equations.

Based on the theory outlined in Section IV,A,2,b, three nonlinear observers are designed and combined to an on-line operated dedicated observer scheme (see Section IV,A,1,a). Fig. 16 shows the dynamic behaviour of the residual r_3 in case the tank 3 leaks from $t = 8s$ to $t = 30s$.

In Fig. 17 the residual r_2 in case the pipe between the tanks 2 and 3 is plugged for $t > 3s$ is displayed.

It can be seen that both types of faults can be detected by means of a simple constant threshold logic [67].

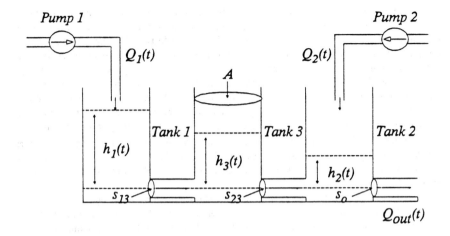

Fig. 15. Three-tank System (Wünnenberg [67])

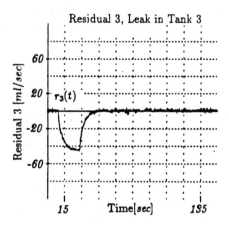

Fig. 16. Residual r_3 (Wünnenberg [67])

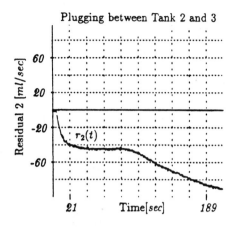

Fig. 17. Residual r_2 (Wünnenberg [67])

B. ROBOT MANUTEC R3

The robot Manutec R3 is a robot with six axes. The arms are connected by rotational joints. Each arm is driven by an electric motor and a gear box. The angles and velocities are measured by an encoder on each motor axis. Fig. 18 shows a draft of the robot Manutec R3 [48].

Based on the theory outlined in Section IV,A,2,b, an observer scheme consisting of six nonlinear observers, one for each axis, is implemented on a personal computer. The measured data from the robot is processed off-line. Fig. 19 shows the residual r_3 of the third observer, when a fault in the gear box of axis three occurs. The fault happens at $t = 1.1s$ while the robot arm performs an over-head manoeuvre with maximum velocity.

It can be seen that the residual clearly reflects the fault. The non-zero value of the residual during the time before the fault occurs is due to the unmodelled mechanical friction. The numerical value corresponds to the values reported in [48]. This effect can be eliminated easily, once the mechanical friction is modelled mathematically. As an alternative to modelling the friction explicitly, one can utilize an adaptive threshold logic, where the threshold is determined as a function of the velocity. Fig. 20 shows the fourth residual corresponding to an instrument fault in the first axis for $t > 1.1s$. Again, the fault can be detected with a constant threshold logic [67].

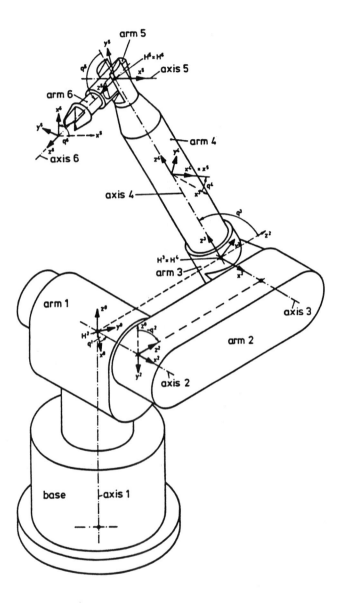

Fig. 18. Robot Manutec R3 (Otter and Türk [48])

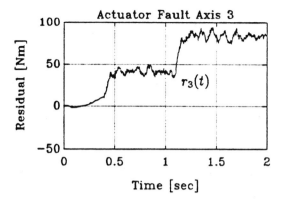

Fig. 19. Residual r_3 (Wünnenberg [67])

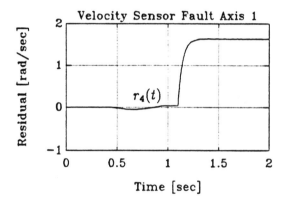

Fig. 20. Residual r_4 (Wünnenberg [67])

C. DC-MOTOR AND CENTRIFUGAL PUMP

This example illustrates experimental results obtained by the parameter identification method discussed in Section IV,B,1. The process is displayed in Fig. 21 and consists of a speed controlled DC-motor and a centrifugal pump [36].

In order to adjust the FDI algorithm initially to the process, the normal state of the system must be determined by a training run [36].

In the first experiment shown here, the contact surface of the motor brushes is reduced by 50% [36].

Fig. 22 shows the estimated physical parameters R_2 (resistance) and $\bar{\Psi}$ (flux linkage).

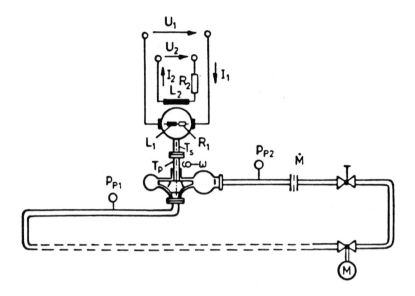

Fig. 21. DC-motor and centrifugal pump (Isermann [36])

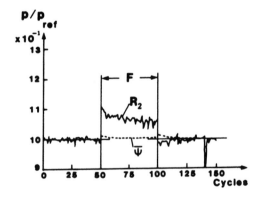

Fig. 22. Estimated physical parameters R_2 and $\bar{\Psi}$ (Isermann [36])

Fig. 23 displays a statistical comparison measure of the physical parameters in the faulty versus the unfaulty case which is sensitive to changes of the parameters. This signal is used as a residual. It is defined via the variances of the process parameters [36].

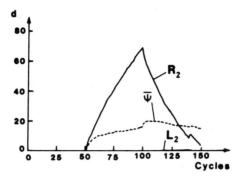

Fig. 23. Comparison of the faulty versus the unfaulty parameters (Isermann [36])

After a certain period of time, these signals clearly reflect the parameter change and therefore the fault. They can be evaluated by a constant threshold logic.

For the second experiment a reduction in the flow of cooling air is inflicted on the system. The results are illustrated in Fig. 24 and Fig. 25 [36].

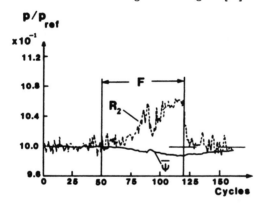

Fig. 24. Estimated physical parameters (Isermann [36])

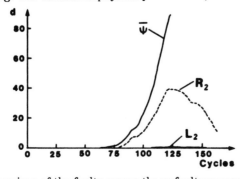

Fig. 25. Comparison of the faulty versus the unfaulty parameters (Isermann [36])

Again, the fault can be detected by applying a constant threshold.

D. STEAM GENERATOR

Based on a linearized model of a U-tube steam generator several FDI methods, i.e. the detection filter (see Section IV,A,1,b), the parity space approach (see Section IV,A,1,c) and the generalized likelihood ratio technique (see Section V,A) are combined to the instrument fault-diagnosis scheme shown in Fig. 26 [53].

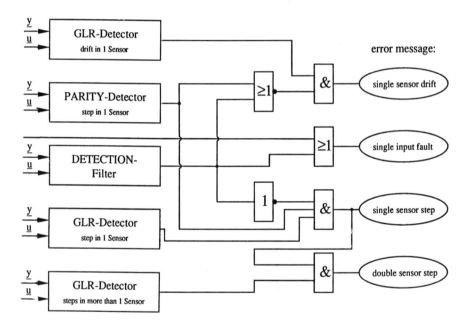

Fig. 26. Scheme for Instrument-FDI in steam generators (Prock [53])

It has been shown in [53] that the performance of the different schemes depends strongly on the fault type, i.e. whether the fault occurs suddenly (abrupt fault) or is slowly developing (incipient fault).

In Fig. 27 the residual generated by the detection filter after a 100% step fault of the input feed water temperature for $t > 133s$ is displayed [53].

The fault is detectable by means of a constant threshold logic. Fig. 28 shows the performance of the general likelihood ratio technique after a 1% drift error has been inflicted on the pressure sensor for $t > 100s$ [53].

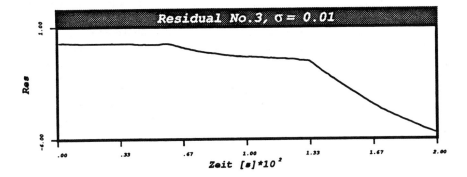

Fig. 27. Detection filter residual (Prock [53])

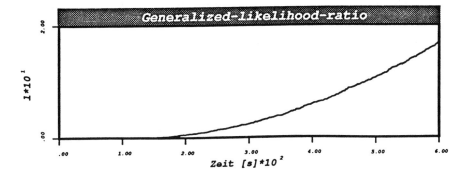

Fig. 28. General likelihood ratio (Prock [53])

IX. CONCLUSIONS

In this contribution we have discussed the analytical redundancy approach to FDI in automatic processes. Various techniques and procedures have been described briefly or in some detail.

The development has now reached a stage where practical applications must follow in order to support the theoretical results and to achieve acceptance by the industry. For several techniques this has already been done on an experimental stage. The experimental results are very promising, which encourages further applications.

It should be pointed out however, that the success of FDI by analytical redundancy in terms of reliability (i.e. the avoidance of false alarms and the security of the detection of faults) strongly depends on the accuracy of the models the algorithms are based on. This point is closely connected to the question of robustness of the algorithms to modelling errors or model-plant mismatches. Recently a lot of progress has been made in this field. We name here particularly the linear as well as the nonlinear unknown input observer approaches and the frequency domain design methods.

The available degrees of freedom for a robustness enhancement are strongly related to the number of independent measurements. This also affects the possibilities for the isolation of different faults.

If the available models are too inaccurate for an application of observer-based or parameter estimation methods alone (like for instance in the chemical industry), one can combine these techniques with a knowledge-based approach in order to exploit all available information for the generation of analytical redundancy.

Considering the development of computer technology and artificial intelligence, a combination of the different methods is bound to become a powerful tool for FDI in automatic processes in the future.

X. REFERENCES

1. M. Athans, R.H. Whitlind and M. Gruber, "The Stochastic Control of the F-8 Aircraft Using Multiple Adaptive Control (MMAC)", *IEEE Trans. Autom. Contr.* **AC-22** (1977).

2. M. Basseville, "Detecting Changes in Signals and Systems - A Survey", *Automatica* **24**, pp. 309-326 (1988).

3. M. Basseville and A. Benveniste (Eds.), "Detection of Abrupt Changes in Signals and Dynamical Systems", *Lecture Notes in Control and Information Sciences* **77**, Springer-Verlag (1985).

4. R.V. Beard, "Failure Accomodation in Linear Systems through Self-Reorganisation", Dept. MVT-71-1, Man Vehicle Laboratory, Cambridge, Massachusetts (1971).

5. L. Cesari, "Asymptotic Behavior and Stability Problems in Ordinary Differential Equations", Springer-Verlag (1971).

6. J. Birk and M. Zeitz, "Extended Luenberger Observer for Non-Linear Multivariable Systems", *Int. J. of Control* **47**(6), pp. 1823-1836 (1988).

7. E.Y. Chow and A.S. Willsky, "Analytical Redundancy and the Design of Robust Failure Detection Systems", *IEEE Trans. Autom. Contr.* **AC-29**, pp. 603-614 (1984).

8. B. Carlsson, M. Salgado and G. Goodwin, "A New Method for Fault Detection and Diagnosis", *Technical Report EE8842*, Dept. of Electr. and Comp. Eng., U. of Newcastle, New South Wales, Australia (1988).

9. R.N. Clark, "A Simplified Instrument Failure Detection Scheme", *IEEE Trans. Aerospace Electron. Syst.* **AES-14**, pp. 558-563 (1978)

10. R.N. Clark, "Instrument Fault Detection", *IEEE Trans. Aerospace Electron. Syst.* **AES-14**, pp. 456-465 (1978)

11. R.N. Clark, D. Fosth and V.M. Walton, "Detection Instrument Malfunctions in Control Systems", *IEEE Trans. Aerospace Electron. Syst.* **AES-11**, pp. 465-473 (1975).

12. J.C. DeLaat and W.C. Merrill, "A Real-Time Implementation of an Advanced Sensor Failure Detection, Isolation and Accomodation Algorithm", prepared for the 22nd *Aerospace Sciences Meeting*, Reno, Nevada (1984).

13. M. Desai and A. Ray, "A Fault Detection and Isolation Methodology", *Proc. of the 20th Conference on Decision and Control*, pp. 1363-1369 (1981).

14. X. Ding and P.M. Frank, "Komponentenfehlerdetektion mittels auf Empfindlichkeitsanalyse basierender robuster Detektionsfilter", *at Automatisierungstechnik* **38**, pp. 299-306 (1990).

15. X. Ding and P.M. Frank, "Fault Detection and Identification via Frequency Domain Observation Approaches", *IMACS/IFAC Symp. on Computing and Applied Mathematics*, Brussels (1990).

16. X. Ding and P.M. Frank, "Fault Detection via Factorization Approach", *Syst. Contr. Lett.* **14**, pp. 431-436 (1990)

17. X. Ding and P.M. Frank, "Frequency Domain Approach and Threshold Selector for Robust Model-Based Fault Detection and Isolation", submitted to *IFAC/IMACS Symp. SAFEPROCESS* (1991).

18. X. Ding and P.M. Frank, "Robust Model-Based Fault Detection and Isolation for Uncertain Dynamical Systems", submitted to *IMACS-MCTS* (1991).

19. A. Emani-Naeini, A. Muhammad and S.M. Rock, "Effect of Model Uncertainties on Failure Detection and Threshold Selector", *IEEE Trans. Autom. Contr.* **AC-33**, pp. 1106-1115 (1988).

20. S. Engell and D. Konik, "Zustandsermittlung bei unbekanntem Eingangssignal", *Automatisierungstechnik*, Teil 1: **34**(1), pp. 38-42, Teil 2: **34**(6), pp. 247-251 (1986).

21. H. Flanders, "Differential Forms with Applications to Physical Sciences", Academic Press (1963).

22. B.A. Francis, "A Course in H_∞ Control Theory", Springer-Verlag (1987).

23. P.M. Frank, "Fault Diagnosis in Dynamic Systems via State Estimation - A Survey", in *S. Tzafestas et al. (Eds.): System Fault Diagnostics, Reliability and Related Knowledge-Based Approaches* **1**, pp. 35-98, Dr. Reidel Publ. Comp., Dordrecht (1987)

24. P.M. Frank, "Advanced Fault Detection and Isolation Schemes Using Nonlinear and Robust Obsrvers", presented at 10th *IFAC World Congress*, Munich (1987).

25. P.M. Frank, "Fault Diagnosis on the Basis of Dynamic Process Models", presented at 12th *IMACS World Congress on Scientific Computation*, Paris (1988).

26. P.M. Frank, "Fault Diagnosis in Dynamic Systems Using Analytical and Knowledge-Based Redundancy - A Survey and Some New Results", *Automatica* **26**, pp. 450-472 (1990).

27. P.M. Frank and L. Keller, "Sensitivity Discriminating Observer Design for Instrument Failure Detection", *IEEE Trans. Aerospace Electron. Syst.* **AES-16**, pp. 460-467 (1980).

28. P.M. Frank and L. Keller, "Entdeckung von Instrumentenfehlanzeigen mittels Zustandsschätzung in technischen Regelungssystemen", *Fort-schrittsberichte der VDI Zeitschriften, Reihe 8* **80**, Düsseldorf (1984).

29. P.M. Frank, "Fault Diagnosis in Dynamic Systems Using Software Redundancy", *Invited Special Lecture at the 2nd Makuhari Conference on High Technology - Information Science and Computer*, Technology-, World Techno-Fair, Chiba, Japan (1991).

30. P.M. Frank, B. Köppen and J. Wünnenberg, "General Solution of the Robustness Problem in Linear Fault Detection Filters", *European Control Conference*, Grenoble (1991).

31. W. Ge and C.Z. Fang, "Detection of Faulty Components via Robust Observation", *Int. J. Control* **47**(2), pp. 581-599 (1988).

32. J.J. Gertler, "Survey of Model-Based Failure Detection and Isolation in Complex Plants", *IEEE Contr. Syst. Magazine* **3-11** (1988).

33. D. Hengy and P.M. Frank, "Component Failure Detection Using Local Second-Order Observers", *2nd European Workshop on Fault Diagnostics, Reliability and Related Knowledge-Based Approaches*, Manchester (1987).

34. D.M. Himmelblau, "Fault Detection and Diagnosis in Chemical and Petrochemical Processes", Elsevier Scientifique Publishing Company, N.Y. (1978).

35. R. Isermann, "Process Fault Detection Based on Modelling and Estimation Methods - A Survey", *Automatica* **20**, pp. 387-404 (1984).

36. R. Isermann, "Process Fault Diagnosis Based on Dynamic Models and Parameter Estimation Methods", in *R.J. Patton, P.M. Frank and R.N. Clark (Eds.): Fault Diagnosis in Dynamic Systems, Theory and Applications*, pp. 253-291, Prentice Hall (1989).

37. A. Isidori, "Nonlinear Control Systems", 2nd Edition, Springer-Verlag (1989).

38. H.L. Jones, "Failure Detection in Linear Systems", Ph.D. Thesis, MIT, Cambridge, Massachusetts (1973).

39. M. Kitamura, "Detection of Sensor Failures in Nuclear Plant Using Analytic Redundancy", *Trans. Am. Nucl. Soc.* **34**, pp. 581-583 (1980).

40. K. Kroschel and A. Wernz, "Sensor Fault Detection and Localization Using Decorrelation Methods", *Proc. 4th Eurosensor*, Karlsruhe (1990).

41. X.C. Lou, A.S. Willsky and G.L. Verghese, "Optimally Robust Redundancy Relations for Failure Detection in Uncertain Systems", *Automatica* **22**(3), pp. 333-344 (1986).

42. M.A. Massoumnia, "A Geometric Approach to Failure Detection and Identification in Linear Systems", Ph.D. Thesis, MIT, Cambridge, Massachusetts (1986).

43. R.K. Mehra and I. Peshon, "An Innovations Approach to Fault Detection and Diagnosis in Dynamic Systems", *Automatica* **7**, pp. 637-640 (1971).

44. W.C. Merrill and J.C. DeLaat, "A Real-Time Simulation Evaluation of an Advanced Detection, Isolation and Accomodation Algorithm for Sensor Failures in Turbine Engines", *Proc. American Control Conference*, pp. 162-169 (1986).

45. W.C. Merrill, "Sensor Failure Detection for Jet Engines", *Control and Dynamic Systems* **33**(3), pp. 1-34, Academic Press (1990).

46. R. Onken and N. Stuckenberg, "Failure Detection in Signal Processing and Sensing in Flight Control Systems", *Proc. IEEE Conf. Dec. Contr.*, San Diego, pp. 449-454 (1979).

47. J. O'Reilly, "Observer for Linear Systems", Academic Press, London (1983).

48. M. Otter and S. Türk, "The DFVLR Models 1 and 2 of the Manutec R3 Robot", *DFVLR-Report, DFVLR Institut für Dynamik der Flugsysteme*, Oberpfaffenhofen, Germany (1988).

49. R.J. Patton, P.M. Frank and R.N. Clark (Eds.), "Fault Diagnosis in Dynamic Systems, Theory and Applications", Prentice Hall (1989).

50. R.J. Patton, S.W. Willcox and J.S. Winter, "A Parameter Insensitive Technique for Aircraft Sensor Fault Analysis", *AIAA Journal of Guidance Control and Dynamics*, pp. 359-367 (1987).

51. R.J. Patton and S.M. Kangethe, "Robust Fault Diagnosis Using Eigenstructure Assignment of Observers", in *R.J. Patton, P.M. Frank and R.N. Clark (Eds.): Fault Diagnosis in Dynamic Systems, Theory and Applications*, pp. 99-154, Prentice Hall (1989).

52. I.E. Potter and M.C. Suman, "Thresholdless Redundancy Management with Arrays of Skewed Instruments", *Integrity in Electronic Flight Control Systems, ARGADOGRAPH-224*, pp. 15-11 to 15-25 (1977).

53. J. Prock, "Signalvalidierung mittels analytischer Redundanz", *Report of Gesellschaft für Reaktorsicherheit (GRS)mbh*, No GRS-A-1482 (1988).

54. R. Seliger and P.M. Frank, "Robust Component Fault Detection and Isolation in Nonlinear Dynamic Systems Using Nonlinear Unknown Input Observers", *IFAC/IMACS Symp. SAFEPROCESS*, Baden-Baden (1991).

55. R. Seliger and P.M. Frank, "Fault Diagnosis by Disturbance Decoupled Nonlinear Observers", submitted to 30th *Conference on Decision and Control*, Brighton (1991).

56. S.G. Tzafestas, "System Fault Diagnosis Using the Knowledge-Based Methodology", in *R.J. Patton, P.M. Frank and R.N. Clark (Eds.): Fault Diagnosis in Dynamic Systems, Theory and Applications*, pp. 509-572, Prentice Hall (1989).

57. N. Viswanadham and R. Srichander, "Fault Detection Using Unknown-Input Observers", *Control Theory and Advanced Technology, MITA Press* 3(2), pp. 91-101 (1987).

58. N. Viswanadham, V.V.S. Sarma and M.G. Singh, "Reliability of Computer and Control Systems", *North Holland Systems and Control Series* 8, North Holland (1987).

59. N. Viswanadham, J.H. Taylor and E.C. Luce, "A Frequency Domain Approach to Failure Detection and Isolation with Application", *Control-Theory and Advanced Technology* 3, pp. 45-72 (1987).

60. B.K. Walker, "Recent Developments in Fault Diagnosis and Accomodation", presented at *AIAA Guidance and Control Conference*, Gatlinburg, TN (1983).

61. K. Watanabe and D.M. Himmelblau, "Instrument Fault Detection in Systems with Uncertainties", *International Journal of System Sciences* 13(2), pp. 137-158 (1987).

62. K. Watanabe and D.M. Himmelblau, "Fault Diagnosis in Nonlinear Chemical Processes", *AICHE J.* 29, pp. 137-158 (1987).

63. J.L. Weiss, K.R. Pattipati, A.S. Willsky, J.S. Eterno and J.T. Crawford, "Robust Detection/ Isolation/ Accomodation for Sensor Failures", *NASA Contr. Rep. 174797*, Lewis Res. Cent. NAS 3-24078, Alphatech Inc., Burlington, Massachusetts (1985).

64. D.N. Wilbers and J.L Speyer, "Detection Filters for Aircraft Sensor and Actuator Faults", *Proc. ICCON '89 International Conference on Control and Applications*, Jerusalem (1989).

65. A.S. Willsky, "A Survey of Design Methods for Failure Detection in Dynamic Systems", *Automatica* 12, pp. 601-611 (1976).

66. J. Wünnenberg and P.M. Frank, "Sensor Fault Detection via Robust Observers", in *S. Tzafestas et al. (Eds.): System Fault Diagnostics, Reliability and Related Knowledge-Based Approaches* 1, pp. 147-160, Dr. Reidel Publ. Comp., Dordrecht (1987).

67. J. Wünnenberg, "Observer-Based Fault Detection in Dynamic Systems", Ph.D. thesis, University of Duisburg, Dept. of Measurement and Control, *Fortschrittberichte VDI* 8(222), Düsseldorf (1990).

68. J. Wünnenberg and P.M. Frank, "Dynamic Model-Based Incipient Fault Detecton Concept for Robots", *Proc. 11th IFAC World Congress*, Tallinn (1990).

69. J. Wünnenberg, "Implementierung des 'Dedicated Observer Schemes (DOS)' zur Instrumentenfehlerentdeckung (IFD) auf einem Intel 8086/8087 Mikrorechner und Erprobung am Labormodell 'Invertiertes Pendel'", Diploma thesis, University of Duisburg, Dept. of Measurement and Control (1984).

70. M. Zeitz, "Nichtlineare Beobachter für chemische Reaktoren", *Fortschrittsberichte der VDI Zeitschriften* 8(27), Düsseldorf (1977).

CATFEM - Computer-Assisted Tomography and Finite Element Modeling

P.M. Finnigan, A.F. Hathaway, W.E. Lorensen, I.J. Connell, V.N. Parthasarathy

GE Corporate Research and Development Center
Schenectady, NY 12301

and

J.B. Ross

GE Aircraft Engines
Cincinnati, Ohio 45215

ABSTRACT

Historically, x–ray Computed Tomography (CT) has been used for visual inspection of cross–sectional data of an object. It has been successfully applied in the medical field as a non–invasive diagnostic tool and for industrial applications for quality evaluation. This chapter presents a conventional look at CT, and in addition, details revolutionary approaches to the use of computed tomography data for engineering applications, with emphasis on visualization, geometric modeling, finite element modeling, reverse engineering, and adaptive analysis. The concept of a discrete solid model, known as a *digital replica*^{TM}, is introduced. The digital replica possesses many of the same attributes intrinsic to a conventional CAD solid model, and thus, it has the potential for broad applicability to many geometry–based applications.

This chapter discusses three–dimensional imaging techniques for the CT slice ensemble using surface reconstruction. Such capability provides the user with a way to view and interact with the model. Other applications include the automatic and direct conversion of x–ray computed tomography data into finite element models. A system that generates both 2–D and 3–D finite element models using the automatic mesh generators, QUADTREE and OCTREE, is presented. These mesh generators are founded on recursive spatial decomposition, thereby enabling a unique integration with the digital replica. The notion of reverse engineering a part is also presented; i.e., the ability to transform a digital replica into a conventional solid model. Other technologies that support analysis automation, including geometry–based problem formulation and adaptive analysis, along with a system architecture, are also described.

This chapter provides sufficient background on CT to ease the understanding of the applications that build on this technology, however, the principle focus is on the applications themselves.

289

I. INTRODUCTION

Computed Tomography (CT) or Computer–Assisted Tomography (CAT) is an imaging technique that generates pictures of cross–sectional slices of a part or patient. The early application and acceptance of CT was in the medical arena, however, this technology is now being used in industrial applications with great success. There are several differences between medical scanners and industrial scanners, that have been driven by the requirements of the respective applications. For example, in medical applications, it is important to create good quality images, but at the same time, the patient should avoid prolonged exposure to x–rays. Since industrial applications are not x–ray limited, industrial scanners typically have higher energy x–ray sources, produce higher resolution images, and can handle very dense materials. Thus, the advancement in industrial CT technology has led to a revolution in the way in which Non–Destructive Evaluation (NDE) techniques are used for quality control. Until recently, industrial CT data has been used primarily as a means for visual inspection to ascertain part quality through flaw detection. High quality images leading to accurate interpretation has been the goal, and to a large extent, this objective has been met. However, it has long been recognized that the potential of CT extends well beyond its original intent of 2–D image generation. The ensemble of CT slices intrinsically houses geometrical and relational information that can be exploited and used in novel ways in support of applications which range from the most prevalent of 3–D image reconstruction and volumetric rendering, to the less obvious applications of finite element modeling and reverse engineering. These applications begin where conventional CT ends. That is, the output of a CT scanner is the slice ensemble, and the interpretation of the ensemble forms the basis on which a growing suite of applications can be built.

The primary focus of this chapter is to explore some of the current and emerging applications that emphasize the importance of, and somewhat unconventional use of, CT data. One such engineering application is finite element modeling in general, and automatic mesh generation in particular. The generation of the finite element mesh continues to be the bane of the analyst and the principal bottleneck in the finite element modeling process. This problem is becoming somewhat ameliorated with the introduction of fully automatic mesh generators. Despite progress, commercial systems still lack robustness and have serious limitations, especially for geometrically complex industrial components. Conventional approaches to automatic meshing of a continuum require a solid model; i.e., a geometric representation that can ascertain if a point in space is inside, outside, or on the object. Unfortunately, *a priori* solid models are not always available for purposes of meshing. The most apparent reason for this is that computer–aided design (CAD) and computer–aided engineering (CAE) are not well integrated. Furthermore, wireframe and surface modeling technology still dominates CAD systems. Thus, there is typically a need to create a solid model as part of the CAE process.

Even in the presence of a solid model, there are still barriers that must be overcome. First, the as–manufactured part may not be the same as the as–designed definition. This may be because of manufacturing tolerances in the system, material shrinkage, or part warpage caused by residual stresses. Thus, when analysis is performed on the designed part, the physical part dimensions may be sufficiently different that the analysis is suspect. Second, although analysis is ideally a scheduled task within the design/analysis process, before manufacturing or test, this is not always the case. In many instances, analysis is not performed at all, often resulting in premature failures, after a component is in test, or worse yet, in service. In other cases, the lead time to perform the requisite 3–D analysis is so long that the component may already have been manufactured before the results of the analysis are known. Third, if a field failure does occur, then analysis of the failed component is mandated. Nominal dimensions may not suffice here. These scenarios point to a dramatic need for rapid turnaround of analysis of the physical component. Computed tomography offers much promise in helping to address these issues. The ensemble of CT slices, along with a set of geometrical operators, can be thought of as another form of solid model, a discrete solid model, rather than a continuous analytic solid model. This new form, henceforth referred to as a *digital replica*™, opens up many new possibilities in the CAE arena.

This chapter describes a system that automatically converts CT data into 2–D and 3–D finite element models. The methods presented will not replace conventional approaches to finite element modeling, but rather will complement existing systems. The work is based on two concepts. The first, a digital replica operates on discrete spatial data and provides functionality similar to classical solid modeling systems. The second concept is fully automatic mesh generation based on recursive spatial decomposition, known as QUADTREE in 2–D and OCTREE in 3–D. These mesh generation approaches naturally map to the discrete data associated with the ensemble of CT data, and thus, it is through the unique coupling of these two concepts that the system becomes possible and practical. It is anticipated that through discussions that highlight the non–standard use of CT, such as finite element modeling, other applications will emerge.

II. COMPUTED TOMOGRAPHY

Computed Tomography (CT), initially developed to improve medical diagnosis, is now being widely used for Non–Destructive Evaluation (NDE) of industrial components. In addition to its qualitative flaw detection capability, CT also has a quantitative dimensional measurement capability that enables CT to be used as a tool for measuring point–to–point, wall thickness, contour shapes, and complete surface geometry. This CT metrology process is fast, giving x–ray CT a unique capability to feed back complex dimensional data from physical parts to engineering. This nicely complements the well developed flow of geometric data from engineering to manufacturing.

CT images are the fundamental source of data for all of the higher order geometry based processes (e.g., reverse engineering and finite element modeling) described in the remainder of this chapter. Today's industrial CT systems can be applied to a wide variety of metallic and non–metallic materials to produce CT images, but image quality is a significant issue. High quality CT images yield accurate geometric data that can then result in better engineering analyses. Poor quality images may result in missing, or worse yet, erroneous data. CT image quality is influenced not only by scanner characteristics such as energy level and resolution, but also by part characteristics, including material density and part size and shape.

This section provides a basic overview of CT technology so that potential users may better understand the image generation process and the domains of applicability of this technology.

A. Background

The computed tomography process generates a cross–sectional view or image of an object by combining a large number of x–ray views taken around the object.

The first commercial x-ray CT system was developed in the early 1970s by G. Hounsfield at EMI, Ltd., in England [1]. Although some industrial testing was done with early CT systems, the primary application was medical diagnostic imaging. Since that time, the value of CT images for medical diagnosis has been firmly established and a variety of high quality systems are now on the market. Medical CT images show cross–sections of the body, with individual organs, tissues, and bones clearly visible. The CT image is a two–dimensional map of linear x–ray attenuation coefficients in the image plane. Highly absorbing substances such as bone appear white and low absorbing substances such as skin appear dark grey. Image display controls can vary the brightness and contrast of the displayed image and reveal subtle differences in image intensity that may correspond to some disease condition.

Medical CT systems optimize the x–ray source, detector, and CT reconstruction characteristics to produce high quality images of human tissue and bone and to limit the amount of x–ray exposure to the patient. Such systems have proven to be useful for industrial applications involving low density materials, carbon–based composite materials, glass, ceramics, plastics, wax, and aluminum alloys.

In 1980, the U.S. Air Force funded GE Aircraft Engines to develop a high resolution industrial CT system for inspection of turbine blades and other small jet engine components. This system featured a higher energy x–ray source and a custom built high resolution detector that could penetrate a wide range of industrial materials such as titanium, steel, and high density nickel–based alloys used in jet engine components. Industrial CT scanners consist of five major components:

    1) x–ray source to generate an x–ray beam
    2) x–ray detector to form individual views of an object
    3) precision manipulator to generate multiple view angles
    4) computing system to combine the multiple views into a CT image
    5) display system to enable visualization and manipulation of CT images

Figure 1 shows a typical CT system used for industrial applications. The x–ray source

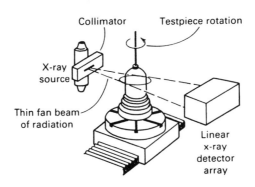

Figure 1. Industrial CT system.

generates a flat fan beam that passes through the object being scanned to the detector. The detector is typically a linear array of closely spaced x–ray sensors connected to sensitive electronics and analog–to–digital converters. For each position of the object, the system delivers a one–dimensional array of digital values corresponding to the x–ray attenuation along each ray passing through the object from the source to the detector. These line images are called projections and are used in two different imaging modes, DR and CT.

B. DR and CT Image Generation

One of the most significant aspects of x-ray CT systems is that they produce digital images that can be processed by computer techniques to enhance details, quantify defects, or measure geometric features. CT systems typically operate in two image generation modes. By moving the object vertically through the x-ray beam, a raster image of the object, known as a Digital Radiograph (DR), can be constructed from the one-dimensional projections. Figure 2 shows a digital radiograph of a jet engine turbine

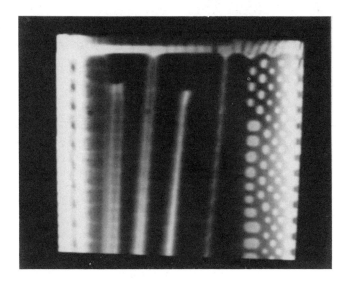

Figure 2. Digital radiograph of a turbine blade.

blade. The DR image shows many internal details and is useful for detecting defects such as voids and microshrink that appear as localized dark areas. The CT image generation process is more complex. The object is rotated about its vertical axis in the presence of the x-ray beam and the projections for 500 to 1500 views are stored in the computer system. The individual one-dimensional projections are combined in the computer using a CT reconstruction algorithm to form the two-dimensional CT image. Figure 3a shows a typical high quality CT image of a jet engine turbine blade airfoil. It clearly reveals the gross internal structure of the blade, as well as small features, such as the laser drilled cooling holes at the leading edge. Blade dimensions including exterior wall thicknesses, cavity-to-cavity wall thicknesses, and cavity radii can be quantitatively measured. In contrast to the sharp image shown in Figure 3a, Figure 3b shows a blurred image of a cross-section of the shank region of the same turbine blade. Because of the greater amount of metal along the x-ray path, and thus greater radiation absorption, image quality is inferior.

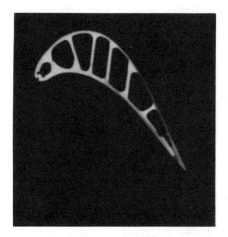 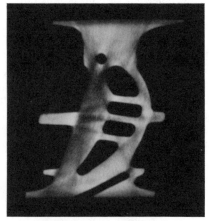

Figure 3a. Good quality CT image
of airfoil section of blade.

Figure 3b. Poor quality CT image of
shank region of blade.

The CT reconstruction process produces a two–dimensional image, with sizes ranging from 128x128 to 1024x1024 or larger, depending on the object's size and the system's spatial resolution. For a system with a spatial resolution of 0.005" per pixel, a 1024x1024 image covers an area of 5.12" x 5.12". Each pixel in the image has a value proportional to the average linear x-ray attenuation coefficient at the corresponding point in the object. The linear attenuation coefficient is approximately proportional to the physical density of the material but is also a function of the spectral distribution of the x–ray beam.

C. 3–D CT Image Generation

The process of reconstructing a series of one–dimensional projections results in a two–dimensional CT image. A three–dimensional CT image can also be generated where each voxel (volume element) represents the linear attenuation coefficient at that point in the volume. Figure 4 is a computer generated rendering of a 3–D CT data set of an automotive piston.

3–D CT images can be produced by two methods;
    1) Collecting a stack of 2–D CT images.
    2) Reconstructing a series of two–dimensional projections to form a 3–D CT image.

Method 1 can be performed on any CT system and is a standard feature of most modern CT systems. Method 2 requires an area detector and specialized 3–D reconstruction software. Based on this technology, a new generation of CT systems is under development which will enable much more rapid generation of 3–D CT images.

D. 3–D CT Image Metrology

The objective of image metrology is to obtain geometrical information about the scanned object such as the location of a point on a model or the surface normal at a point. Ideally, the CT reconstruction process should produce clear, sharp images that correlate closely with object density and exhibit a one–to–one correspondence with the physical object. Density and wall thickness measurements could be made from the image, and

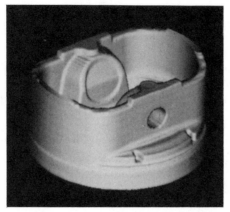

Figure 4. 3–D rendering of CT data set for automotive piston.

simple edge detection algorithms based on thresholding could be used to determine the object's boundary.

In real systems, several factors combine to degrade the images by blurring part edges and by introducing CT reconstruction artifacts that further distort the edges. These factors are strongly influenced by the size, shape, and material of the scanned object and produce complex distortions in the image intensity values. Attempts to infer density, edge position, and object boundaries may result in significant errors if image artifacts are ignored.   Today, successful CT systems utilize calibration procedures and image compensation algorithms to minimize image artifacts, and to enhance image metrology. A discussion of some of the more important artifacts affecting image metrology are described in the paragraphs that follow.

Image Blurring: Typical x–ray sources used in CT systems have a relatively large focal spot that compromises geometric sharpness. This effect can be minimized by moving the part away from the source, but this reduces the signal at the detector. Figure 5 shows a

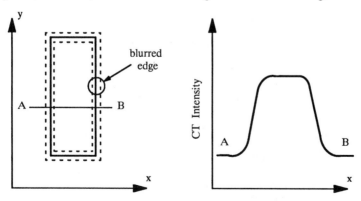

Figure 5. Typical edge profile for CT image.

typical edge profile from a CT image. If the edge is uniformly blurred, the location of the physical part edge can be determined from the 50% intensity point. In practice, other artifacts distort the edge profile and a more reliable indicator of edge location is the point of maximum slope, or inflection point, of the profile.

Beam Hardening: X–ray tubes generate a range of energies that are selectively absorbed by part materials. Low energies have the highest absorption. Consequently, higher energy x–rays are more effective than lower energy x–rays in passing through thick materials. Thus, the preferential transmission of higher energy x–rays is referred to as beam hardening. The relationship between image intensity and part thickness is nonlinear. Calibration procedures on wedge shaped specimens result in a beam hardening correction that linearizes the intensity vs thickness function. Figure 6 shows an

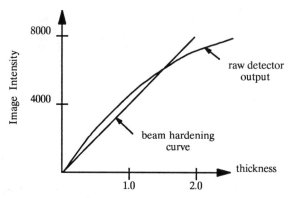

Figure 6. Intensity vs. thickness plot.

example of an intensity vs. thickness plot for rene–80, a high strength nickel alloy used in jet engine turbine blades, and the linearized intensity vs. thickness curve. Beam hardening artifacts result in non–uniform intensity distributions in CT images, thereby distorting edge location. With no beam hardening correction, CT images of rectangular shapes become "cup shaped". The effect is shape dependent and is most pronounced on parts with high aspect ratios.

Detector Rings: Linear detector arrays are preferred over single detectors to reduce CT data acquisition time. Array detectors are calibrated to equalize the gain across the detector elements by measuring the detector signal with some known material in the beam path, typically air or water. Nonlinearities in the response to x–rays may cause non–uniform detector output from the same x–ray excitation. This results in a series of rings of varying intensity, centered on the CT center point. In 3–D data sets generated from a stack of slices, the rings produce concentric cylinders. Detector rings can produce deviations in edge locations in those areas where the rings intersect the object's boundaries. Figure 7 shows a CT image with both beam hardening and detector ring artifacts. Beam hardening errors produce the bowing out of the horizontal edges.

Smearing: A fundamental result of the CT reconstruction process is that edge definition accuracy varies with the part geometry. Consider the rectangular shape shown in Figure 8a. The long edges are defined by rays passing through the long block dimension, while the ends of the block are defined by rays passing through the short block dimension. Attenuation will be less in the short dimension than in the long dimension. The result is that long edges tend to be smeared out when compared to short edges. This

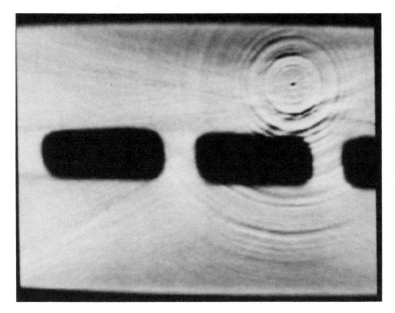

Figure 7. CT image with detector rings and beam hardening artifacts.

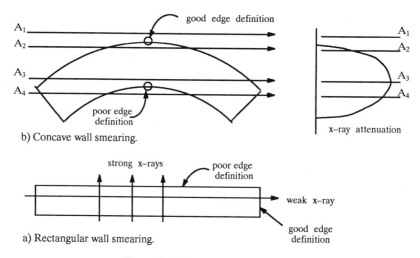

b) Concave wall smearing.

a) Rectangular wall smearing.

Figure 8. Edge accuracy vs. geometry.

effect becomes significant when the x-ray path length approaches the maximum for the material type. Another example of this effect is that concave walls are never as sharp as convex walls because the path lengths are always longer for concave walls, as shown in Figure 8b.

Edge Extension Grooves result from errors in the CT reconstruction process. They appear in CT images as narrow dark lines cutting through part walls and may be mistaken

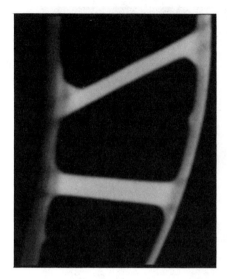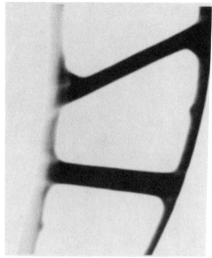

Figure 9. Edge extension grooves.

for defects. Edge grooves result when a narrow wall joins a relatively long straight edge. Figure 9 shows edge grooves in the outer wall of a turbine blade airfoil. The straight edges of the interior ribs are the source of the edge grooves.

Lack of Penetration: X-ray absorption places a limit on the size of parts that can be successfully scanned by the CT process. High quality CT images require an adequate signal–to–noise ratio at the detector. If a part is too large, i.e., the x–ray path in the part is too long, the detector signal will be noisy, resulting in poor image quality. Some improvement can be obtained by increasing signal integration time, however, there is a fundamental limitation on part size that can only be overcome by increasing source energy.

1. Edge detection methods

The goal of edge detection is to extract an ordered list of x,y,z points from the boundary of an object embedded within the CT data set. Commercial image processing packages are a good source of edge detection algorithms. Two major approaches have proven useful with CT images, thresholding with bi–linear interpolation and Laplacian–of–Gaussian.

The simplest method is thresholding combined with sub–pixel interpolation. In this method, pixel intensities are compared with a specified reference threshold value to determine if the pixel is inside, outside, or on the boundary. If the pixel is on the boundary, a bi–linear interpolation is performed to locate the point within the pixel where the boundary passes. Failure to interpolate the data results in a jagged boundary that does not fully exploit the accuracy of the CT data. Figure 10 shows a schematic of an edge profile, along with the resulting edge contour for a turbine airfoil cross–section.

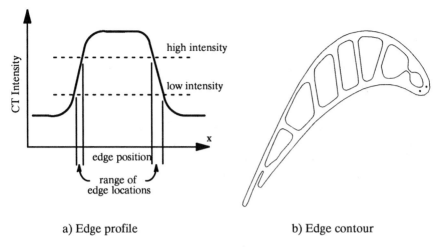

a) Edge profile	b) Edge contour

Figure 10. Threshold–based edge detection.

Threshold–based edge detection works well when image quality is high. The accuracy with which an edge is located is a function of the reference threshold. The optimum threshold value can be established by measuring the part being scanned, comparing that against the calculated edge locations, and adjusting the reference threshold value to achieve a best fit. Beam hardening, and penetration–related artifacts, which selectively alter absolute image intensity, may contribute significant errors to edge location. All artifacts increase with object size so the magnitude of edge errors increases with part size.

The inflection point of the intensity profile is a better measure of edge location because it is insensitive to local variations in image brightness. An improved method of edge detection uses the Laplacian–of–Gaussian (LoG) [2] operator to find the inflection point in the edge intensity profile. In addition to being less sensitive to image quality, this method eliminates the need for threshold determination. Figure 11 is similar to Figure 10, with the exception that it is based on LoG edge detection vs. thresholding.

2. Practical limits of current technology

Industrial CT systems, capable of scanning a wide variety of part sizes, are currently available from several suppliers. These systems generally use a 320 kV or 420 kV x–ray source. Higher energies can be obtained by using a linear accelerator or a cobalt–60 source. Detectors are generally based on xenon gas, scintillator plus photomultiplier, or scintillator plus photodiode array.

At GE Aircraft Engines, three CT scanners are available: GE Medical Systems CT9800 scanner, a custom built medium volume scanner, and a large volume scanner. The CT9800 system utilizes a 150 kV source and a xenon detector, although a solid state detector is now available. Both custom built systems utilize a 420 kV source and a high resolution multi–element xenon gas detector. Image pixel size varies from 0.020" for the CT9800 to 0.005" for the medium volume system.

Part size is generally limited more by x–ray penetration than by the scanning volume of the CT system. A 420 kV source can penetrate 8" of aluminum, 4" of titanium, 3" of

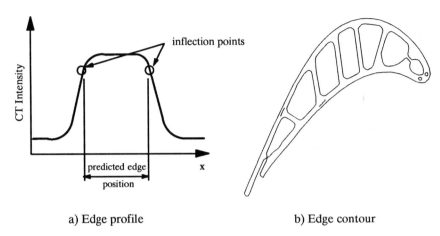

a) Edge profile   b) Edge contour

Figure 11. LoG–based edge detection.

stainless steel, and 2" of high nickel alloy. Typically, good image quality is possible on turbine blades with path lengths up to 2".

III. GEOMETRY

The rigorous use of geometry is becoming an increasingly important part of the engineering design/analysis/manufacturing process. This has been stimulated by the evolutionary advancements in geometric modeling, which has seen this technology move from wireframe, to surface, and now to solid models, whereby each step subsequently provided more complete representational information that could then be exploited by various CAD/CAM/CAE applications. A solid model, of course, provides the most complete information, in fact, it constitutes an unambiguous representation of a part. Simply stated, a solid model can determine if a point in space is inside, outside, or on the part. Neither surface nor wireframe models can answer that question. The applications addressed in this chapter are primarily interested in solid modeling, however, there are elements of surface and wireframe modeling as well.

A. Solid Modeling

There are two conventional solid modeling representational schemes; Constructive Solid Geometry (CSG) and Boundary Representation (B–rep). With CSG, an object is defined in terms of a number of primitives that can be scaled, positioned, or combined using Boolean operations. The CSG representation of an object is an ordered binary tree where the root of the tree represents the final object, the leaf or terminal nodes, are the primitives, and the non–terminal nodes are either Boolean operators or rigid body motions. The CSG tree can then be evaluated for purposes of producing a B–rep. A B–rep model is one that is described in terms of geometry (i.e., points, curves, and surfaces) and topology. Topology provides the relational information between individual geometric entities, as well as the trimming information for curves and surfaces. Geometric entities plus topology provide a complete and unambiguous representation for a solid. The solid models' topology discussed in this chapter are two–manifold. A two–manifold solid is one where each model edge is used by, at most, two model faces. Figure 12 shows a B–rep's model data that consists of a conventional hierarchical topological model and the associated geometry. In addition, the relationship between topology and geometry is depicted. A set of working definitions is provided in the following paragraphs.

Geometric Entities: There are three distinct geometric entity types: points, curves, and surfaces.

- A *point* is a location in space defined by a vector of dimension three.
- A *curve* is the trajectory of a single point through three–dimensional space.
- A *surface* is a two–dimensional locus of points within three–dimensional space.

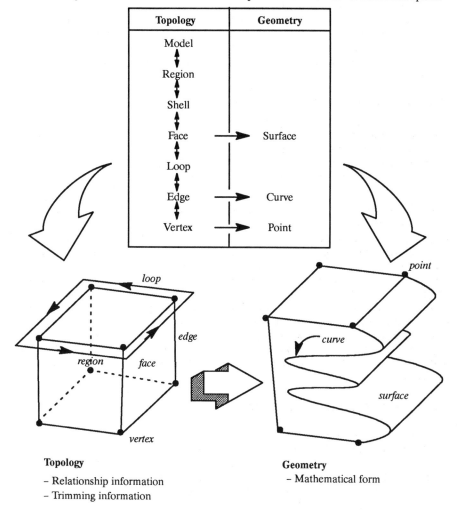

Topology
- Relationship information
- Trimming information

Geometry
- Mathematical form

Figure 12. Relationship between geometry and topology.

Topological Entities: For a two–manifold solid model, there are six distinct topological entity types. Topological entities are hierarchical in nature, with higher order entities being comprised of lower order entities.

- A *vertex* is a topological representation of a point in three–dimensional space. A vertex bounds an edge and forms the joining element between edges. A geometric point is associated with a vertex.
- An *edge* is the topological equivalent of a curve and consists of a starting and ending vertex, which defines the direction of edge traversal. An edge is used as part of the definition of a loop. A geometric curve is associated with an edge.
- A *loop* is an ordered, closed, connected, non–self–intersecting collection of edges. Loops define the boundaries of faces.
- A *face* is defined by one or more loops, one of which defines the outer face boundary and additional loops define the holes. The face forms part of the shell. A geometric surface is associated with a face.
- A *shell* is a set of faces that forms an oriented boundary for a region.
- A *region* is one or more shells and represents a volume of space.

B. TAGUS (Topology And Geometry Utility System)

Ideally, a geometric modeling system should have an open architecture that provides external applications with access not only to the geometric model, but also to the low level geometric operators and evaluation tools inherent in the modeling system. However, most commercial geometric modeling systems have closed architectures that limit the access to their system to the documented user interface, and possibly a model structure retrieval system.

To address this limitation of these systems, GE Corporate Research and Development (GE–CRD) has developed a geometric modeling utility system known as TAGUS (Topology And Geometry Utility System). The TAGUS system, shown in Figure 13, translates evaluated B–rep data from a variety of commercial modeling systems, such as GEOMOD™, CATIA™, or UNIGRAPHICS™, into its neutral representation, and through its operator–driven interface, bridges the gap between closed modeling systems and geometry–dependent application programs, such as automatic mesh generation, reverse engineering, or adaptive analysis. TAGUS augments geometric modeling systems by providing application programs with access to and manipulation of geometry and topology, regardless of the initial modeling environment. The TAGUS library is designed to provide application programmers with tools needed for construction, modification, deletion, queries, and other manipulation facilities of models for design, manufacturing, or analysis.

IV. THE DIGITAL REPLICA

A digital replica is a reproduction of a physical object in digital form. The digital replica consists of two components: 1) a digital representation of the object and 2) operators that act on the representation to answer geometric and topologic questions about the data. There are two primary motivations for creating a digital replica: analysis and visualization. Analysis applications include mass property calculations, fluid flow and structural analysis. Visualization is used for inspection, geometric manipulation and simulation. Analysis and visualization often require different operators that are appropriate for the application. For example, an operator that calculates the intersection of a line with a solid contained within the digital replica may be used in a finite element

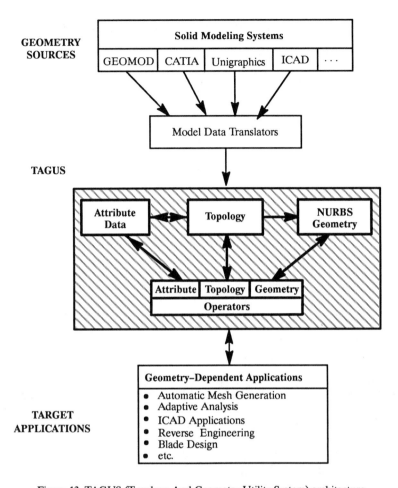

Figure 13. TAGUS (Topology And Geometry Utility System) architecture.

mesh generator but may not be useful in creating a three–dimensional image of the solid. Also, analysis and visualization may store the digital replica using different representations, possibly trading off efficiency for functionality. Although a polygonal representation of the surface of a solid suffices for display, inside/outside tests of points in three dimensions are not efficiently performed on closed polyhedra consisting of hundreds of thousands of triangles. Since our applications include both analysis and visualization, we use different representations and operators for each. This section describes the digital replica used for analysis and the next section describes two representations for visualization.

A. Terminology

The output of a scanning system is an intensity at each *sample point*. A two–dimensional set of intensities is called a *slice* and a set of slices taken at different locations is called a *volume*. A set of eight adjacent intensities, four from two adjacent slices is called a *voxel*, or volume element, Figure 14. The value of the intensity at a sample point depends on the

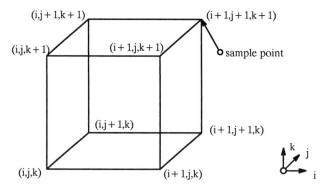

Figure 14. A voxel.

material being sampled and the type of quantity measured by the scanning device. Recall that for x–ray computed tomography, the measured quantity is attenuation and is related to the amount of the x–ray beam that is absorbed, which in turn is related to the density of the material. For example, air absorbs almost none of the x–ray beam while metal absorbs an amount based on its density and thickness. This property of selective absorption of the beam permits us to distinguish between air and different types of metals by looking at the intensity at a sample point.

Figure 15a. CT image of piston.   Figure 15b. Thresholded image of piston.

1. Inside/Outside

Using a technique called thresholding, we can classify each point in the volume, whether it lies inside or outside a given material, by comparing the sample intensity with a fixed range of intensities. For our purposes here, we assume the object consists of one material surrounded by air, so we need not test for a range of values, but rather for a given intensity above a user–specified threshold. Figure 15a shows a typical CT slice through an automotive piston, while Figure 15b shows the same CT slice using thresholding, where the bright sample points are metal and the dark sample points are air.

2. Interpolation

The digital intensities produced by the scanner classify each sampled point according to the material that is present at the point. But analysis applications need to know the inside/outside classification at arbitrary points in 3-D space. A 3-D interpolation scheme must be provided to calculate the intensity at points other than the sampled points. Because in industrial scanners, the size of the voxels is small compared to the major features in the objects, linear interpolation suffices to calculate intermediate intensities. Figure 16 shows schematically the tri–linear interpolation process applied to a voxel.

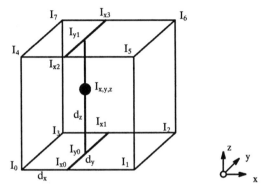

Figure 16. Tri–linear interpolation.

Three successive linear interpolations calculate the intensity at any point from the intensities at the eight sample points:

$$I_{x0} = I_0 + d_x * (I_1 - I_0)$$
$$I_{x1} = I_3 + d_x * (I_2 - I_3)$$
$$I_{x2} = I_4 + d_x * (I_5 - I_4)$$
$$I_{x3} = I_7 + d_x * (I_6 - I_7)$$
$$I_{y0} = I_{x0} + d_y * (I_{x1} - I_{x0})$$
$$I_{y1} = I_{x2} + d_y * (I_{x3} - I_{x2})$$
$$I_{xyz} = I_{y0} + d_z * (I_{y1} - I_{y0})$$

3. Surface Normal

Another important issue in geometric modeling is the calculation of the normal to a surface. In the digital replica, the normal is calculated from the gradient of the intensities. The gradient in a sampled volume is estimated by taking the central differences between the intensities surrounding the point.

$$g_i = (I_{i+1, j, k} - I_{i-1, j,k}) / 2$$
$$g_j = (I_{i, j+1, k} - I_{i, j-1,k}) / 2$$
$$g_k = (I_{i, j, k+1} - I_{i, j,k-1}) / 2$$

Tri–linear interpolation uses the voxel corner gradients to find the gradient at any x, y, z.

B. Digital Replica Geometric Operators

Several geometric operators have been developed for the digital replica that are useful for a variety of analysis applications. These operators and the representation described above form the core of a Digital Replica Geometric Modeling Utility, DRGMU. The DRGMU hides the representation from the application program by only allowing access to the representation through the operator procedure calls. This information hiding is as important in the digital replica as it is in other large software systems, protecting the application code from changes in the internal representation. The following operators are useful in finite element mesh generation.

1. DetermineUniverse

Purpose: Determine the universe of the model. The universe is a cube. Its center is at the intersection of one of the three mutually perpendicular bisectors that is geometrically closest to the center of the model. For efficiency reasons, the size of the world will be such that an octree decomposition of the world will lie along the naturally recurring scan planes.

Approach:
a) Calculate the resolution of the volume from the input data set.
b) Find the maximum resolution in x, y and z.
c) Define the range of the data as the closest power of two to the maximum resolution.
d) Return the center of the data and the range.

2. LocateVoxel(x,y,z)

Purpose: Given an x, y, z, return the voxel that contains it.

Approach:
a) Searching for the voxel that contains a point is trivial since the sample points in a volume are stored in arrays and the indices of the array address the sample points directly. Recall that a voxel consists of eight sample points, four each from two adjacent slices. The sample point address (i, j, k) of the first voxel corner is (*floor* (x), *floor*(y), *floor*(z)), where *floor* returns the integral part of a floating point number.
b) Return the eight voxel corners:

0: (i, j, k),	1: $(i + 1, j, k)$,
2: $(i + 1, j + 1, k)$,	3: $(i, j + 1, k)$,
4: $(i, j, k + 1)$,	5: $(i + 1, j, k + 1)$,
6: $(i + 1, j + 1, k + 1)$,	7: $(i, j + 1, k + 1)$

3. Interpolate(x,y,z)

Purpose: Given a point in space, calculate the intensity in the digital replica.

Approach:
a) LocateVoxel containing x, y, z.
b) Calculate point's offset within the voxel:
$$dx = x - \text{floor}(x)$$
$$dy = y - \text{floor}(y)$$
$$dz = z - \text{floor}(z)$$

c) Interpolate along the four x edges of the voxel:
$$I_{x0} = I_0 + d_x * (I_1 - I_0)$$
$$I_{x1} = I_3 + d_x * (I_2 - I_3)$$
$$I_{x2} = I_4 + d_x * (I_5 - I_4)$$
$$I_{x3} = I_7 + d_x * (I_6 - I_7)$$
d) Interpolate I_x along the y edges:
$$I_{y0} = I_{x0} + d_y * (I_{x1} - I_{x0})$$
$$I_{y1} = I_{x2} + d_y * (I_{x3} - I_{x2})$$
e) Interpolate between slices:
return $I_{y0} + d_z * (I_{y1} - I_{y0})$

4. InOutTest (x,y,z)

Purpose: Given a point in space, determine whether the point is inside, outside, or on the boundary of the model.

Approach:

a) Calculate the intensity at the point using Interpolate.
b) If the intensity is equal to the surface intensity, return (ON). If the intensity is below the surface intensity, return (OUT). If the intensity is above the surface intensity, return (IN).

5. DetermineNormal(x,y,z)

Purpose: Given a point in space and on the model, determine the outward normal to the model.

Approach:

a) LocateVoxel containing x, y, z.
b) Calculate the normals (n_x, n_y, n_z) at each voxel corner using central differences.
c) n_x = Interpolate (normals_x, x, y, z)
   n_y = Interpolate (normals_y, x, y, z)
   n_z = Interpolate (normals_z, x, y, z)
d) Normalize the normal n_x, n_y, n_z.

6. IntersectOctantEdgeWithModel

Purpose: Given two points defining an octant edge, determine the point(s) of intersection.

Approach:

a) Find the voxels containing the two points.
b) For each line segment on the line, check for a transition from inside to outside or outside to inside.
c) If there is a transition, locate the surface intersection using linear interpolation. Separate logic is used for intersections along x, y, and z. For x intersections,
$$t = (\text{SurfaceValue} - I_i) / (I_{i+1} - I_i)$$
$$x = i + t$$
d) Return all intersections.

7. IntersectRayWithModel (p0, direction)

Purpose: Given a ray defined by a point and a direction, determine the distance to the surface along the ray.

Approach:
  a) Determine whether p0 is in or out with <u>InOutTest</u>.
  b) Step from p along direction, checking in/out of
$$p1 = p0 + \text{direction} * \text{step\_size}$$
  c) If there is a transition from in to out or out to in, find the transition point using a binary search. If there is no transition, continue until the maximum number of steps is reached or the ray exits the model.

8. IntersectLineWithModel (p1, p2)

Purpose: Given two points defining an arbitrary line, determine the point(s) of intersection.

Approach:
  a) Calculate a direction from p1 to p2.
  b) Start at p1
  c) Travel to p2 find the closest intersection point using <u>IntersectRayWithModel</u>.
  d) Store intersection and continue c), starting at intersection.

9. PopPointToModel

Purpose: Given a point in space and close to the model, pop the point directly onto the model.

Approach:
  a) Find the normal at the point with <u>DetermineNormal</u>.
  b) Fire a ray in two opposite directions along the normal using <u>IntersectRayWithModel</u>.
  c) Return point that is closest to original point.

V. VISUALIZATION

Although visualization places different demands on the digital replica than does modeling, many of the same principles apply. In particular, the notions of inside/outside, interpolation, and surface normal calculation are used in many visualization algorithms. Since many visualization applications do not need to understand the topology of the models residing within the digital replica, these algorithms often act on individual voxels.

Ordinary CT images can be displayed on any computer system that can show a range of grey levels. Efficient display of 3-D CT data sets requires special techniques. A variety of display techniques are available and most systems offer several. The most useful current methods are described in the following paragraphs. Kaufman [2] contains a collection of reprints and original papers that describe many volume rendering techniques.

Three-dimensional volume visualization algorithms fall into two categories: algorithms that create images directly from the volume data and those that first create an intermediate representation before display.

A. Volume Rendering

Projection and ray tracing algorithms process the volume data for each change of viewing parameters. Projection techniques include:

- Maximum intensity projection – rays are cast through the volume. The most intense sample point encountered is displayed [3]. This technique is fast and relatively insensitive to background noise. The images are best presented in a movie sequence.
- Sum projection – similar to maximum intensity projection, the intensity of the sample points are accumulated as the ray passes through the volume.
- Compositing – slices in the volume are treated as transparent gels that are composited during the projection step. Each new view of the volume requires a resampling of the volume data [4].
- Ray tracing – rays are cast through the image plane into the rotated volume. As a ray passes through a voxel, the voxel contributes a color and opacity that depends on the materials present in the voxel. [5]

B. Volume Modeling

A variety of algorithms first create a model before creating a 3–D image. The algorithms vary in the type of primitive used to create the surface. Once a model is extracted, primitive-specific rendering techniques can produce images from a variety of angles with varying light sources and surface textures.

Initial approaches by Keppel[6] and Fuchs[7] created contours of material boundaries one slice at a time. Using the contours from adjacent slices, these algorithms stitched the contours together to form triangular surfaces. These contour-based algorithms could not reliably and automatically handle adjacent slices that contained different numbers of contours. Another approach, developed by Herman and colleagues[8], creates surfaces from cuberilles, a dissection of space by three orthogonal sets of parallel planes.

Two successful approaches to surface generation from 3–D slices use a divide–and–conquer technique to reliably produce triangular surfaces. The Marching Cubes algorithm by Lorensen and Cline[9], creates triangles from 3–D sets of slices, while Dividing Cubes creates points with surface normals[10]. Both algorithms solve the surface construction problem one voxel (or cube) at a time. Just as in the digital replica, a cube is formed from eight pixel values, four from each of the two adjacent slices.

1. Marching Cubes

Using the notion that high sample values exist within a solid body and low values exist outside the body, the Marching Cubes algorithm classifies each of the eight vertices of the cube as either inside or outside. The surface cuts any cube edge that has one vertex above the surface value and one below. This means there are only 2^8 or 256 possible cube configurations. Because of the symmetry of the cube, the fifteen cases shown in Figure 17 can be used to derive all 256 cases. Marching Cubes builds a table of edge intersections for each of the 256 cases and sequentially finds if and where a surface intersects each cube. To enhance the visualization, the algorithm calculates a normal for each edge intersection, using the gradient of the original CT data. With the triangles and normals, conventional computer graphics techniques and hardware can render a view of the 3–D triangle model from any vantage point. Figure 18 shows a surface model of a jet engine turbine blade, automatically generated from 300 CT slices, 512 by 512 each. The surface model of the turbine blade illustrates the detail available with the digital CT slices. This approach generates many triangles (1.6 million for the turbine blade), so it is used only for creating surfaces for viewing.

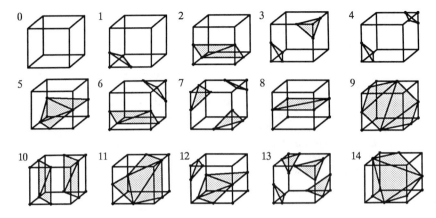

Figure 17. Cases for Marching Cubes.

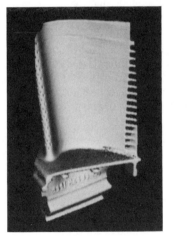 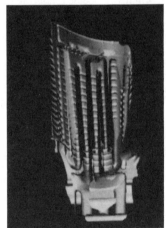

Figure 18. 3–D surface model of a blade.

2. Dividing Cubes

Since many of the triangles generated by Marching Cubes only contain a few pixels, an alternative surface construction algorithm, Dividing Cubes, was developed. This algorithm classifies each cube as to whether it is a surface cube or not. Surface cubes have one or more vertices that have a higher or lower density than at least one other vertex. When a surface cube is detected, Dividing Cubes subdivides that surface cube into multiple cubes. Each of these smaller cubes is checked to see if the surface passes through it using the same criterion as for the larger cubes. The amount of subdivision depends on the desired resolution of the final projected images. Any small cubes that intersect the surface are saved and normals to the surface at these "points" are calculated. Image quality from the Dividing Cubes algorithm is comparable to Marching Cubes.

3. Comparison of the Two Methods

Surfaces created by Marching and Dividing Cubes produce similar images since each locates the surface using linear interpolation and uses the original data to derive surface normals. Because they contain triangles, models created with Marching Cubes can be displayed with conventional polygon rendering algorithms and hardware. Dividing Cubes produces a more specialized primitive, a point plus a normal, that requires special hardware or algorithms to render; but, since it works in fixed point arithmetic, even software rendering is fast. The number of triangles that Marching Cubes produces is related to the size of the digital replica, and remains constant for an increase in display resolution. On the other hand, the number of points in a Dividing Cubes surface depends on both the surface area and the resolution of the final display; only enough points to project a filled image are required. Thus, Marching Cubes is efficient for small digital replicas while Dividing Cubes is more efficient for large volumes.

C. Examples

Any visualization environment should provide a variety of techniques to the user. Although the surface–based algorithms described previously satisfy many display requirements, other techniques can help explore and understand the complex models that exist within a digital replica.

1. Oblique reformating

A cube of data is displayed and manipulated on a graphics workstation. Cube faces can be moved through the data to show internal planes of the object. An oblique cutting plane can be positioned to show any internal plane. Figure 19 shows an example of interactive reformating by Sun Microsystems' SunVision software.

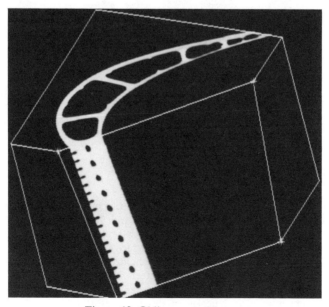

Figure 19. Oblique cut plane.

2. Opaque volume rendering

The entire 3–D data set is rendered using ray tracing: a light source location is selected and rays are fired at the 3–D data set. Voxels with intensity above a threshold reflect the ray in a direction determined by neighboring voxel values. A realistic image of the object results with all solid volumes rendered as solid. Figure 20 is an example done using Sun Microsystems' SunVision software.

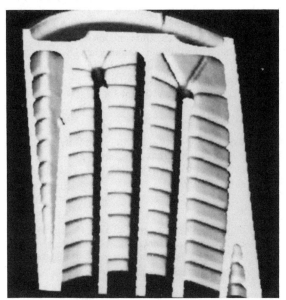

Figure 20. Volume rendering.

3. Translucent volume rendering

The entire 3–D data set is rendered by ray tracing: a classification table is established that reads CT intensity and computes voxel color and opacity. If the volume contains multiple materials that appear as different CT intensities, the classification table can be set up to selectively view any material. Figure 21 is an example done with Sun Microsystems' SunVision. This object was an electrical transformer with a ferrite core, copper windings, and lead/tin solder.

VI. REVERSE ENGINEERING

In general terms, the concept of reverse engineering involves starting with a product, and moving back up the design and production tree to extract required information necessary to perform specific functions. Of course, reverse engineering can mean many things. To an electronics engineer it may mean the duplication of an electronic component without having a schematic; to a biochemist it may mean the production in the laboratory of an antibody with certain characteristics. For the purposes of the present work, reverse engineering is considered to be the automatic conversion of a scanned object, into a conventional analytical geometric format, without *a priori* knowledge of the object's topology, or of the curves and surfaces comprising the object. To this end, a

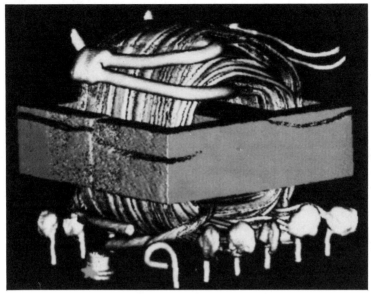

Figure 21. Translucent volume rendering.

system known as CATRE (Computer-Assisted Tomography and Reverse Engineering) is currently under development, and is shown in Figure 22.

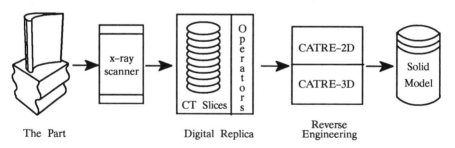

Figure 22. CATRE system architecture.

A conventional geometric model produced from a CT ensemble is useful for a number of applications including:

- For the purposes of nondestructive evaluation, comparisons of "measurements" of production parts and designed parts are frequently required. With a geometric model based on scanning of the physical part, measurements of radii or flaw sizes can be obtained. Analyses of the flawed parts may then be performed to evaluate the effect of the flaw on part life.

- In the design process for jet engines, many components may be derived from previous designs, with sometimes only small variations. An existing part could be scanned, and through reverse engineering, a representation could be made of the part. Design modifications could then be applied to that model, and a casting mold made for production of the new part.

- In the medical field, the lifetime of a prosthesis can be significantly extended if the fit can be improved. From a CT scan of the area of interest, a geometric model can be produced through reverse engineering. Using this model, a design for a prosthesis that provides a good match with, for example, the patient's femur, can be made. This can then be used to aid in the manufacture of a custom prosthesis.

Reverse engineering opens up the world of the digital replica to applications built on top of conventional geometric modeling systems. For CATFEM-2D, discussed in Section VII,E, accessing discrete point data from slices through a digital replica, and using them to produce a 2-D piecewise linear boundary representation, is sufficient. It is possible that other applications could be retrofitted to directly accept the digital replica data. OCTREE was well suited to this modification for use in CATFEM-3D. However, in general, extensive modification of existing applications may not be desirable, and the preference would be to transform the object data from the discrete geometry representation into a more conventional solid model form. Using a reverse engineering process, a boundary representation (B-rep) may be determined from a digital replica, and geometry-dependent applications may function directly off the B-rep, without requiring major modifications.

The solid modeling representation defined in Section III provides a framework for the reverse engineering process. It has a general form that can capture the scanned part definition, but at the same time provides a structure rigorous enough to be transferred into a CAD system.

The CT image slices provide discrete data consisting of an array of locations and image intensities. The digital replica provides a discrete collection of points lying on, within, or close to the boundary of an object. This data will inherently be noisy, and inevitably, the quality of the CT image as well as the specific algorithms used, will have a significant effect on the success of the reverse engineering process. An important factor to note is that the points derived from the digital replica should not be considered to be invariant for the purpose of producing a geometric model. They will not lie precisely on the boundary of the real object. The reverse engineering process should treat these points merely as approximations to the true location of the object's boundary, within some tolerance.

A. CATRE-2D

The major reverse engineering problem is the identification of the model topology embedded within the digital replica and the extraction of the geometric entities associated with the topology. In two dimensions the problem seems tractable, and is an extension of the way QUADTREE functions. From a slice through the digital replica, a set of boundary points can be generated. Because of the organization of the digital replica data, this point set can be easily separated into an outer loop of points, together with any inner loops. At this stage, a geometric model could be formed from the piecewise linear segments forming the loops. This is the type of geometric model used at present by CATFEM–2D, and in itself, has some value. It certainly satisfies the criteria of reverse engineering in that it is a valid geometric representation of the object. One major disadvantage is the number of points per section. Fifteen hundred to two thousand points per cross-section are typical. This is not very manageable, and thus, reverse engineering should take this process a step further. By using curve fitting techniques, it is possible to selectively thin these points, keeping the points that logically constitute vertices, and then fitting curves, (lines, arcs or splines), to the thinned points between the vertices. Figure 23 illustrates this procedure .

As an example, Figure 24a shows the contour point data extracted from a CT image of a precision gauge block 350 mils x 350 mils in size. A Laplacian-of-Gaussian operator[11], was used to extract the points. Since the block was small and there were negligible beam-hardening effects, the image was of high quality, however, the LoG operator still rounded the corners. This, and other error effects of the LoG operator, is discussed by Berzins[12]. Figure 24b shows a piecewise linear geometric representation of the block. This was produced by selecting every fifth point in the contour, and shows how by taking this simplistic, albeit valid approach to reverse engineering, critical information is lost. In

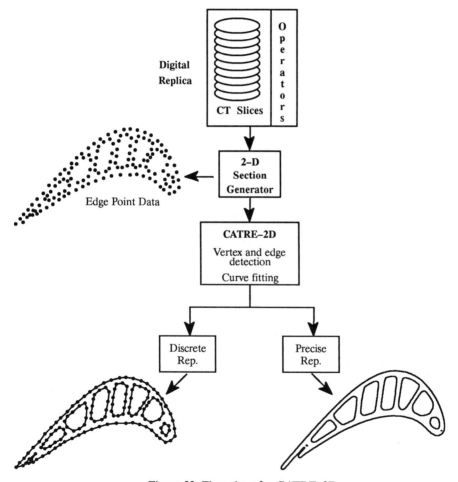

Figure 23. Flow chart for CATRE–2D.

this case, the bottom left corner is eliminated. Figure 24c displays a geometric representation of the block, produced from the same contour points, and is clearly a significant improvement over 24b. A best-fit technique was used to search for lines. The search was totally automatic, only requiring the input of a tolerance, a half-pixel for this example. All the contour points were utilized and the lines were fitted to all the points within the given tolerance.

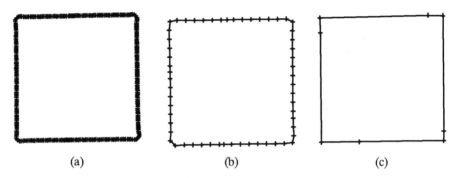

(a)   (b)   (c)

Figure 24. a) Point contour of gauge block; b) Piecewise linear model;
c) Best–fit linear model.

Of course a block is a particularly simple object to analyze. Figure 25 shows a more complicated set of contours extracted from a CT slice cut through a jet engine blade. These were also obtained using a LoG edge-detection operator. This particular slice shows the platform region of the blade (outer contour), and the cooling holes through the shank (the inner contours). The contour point data depicted in Figure 25a has 1785 points in the outer contour, and a total of 2672 points in all contours. Figure 25c shows a geometric representation of the part obtained after a simple sequence of operations were performed. First, a best-fit search for lines was carried out to thin the points, then a best-fit search for circular arcs was made on these thinned points. The fit of the arcs also checked that the geometry remained within a pixel of the original contour data.

The examples in Figures 24 and 25 utilize a simple localized best-fit approach to reverse engineering, although a more global approach would be preferable. This would entail fitting curves to points such that locally a curve may not supply the best-fit to its associated points, globally the loop of curves does provide the best-fit solution to the complete contour of points. However, these examples are sufficient to illustrate the possibilities of reverse engineering, and suggest some   requirements of a 2-D reverse engineering system:

- The geometric model must be closed.

- If there are multiple contours in a model, they must not intersect.

- The geometric model must maintain a tolerance with the original contour data, but not treat any point as invariant.

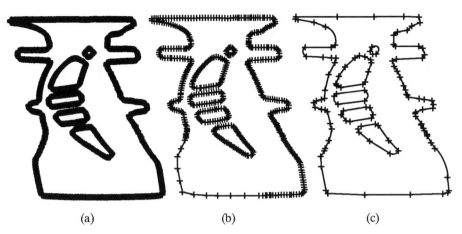

Figure 25. a) Point contour of blade; b) Piecewise linear model; c)
Best–fit geometric model.

- Some initial analysis should be carried out to determine a good place to start the fitting procedure. Ideally the search should begin in an area where the points fit a straight line, not an area where the gradient is changing rapidly. The resulting geometric model should then be independent of the starting location of the original contour of points.

B. CATRE-3D

In three dimensions, the problem is complicated by the question of surfaces and faces. In two dimensions, the only possible surface is the plane of the slice. In three dimensions however, no such shortcuts can be made regarding the topology and geometry, and therefore the scope of the variables is much greater. A simplistic approach to handling the transformation of a three-dimensional digital replica into a geometric model would be to utilize the two-dimensional approach directly. Each slice could be treated separately and then the resultant collection of two-dimensional profiles lofted to define the surfaces and faces. This is a possible solution for simply connected bodies, however, the system would rapidly degenerate for complex parts, where the topology between the slices is not constant, for example a jet-engine blade consisting of an airfoil, platform and dovetail. It is therefore desirable to define a general methodology that is independent of part complexity and scan orientation.

A more general approach would be to modify the algorithm used in Marching Cubes (discussed in Section V,A,1). Marching Cubes constructs a surface representation of the object, in the form of triangular facets, from the digital replica. Similar to the 2-D concept of piecewise linear curves, these triangles are used to create a piecewise planar surface. With proper orientation of the facets, a valid geometric model is formed, however, in most practical cases the number of facets will be large, typically being composed of a million or more triangles. For purposes of reverse engineering, the digital replica can be utilized slightly differently.

Section IV described how the digital replica can be considered as a collection of voxels, with each voxel containing geometric information. Each voxel can be classified as being inside, outside, or on the boundary of the object. Also, the digital replica contains information regarding neighboring voxels, i.e., given a voxel, what are the 26 (or fewer if it is located at the extremes of the image) adjacent voxels? This means that surface information implicitly contained within the voxels can be tracked throughout the entire digital replica. This will then allow point sets of boundary information to be isolated. These points can be separated into an outer shell of the object, together with any inner shell(s). Once the shells have been determined, the edges used to define loops and faces need to be detected. A possible starting point would be to track the faceted model's edges where the dot product of the normals of adjacent facets exceeds some specified value. With the topology information complete, the underlying curves and surfaces can be fit to the topology. Figure 26 illustrates the procedure.

This section has provided a general overview of some of the possibilities of reverse engineering, and has noted some ways to set up a reverse engineering process. Using reverse engineering as a conduit, many types of application could utilize the geometrical information stored within a digital replica. Currently, 2-D reverse engineering is in progress, and is showing much promise, whereas the work in 3-D has not yet begun.

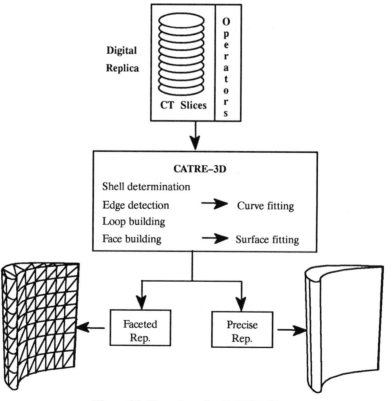

Figure 26. Flow chart for CATRE–3D.

VII. FINITE ELEMENT MODELING AND ANALYSIS

Automation within the mechanical computer–aided engineering arena in general, and the finite element modeling and analysis process in particular, is of growing concern because of the pressure on business to reduce its design cycle time[14]. Today, a large part of computer–aided engineering is devoted to finite element modeling and analysis. The typical FEM/FEA process includes the three components shown in Figure 27; preprocessing, analysis, and postprocessing. This figure also shows an estimate of the typical applied engineering involvement. Clearly, the productivity gains come from improvements in pre– and postprocessing. Historically, this has been done by making extensive use of interactive graphics to automate the manual methods. This approach has been useful in making incremental advances but there is still a long way to go if the final objective is to automate the entire process.

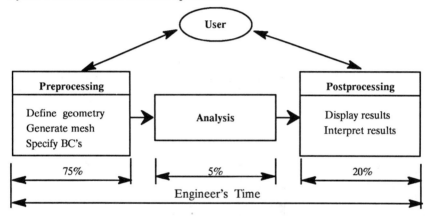

Figure 27. Typical FEM/FEA process.

To make a quantum leap towards automating the entire finite element modeling and analysis process, several underlying technologies must be addressed in a rigorous manner; namely, geometry, fully automatic mesh generation, and adaptive analysis. As described in previous sections, geometry within the CATFEM system is provided by the digital replica, as well as through the reverse engineering process. The following subsections describe an approach to automatic mesh generation, the integration of this approach with the digital replica, adaptive analysis within the CATFEM system, and finally, full automation of the finite element modeling and analysis process using the CATFEM system.

A. Automatic Mesh Generation

Historically, mapped mesh generation techniques have dominated finite element mesh generation. In the world of advanced applications requiring automation under program control to which the CATFEM system is targeted, the need for fully automatic mesh generation is critical.

In the context of this chapter, a fully automatic mesh generator is an algorithmic procedure that generates a valid finite element mesh in a domain of arbitrary complexity, given no input past the original geometric description of the domain and some element discretization information. Many of the fully automatic mesh generation approaches that follow the above definition are limited to 2–D and cannot be easily extended to 3–D. The approach that looks most promising is based on recursively subdividing a geometric model[15][16][17] to the point where element generation can be easily performed. Quadtree/octree decomposition naturally lends itself to fully automatic mesh generation.

B. Approach

Fully automatic mesh generation based on the quadtree/octree algorithms is a three step process: (1) cell decomposition or tree building, (2) element generation, and (3) mesh smoothing or nodal repositioning (Figure 28). Each step proceeds according to information obtained through interrogation of an unambiguous geometric description.

1) Tree Building   2) Element Generation   3) Mesh Smoothing

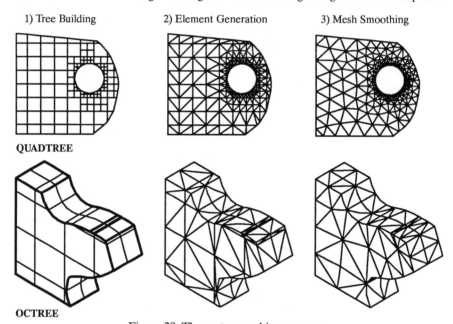

QUADTREE

OCTREE

Figure 28. Three step meshing process.

Tree Building. A quadtree/octree approximation of a geometric model consists of a collection of variably sized, non–overlapping cells whose union approximates the geometric model. The quadtree/octree approximation for a geometry is obtained through recursive subdivision of the physical space enclosing the geometry. Figure 29 shows an example. The approximation process proceeds as follows:

1) Enclose the geometry with a square Cartesian box and subdivide the box into four/eight regular cells, quadrants in 2–D and octants in 3–D.
2) Classify each of the cells with respect to the geometry. When the cell is wholly interior or outside the geometry, the decomposition of that cell ceases. When a cell cannot be so classified, it is further subdivided into four/eight new cells.
3) Continue the decomposition and classification process until the desired resolution is achieved.

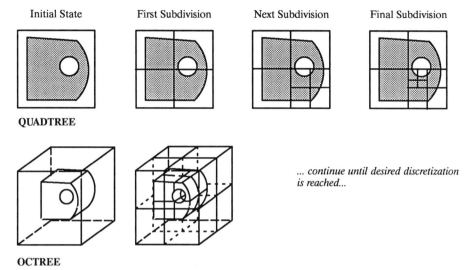

QUADTREE

OCTREE

... *continue until desired discretization is reached...*

Figure 29. Tree building process.

Approximations produced by the decomposition scheme outlined above can be represented by logical trees whose nodes have four/eight offsprings. Each node corresponds to a cell in the recursive decomposition process.

For mesh generation, the classical quadtree/octree approximation must be modified. In its standard form, the boundary of an object is represented as a series of re-entrant steps that introduce artificial discontinuities into the mesh. This situation is resolved by modifying boundary cells such that the true geometry is captured. Furthermore, it is possible to represent the geometry with neighboring cells that have large size differences, and thus, to avoid abrupt changes in mesh density, neighboring cells are forced to have no more than a 2:1 level difference. This is achieved by subdividing the larger octants to the appropriate level.

Element Generation. Once the cell decomposition process has been completed, the original geometry is represented as the union of all cells. In any given cell, the geometric complexity has been greatly reduced, thereby simplifying the element generation process.

Element generation proceeds on a cell–by–cell basis, creating triangular elements for 2-D, and tetrahedral elements for 3-D. Cells that are interior to the geometric model are decomposed into elements using templates as shown in Figure 30. The difficulty in element generation arises in those cells lying on the boundary of the geometric model. Here, a more general approach, based on the Delaunay triangulation[18], is utilized.

The Delaunay triangulation has the property that the n–dimensional simplexes that compose it have circumscribing hyperspheres (in 2-D, circumcircles, and in 3-D, circumspheres) that contain no other point of the triangulation (Figure 31). The Delaunay algorithm is ideally suited to generating finite element meshes because the resulting triangulation has been shown to be optimal.

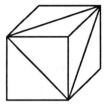 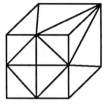

Figure 30. Element generation via templates.

Figure 31. 2-D Delaunay triangulation.

Because the Delaunay triangulation uses only point data, and does not incorporate higher dimensional geometry information, the generated mesh may be topologically incompatible with the original geometry[19]. This situation may be resolved by introducing additional points into the triangulation at the point of incompatibility. For example, the initial Delaunay mesh in Figure 32b is incompatible with the original geometric model; elements *T1* and *T2* "pierce" the geometry (i.e., edge *E*). Point *P* is added and the data is re-triangulated resulting in a compatible mesh shown in Figure 32d.

In the past, the use of Delaunay triangulation techniques to produce computational grids has proceeded across an entire point set. In the quadtree/octree based approach, elements are generated on a cell-by-cell basis. Information is then communicated between cells to insure mesh compatibility. This approach provides a method for error recovery since meshing failures in a cell can be identified and then corrected on a local basis.

Smoothing. Once the mesh has been generated for the geometric model, further improvement in overall mesh quality can be achieved by smoothing the mesh. In smoothing, the topology of the mesh remains unchanged, but the coordinates of the nodes are adjusted to improve element shape. A simple but effective approach is Laplacian smoothing which iteratively repositions the nodes based on the connectivity of a node to its neighbors.

C. Examples

Figure 33 contains examples of both 2-D and 3-D meshes generated by an implementation of the quadtree/octree algorithms. This implementation is capable of producing linear beam, triangular shell, and tetrahedral solid elements. It can generate valid meshes from solid, shell, wireframe, and combination solid/shell/wireframe

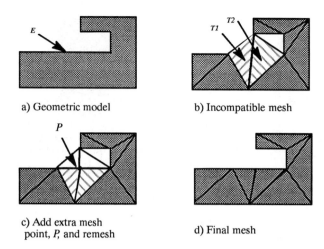

a) Geometric model   b) Incompatible mesh

c) Add extra mesh
point, P, and remesh   d) Final mesh

Figure 32. Resolving geometric incompatibilities.

geometric models. Some of its capabilities include meshing of multiple material models and performing mesh refinement at arbitrary locations within or on the boundary of the geometry. Performance studies show that this algorithm has a linear computational growth rate.

D. Role of Geometry

The role of geometry is extremely important in fully automatic mesh generation because the mesher must correctly capture the geometry to be meshed without user intervention. This requires full access to a geometric representation, and in addition, requires a set of geometric operators that are used to interrogate that representation. The set of operators required by a fully automatic mesh generation algorithm is a function of the algorithm and its implementation.

The set of requisite geometric interrogation operators used in the quadtree/octree implementation are similar to those discussed in Section IV,B, and fall into the following classes:
- Basic queries (e.g., *Determine Universe*)
- Intersection operations (e.g., *In/Out Test, Intersect Line With Model*)
- Parametric evaluations (e.g., *Evaluate Parametric Location*)

In the quadtree/octree approach, the three step meshing process proceeds based on information received from the various geometric interrogations made in a previous step. For example, cell decomposition (i.e., recursive subdivision process) needs to compute intersections. To construct cells that accurately represent the true geometry of the part, operators are used to insure that the intersection results are properly classified with respect to the model. During element generation, points are added to the Delaunay triangulation, requiring the intersection operators. Finally, much of the mesh smoothing process is performed in parametric space, requiring the re-evaluation of many points.

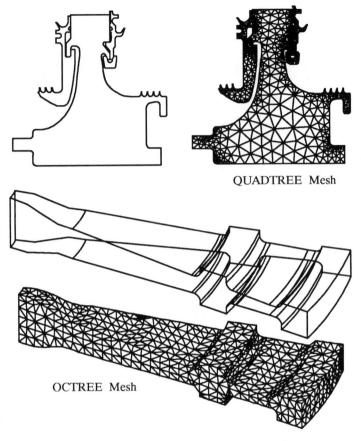

QUADTREE Mesh

OCTREE Mesh

Figure 33. QUADTREE and OCTREE meshes.

E. Integrated Computed Tomography and Finite Element Modeling

The CATFEM (Computer–Assisted Tomography to Finite Element Modeling) system is a combination of hardware and software. Figure 34 depicts the system architecture. Three subsystems comprise CATFEM: the x–ray scanner, the digital replica, and the fully automatic mesh generators. The CATFEM process begins by scanning an object with an x–ray computed tomography system. From the CT scanner comes a stack of contiguous cross sections. This ensemble of data, along with a set of geometrical operators, comprise the digital replica. The digital replica provides a foundation for geometry–dependent applications, such as automatic mesh generation. CATFEM uses two automatic mesh generators, depending on whether 2–D or 3–D solid meshes are desired. QUADTREE generates two–dimensional meshes on arbitrary slices through the data set, whereas OCTREE generates general three–dimensional solid meshes. The resulting finite element meshes can be input to any finite element code with a simple formatting program.

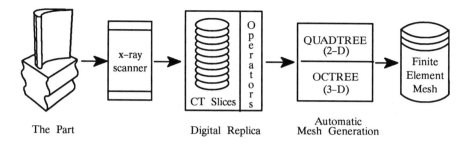

Figure 34. CATFEM system architecture.

The process of converting discrete computed tomography data into a finite element model begins by operating on the raw data received from the x–ray scanner. This approach couples the quadtree/octree algorithms of spatial decomposition and Delaunay triangulation with the CT data, using the geometric operators provided by the digital replica, discussed in Section IV,B. Quadtree/octree cell decomposition of the computational domain uses a top–down recursive subdivision algorithm. If uniform subdivision occurs far enough, then the resulting tree representation is equivalent to the digital replica. In other words, the CT slices that form a set of contiguous cross–sections form voxels (cubes made up of eight pixels) and the quadtree/octree decomposition forms quadrants/octants. The one–to–one correspondence between voxels and octants illustrates the commonality between the digital replica and the quadtree/octree approach to finite element mesh generation. However, since the digital replica is at the resolution of the x–ray scanner, the problem that quickly emerges is that using the entire CT database (every voxel as an octant) could result in the generation of several million finite elements, which renders it impractical. Therefore, the goal is to generate a tree representation at a coarser resolution using the digital replica operators to extract the appropriate geometric detail, resulting in the desired finite element mesh, as depicted in Figure 35. Clearly, the digital replica serves the same function as the conventional CAD geometric model.

Figure 36 shows a schematic for converting CT data to finite element models using the programs QUADTREE and OCTREE. From the figure, one can see that a fundamentally different approach has been adopted depending on whether 2–D or 3–D meshing is specified. The reason for this will become clear in the subsections that follow.

1. CATFEM–2D

A system for automatically generating a 2–D finite element model from a digital replica has been developed and fully tested. CATFEM–2D first converts a 2–D slice of CT data into a 2–D boundary representation, making it similar to our conventional meshing system, which is founded on the use of a CAD model. Therefore, the only geometric interrogation operator required for CATFEM 2–D is to extract a 2–D boundary representation directly from a selected CT slice within the digital replica. Since the data within a CT slice is discrete point data, a more general 2–D boundary representation is created by linking the point data together to form piecewise linear loops. Using this representation, QUADTREE produces a finite element model. Since the meshing is performed on a conventional boundary representation, mesh density information can be

Tree Building   Element Generation   Mesh Smoothing

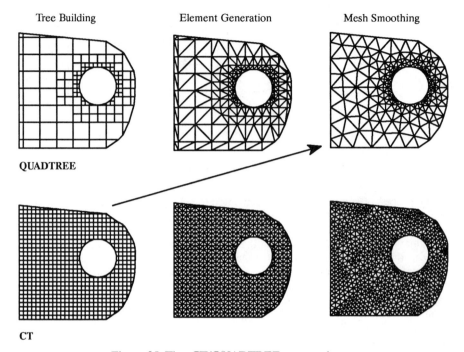

QUADTREE

CT

Figure 35 The CT/QUADTREE connection..

applied to any geometric entity. Figure 37 shows a 2–D boundary representation generated from a CT slice of an aircraft engine airfoil and the resulting QUADTREE–generated mesh, while Figure 38 shows a mesh of a transmission housing.

2. CATFEM–3D

As outlined in the introduction, the primary need for generating a valid finite element model from CT scan data stems from the requirement to conduct appropriate down–stream mechanical analysis of as–manufactured parts. Similar to the CATFEM–2D implementation, a system to generate 3–D finite element models is under development. Following the 2–D architecture, the logical first step would have been to generate an intermediate 3–D boundary representation and later use it to obtain the final 3–D finite element mesh. However, such a strategy was not adopted for two reasons. In contrast to 2–D, 3–D reverse engineered models, even for simple configurations, are at least an order of magnitude more difficult to construct, and the methodology is currently not available (See Section VI,B on 3–D reverse engineering). Second, the close relationship between voxel data and octant data, can be exploited to directly achieve fast decomposition of the computational domain. Moreover, in a non–manufacturing environment, such as biomechanical analysis, a conventional geometric representation of the model being analyzed, e.g. tooth, bone, etc. does not exist. The direct generation of a finite element mesh from the digital replica becomes the preferred solution strategy for such cases. Substantial progress has been made with the generation of 3–D finite element models using this methodology. In what follows, the direct conversion of 3–D CT data to a

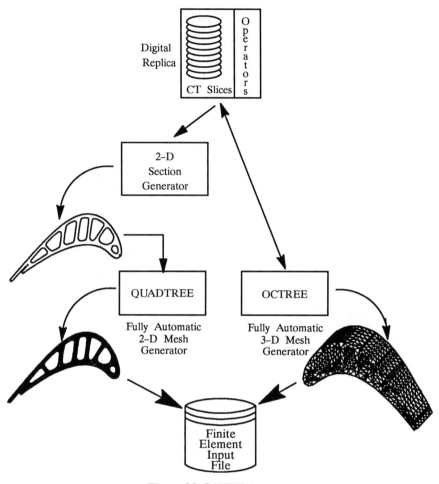

Figure 36. CATFEM system.

finite element model using OCTREE is explained as we walk through the three–step meshing process.

The first step in the process aligns the digital replica with the root octant, as shown in Figure 39. This becomes the bounding–box for the model, thereby guaranteeing that all intersections will be performed along CT slice planes (or scan lines). Therefore, the only data within a cell will lie along the boundary of a cell/octant, i.e., along cell/octant edges.

Next, the tree building process proceeds in the conventional way by using the digital replica intersection operators. The only algorithmic change in the tree building process was in "linking" the intersection points into octant edges. The problem of determining when an edge is inside or outside the model in moving from intersection point to intersection point has been addressed differently. The conventional boundary representation version relied on computing surface normals at each point. To take advantage of the digital replica, a small portion of the algorithm was rewritten to exploit

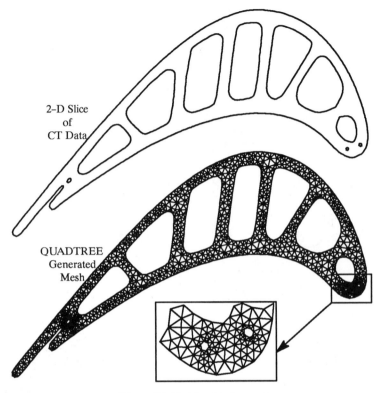

Figure 37. Turbine blade mesh.

the facts that the inside/outside classification of the first point is already known, and the *InOutTest operator* available in the digital replica is more efficient than its conventional counterpart.

After tree building, the element generation process also proceeds as in the conventional approach. The element generation process for a cell derived from a digital replica is generally much simpler since there is never any data interior to the cell. Nevertheless, at reasonably fine levels of discretization of the tree, the large number, and close spacing, of the intersection points along octant edges of the boundary cell, can cause degenerate situations for the Delaunay triangulation algorithm. An algorithmic enhancement that fictitiously redistributes nodes equally along octant edges, eliminates the occurrence of such degenerate point-sets for the mesher.

At coarse levels of discretization, elements in the concave regions of the part "fill in" the model geometry. This is because of the fundamental property of the Delaunay technique, which always yields a mesh of the convex hull of the points[20]. A method to evaluate whether an element is inside or outside the model was implemented. The approach taken performs an in/out test of the centroid of those elements, whose four nodes are classified on the model boundary, followed by an in/out test of a set of pseudo-nodes (usually 4) within the element. Pseudo-nodes are first generated by "pulling-in" the actual nodes of the element along the line joining the centroid and the

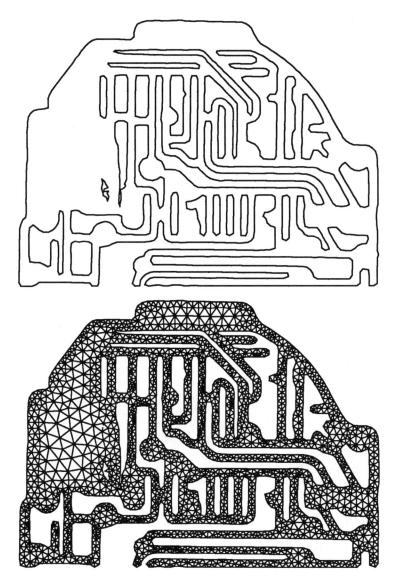

Figure 38. Cross–section of an transmission housing.

node in question. If the result of all five tests are affirmative, then the element is considered as inside the model, otherwise it is outside the model.

A major difference between a conventional boundary representation and a digital replica (in both 2–D and 3–D) is that there is no parametric representation for the surface/edge data. In fact, for 3–D models, there is only one logical surface for the entire model. Inherent in many of the octree algorithms is the concept that data can always be

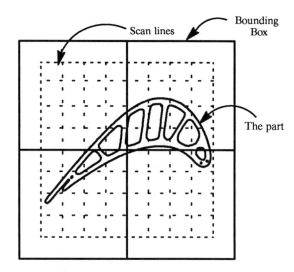

Figure 39. Construction and alignment of bounding–box and CT image.

maintained and manipulated in parametric space for geometric entities lying on the boundary. In some instances, the existing algorithms had to be modified to provide other ways of evaluating and maintaining data on the boundary. One such case is the third step of the octree algorithm, mesh smoothing. The initial approach taken in mesh smoothing was to smooth in world coordinates, allowing points to move off the boundary and then to "pop" the points back onto the boundary. This method yielded only reasonable success. An alternate approach currently being implemented is the use of surface normals at each node. By comparing the normal direction values before and after smoothing, the node can be restrained in its movement, both in terms of distance and direction, to lie within the physical surface boundaries. This method is also expected to ameliorate the difficulties associated with the non–existence of explicitly defined geometric edges in the model, by restraining the finite element nodes from moving across physical edges in the part.

Presently, for CATFEM–3D, the mesh density information can be specified only by the number of levels of refinement desired in the tree for the entire model. Similar to the conventional schemes, the finite element model captures only the part geometry available at the resolution of the leaf octants. Another method currently under consideration would use a reference CAD model of the part in question. Since manufacturing is invariably preceded by first creating a conventional CAD model, such a reference model is usually available. The initial specification of mesh density on the reference model could then be mapped to the digital replica on an octant–by–octant basis. A more sophisticated approach, applicable irrespective of the availability of a conventional geometric model, is adaptive mesh control as will be discussed in the following subsection.

In CATFEM–3D, the results to date have been encouraging. Figure 40 shows representative meshes derived via the CATFEM process.

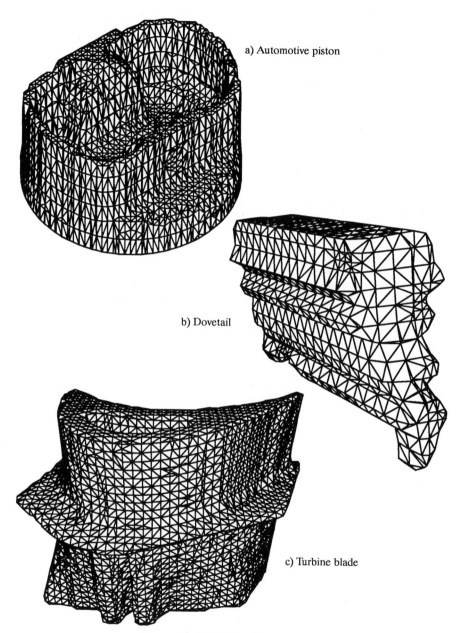

a) Automotive piston

b) Dovetail

c) Turbine blade

Figure 40. CATFEM–3D examples.

F. Adaptive Analysis

The purpose of adaptive analysis is to determine and control the discretization error, thereby ensuring the reliability of the finite element solution. Conventionally, this has been carried out by finite element analysts who judge the acceptability of a mesh and the resulting solution based on previous experience or simple calculations. This iterative process is not only time–consuming and error–prone, but it also requires extensive knowledge of finite element methods. While still an active research topic, adaptive analysis is an important element of an automated analysis system. With regards to CATFEM, adaptive analysis is important to 2–D, but critical to its ultimate success in 3–D, as will become clear in the following paragraphs.

A fundamental difference between the solid model resulting from conventional approaches to CAE and the digital replica is the level of geometric complexity contained within the model. As mentioned earlier, the CAE community typically creates its own solid model for purposes of analysis. Implicit in that model is an *a priori* recognition of which geometric features are important, or potentially important, for the particular analysis in question. With a digital replica, a 100% complete solid model is present. All geometric detail, down to the resolution of the x–ray scanner, is present. What is required is a method that includes important geometry in the analysis, and disregards unimportant geometry. Ideally, this should be done in an automated way, without requiring user interaction, since that would partially defeat the automation gains resulting from CATFEM. Adaptive analysis seems well suited to solve this problem. With adaptive analysis, algorithmic procedures can be used to selectively and incrementally include important features. There are two components to adaptive analysis; mesh improvement and error estimation. Each of these components are briefly described in the following paragraphs.

Three approaches to mesh improvement have been identified for adaptive analysis and can be categorized as follows:

1) r–method: keeping the same number of elements, this approach improves the solutions by re–positioning nodes to more optimal locations

2) h–method: in this method, the accuracy is improved by locally refining elements in areas of high solution gradient.

3) p–method: given a finite element mesh, this approach improves the solutions by increasing the polynomial order of the approximating shape functions used for the particular element type

The other component of adaptive analysis is the ability to obtain an error measure from the finite element solutions. Since errors are not known *a priori*, this has motivated many researchers to introduce the so–called error estimator or error indicator. Mathematically, the error indicates how large the variation of the solution is throughout the entire domain. This suggests that a zero error can be achieved when the error indicator is the same everywhere in the domain. The inclusion of element size in the error indicator implies that this can be achieved by making the size of an element with a large error indicator smaller. This, of course, is the technique used in the h–method.

For conventional CAE, all the approaches described above have met with reasonable success. For example, consider a stress concentration in the fillet of a gear tooth under bending as shown in Figure 41. A concentrated load of 1000 lbs. is used to simulate contact between two gear teeth. Photoelastic results indicate a stress value of 504 psi at the bottom fillet. The finite element models at various stages in the adaptive process using h–refinement are shown. The maximum stress at the bottom fillet obtained from the adapted mesh has been shown to be within 0.5% of the experimental results.

For CATFEM, only the h–method of mesh refinement seems generally appropriate. While the r– and p–methods are ill–equipped to pick up significant additions in geometric

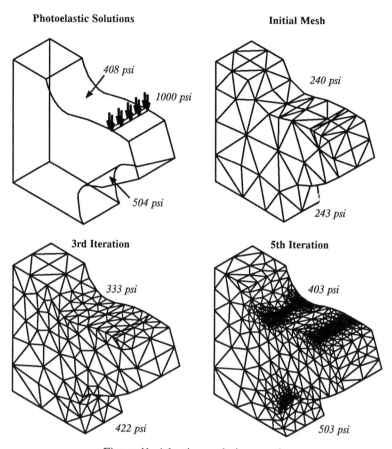

Figure 41. Adaptive analysis example.

complexity, the h–method can capture geometry at the resolution of the data acquisition system, and can do so in a controlled, rational, and algorithmic (as opposed to heuristic) way.

G. Automated Analysis – Putting it all together

Full analysis automation of CATFEM for both single–pass and iterative analyses will require the development and integration of the technologies discussed in the previous sections. Figure 42 shows a conceptual view of such an automated analysis system. There are six major components:

    1) the digital replica
    2) the automatic mesh generators, QUADTREE and OCTREE
    3) the reverse engineering subsystem
    4) the geometry–based problem specifier
    5) finite element analysis code
    6) the adaptive analysis subsystem

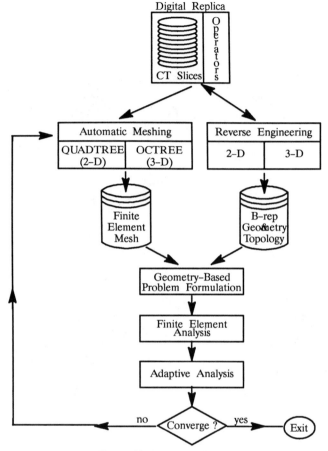

Figure 42. Automated analysis system.

The process assumes the existence of the previously mentioned digital replica. The digital replica serves two purposes. First, it is used by one of the automatic mesh generators (QUADTREE or OCTREE) to generate a 2-D or 3-D finite element mesh. Second, it is used by the reverse engineering application to produce a boundary representation. The B-rep can be used later for high level specification of the problem (e.g., loads, boundary conditions, mesh control) and for automatically mapping that data onto the finite element mesh created by the mesh generators. The finite element code analyzes the data and the results are interpreted by the adaptive analysis subsystem. Here, a determination is made with regards to solution convergence. If the solution has not converged then new element discretization criteria are fed back to the mesh generator and re-analysis is performed. The process continues until convergence is achieved.

VIII. SUMMARY

This chapter has presented an overview of computed tomography, along with conventional and non–conventional applications of this technology, with special emphasis placed on finite element modeling. The output of a CT scanner is an ensemble of 2–D images. These images are typically used for medical diagnosis or part quality inspection. The ensemble has also been used for 3–D image generation. To that end, the Marching Cubes and Dividing Cubes algorithms were developed.

The recognition that the CT ensemble contains information far beyond that which is required for image processing, opens many new possibilities. This information is geometrical in nature and can be exploited in numerous ways. The basis for this exploitation begins with the concept of a digital replica, i.e., the CT slice ensemble along with a set of application–independent geometric operators that provide data access and manipulation capabilities. The digital replica hides the details of the data from the application. The digital replica forms a new type of model, a discrete vs. continuous solid model, which can be used with those geometry–dependent applications requiring the existence of a solid model. This is the case for such applications as reverse engineering and automated finite element modeling and analysis.

The concept of automated reverse engineering of a part was presented. That is, beginning with a part, scan the part to create the digital replica, and from it, create a conventional CAD model, without need for manual intervention. Reverse engineering is important for dimensional assessment of manufactured parts, obtaining derivative designs, producing electronic solid models when only wireframe or surface models exist, and for use with finite element modeling. This type of application emphasizes the geometrical significance of the digital replica. Good progress has been made on a two–dimensional reverse engineering system, CATRE–2D. Work on a 3–D system has yet to begin, however, reasonable algorithmic procedures have been identified and we expect that an automated system with minor manual intervention will be possible.

Algorithms for fully automatic mesh generation were also presented. These algorithms, QUADTREE in 2–D and OCTREE in 3–D, are based on recursive spatial decomposition. Serendipitously, the approaches to automatic mesh generation that GE–CRD is pursuing map naturally to the digital replica, and thus, the integration of these two technologies form a powerful system known as CATFEM (Computer–Assisted Tomography to Finite Element Modeling). The CATFEM–2D system is robust. A CT slice is selected for meshing and a topological face is created and read into QUADTREE. From this point on, QUADTREE does not interact with the digital replica, and in this respect, 2–D fully automatic mesh generation is similar to the way a conventional CAE system would operate.

Good progress has been demonstrated with the CATFEM–3D system. It operates in a fundamentally different way than its 2–D counterpart. No conventional *a priori* geometric information is available to OCTREE throughout the course of the meshing process. OCTREE relies 100% on the digital replica to obtain all geometric information about the part, and to perform specific geometric calculations during the meshing process. OCTREE is a prime example of an application that was originally designed to work off a conventional solid modeling boundary representation, but which has successfully been integrated with a discrete solid model (i.e., the digital replica). The principal effort has focused on replacing B–rep functionality with digital replica capability. Care has gone into keeping the meshing algorithms as independent of the form of the solid model as is possible.

Mesh generation is, of course, just part of the finite element modeling process. There are other technologies that also must be developed and integrated with CATFEM to achieve a fully automated system. The most relevant technologies are adaptive analysis, and geometry–based modeling. An architecture for such a system was also described.

Although this chapter has stressed x–ray as a means for defining the digital replica's data, clearly other modalities, such as magnetic resonance and CMM (Coordinate Measuring Machine) data, are also appropriate. In fact any method that captures continuous spatial data in a regularized (i x j x k points) discrete fashion would be appropriate for this technology. Finally, the unconventional use of CT technology, such as that described in this chapter, is in its relative infancy, but the power of this technology is real, and its potential should be apparent. In the future, one should expect to see additional revolutionary approaches to certain classes of problems using this technology which today are intractable with conventional procedures.

IX. REFERENCES

[1] Dennis, M.J., 1989, "Industrial Computed Tomography," Metals Handbook, Vol. 17, 9th Edition: Nondestructive Evaluation and Quality Control.

[2] Kaufman, A., 1990, Volume Visualization, IEEE Press.

[3] Cline, H.E., Dumoulin, C.L., Lorensen, W.E., Souza, S.P. and Adams, W.J., 1991, "Volume Rendering and Connectivity Algorithms for MR Angiography," *Magentic Resonance in Medicine*, Vol. 18, pp. 384–394.

[4] Drebin, R.A., Carpenter, L., and Hanrahan, P., 1988, "Volume Rendering", Vol. 22, No. 4, pp. 64–75.

[5] Levoy, M., 1988, "Display of Surfaces from Volume Data," *IEEE Computer Graphics and Applications*, Vol. 8, No. 5, pp. 29–37.

[6] Keppel, E., 1975, "Approximating Complex Surfaces by Triangulation," *IBM J. Res. Develop.*, Vol. 19, No. 1, pp. 2–11.

[7] Fuchs, H., Kedem, Z.M. and Uselton, S.P., 1977, "Optimal Surface Reconstruction from Planar Contours," *Communications of the ACM*, Vol. 20, No. 10, pp. 693–702.

[8] Herman, G.T., and Liu, H.K., 1979, "Three–Dimensional Display of Human Organs from Computed Tomograms," *Computer Graphics and Image Processing*, Vol. 9, No. 1, pp. 1–21.

[9] Lorensen, W.E. and Cline, H.E., 1987, "Marching Cubes: A High Resolution Surface Construction Algorithm," *Computer Graphics*, Vol. 21, No. 3, pp. 163–169.

[10] Cline, H.E., Lorensen, W.E., Ludke, S., Crawford, C.R. and Teeter, B.C., 1988, "Two Algorithms for the Three–Dimensional Reconstruction of Tomograms," *Medical Physics*, Vol. 15, No. 3, pp. 320–327.

[11] Marr, D. and Hildreth, E., 1980, "Theory of Edge Detection," *Proc. Roy. Soc. London Ser. B,* **207**, pp. 187-217.

[12] Berzins, V., 1984, "Accuracy of Edge Detectors," *Computer Vision, Graphics and Image Processing,* Vol. 27, 195-210.

[13] Finnigan, P.M., Kela, A., and Davis, J.E., 1989, "Geometry as a Basis for Finite Element Automation," *Engineering With Computers,* No. 5, pp. 147–160.

[14] Shephard, M.S., and Finnigan, P.M., 1989, "Towards Automatic Model Generation", in A.K. Noor (ed.), *State-of-the-Art Surveys in Computational Mechanics, ASME.*

[15] Jackson, C.L., and Tanomoto, S.L., 1980, "Oct-trees and Their Use in Representing Three-Dimensional Objects," *Computer Graphics and Image Processing,* Vol. 14, No. 3, pp. 249–270.

[16] Shephard, M.S., 1988, "Approaches to the Automatic Generation and Control of Finite Element Meshes" *Applied Mechanics Review, ASME,* Vol. 41, No. 4.

[17] Kela, A., 1989, "Hierarchical Octree Approximation for Boundary Representation Based Geometric Models", *Computer-Aided Design*, Vol. 21, No. 6, pp. 355–362.

[18] Cavendish, J.C., Field, D.A., Frey, W.H., 1985, "An Approach to Automatic Finite Element Mesh Generation," *International Journal Numerical Methods in Engineering,* pp. 329–347.

[19] Schroeder, W.J., and Shephard, M.S., 1990, "A Combined Octree/Delaunay Method for Fully Automatic 3-D Mesh Generation," *International Journal Numerical Methods in Engineering,* Vol. 29, No. 1, pp. 37–55.

[20] Sibson, R., 1978, "Locally Equiangular Triangulations," *Computer Journal,* Vol. 21, No. 3, pp 243–245.

DECISION AND EVIDENCE FUSION

IN SENSOR INTEGRATION

STELIOS C. A. THOMOPOULOS

Decision and Control Systems Laboratory

Department of Electrical and Computer Engineering

The Pennsylvania State University

University Park, PA 16802

I. INTRODUCTION

A. Distributed Decision Fusion and Evidence Processing

Sensor integration (or sensor fusion) was defined in [15] as "... the process of integrating raw and processed data into some form of meaningful inference that can be used intelligently to improve the performance of a system, measured in any convenient and quantifiable way, beyond the level that any one of the components of the system separately or any subset of the system components partially combined could achieve." A *taxonomy* for sensor fusion that involves three distinct levels at which information from different sensors can be integrated was proposed in [15]. According to the proposed taxonomy, sensor fusion can be accomplished at three different levels: the *signal* level, the level of *evidence*, and the level of *dynamics*. Each level is identified by the distinct features that the information that is fused carries, which (features) determine in turn the suitable processing methods for combining this information. The signal level in this taxonomy is characterized by the lack of a complete mathematical model that fits the data; the appropriate techniques for fusing information at this level are then correlation and learning through association. At the evidence level, a statistical model (not necessarily precise) describing the observed phenomenon or process is assumed; statistical techniques are then applied to fuse the information from different sensors. At the level of dynamics, a mathematical model that describes the process from which data are collected through some linear or nonlinear transformation of the process state. At this level, analytical tools can be used to fused the information from the sensors in either a centralized or decentralized way. The taxonomy in [15] is analogous to the three level - data, feature, and decision,- analysis that is used in classical pattern recognition problems. This chapter is primarily focused on different theories and approaches related to data fusion at the level of evidence. Data fusion at the evidence level will be referred to as decision or evidence fusion, depending on the semantic attributes associated with the fused information.

Distributed Decision (Evidence) Fusion (or DD(E)F in the sequel) exhibits some interesting characteristics which are not present in centralized, or raw data, fusion. The interesting characteristics relate to the semantic information that the decisions (in the broader sense of the term) convey which is not present, at least explicitly, when raw data is fused. Different theories and results related to Distributed Decision Fusion (DDF) have appeared in the literature the last decade [TeSa 81, Sadj '86, ChVa '86, Srin '86, TVB '87, VTT '88, TVB '88, Demp '68, Shaf '76, Thom '90]. Each theory takes a different stand on the definition on how to measure evidence or combine decisions. The objective of this paper is to investigate the nature of DD(E)F, present some of the dominating theories on DDF and DEF, highlight similarities and differences among them that result from the semantic format of the fused information, and exploit natural topological equivalences between DDF and structures that exhibit learning abilities, such as neural networks.

To avoid concealing some of the issues under structural complexities and keep the discussion focused and as clear as possible we consider the simplest, yet fundamental, DDF topology and problem. We assume a parallel topology in which each sensor receives data from a common volume, Fig. 1. Furthermore, we assume that the sensors are perfectly aligned, so the problem of mismatch does not arise [13]. In this parallel topology we assume the simplest DDF problem with each sensor's data statistically independent from the other sensors. Each sensor performs a local operation on its data and transmits the outcome to the fusion. The fusion collects all the local information from the sensors and produces the global inference. In this framework, we consider both single-level logic decision rules, in which the number of permissible local decisions coincides with the number of hypotheses under test, and multi-level logic decision rules, in which the number of permissible local decisions exceeds the number of tested hypotheses. Most of the results in this paper pertain to the binary hypothesis testing problem. Extensions to multi-hypothesis testing are also included.

In DDF, the outcome of the global processing (fusion) depends on the outcome of the local data processing (sensor level) and the semantic format of the fused information. In the Bayesian context, the outcome of the local processing can be either hard decisions in a single-level logic, or soft decisions in a multi-level logic [16], or it can be the outcome of a simple quantization of the data, if no semantic attributes are attached to the outcome of the local processing [19]. In the context of the Dempster-Shafer's (D-S) theory, the outcome of the local processing is a set of probabilities that relate to the degree of support of the each proposition in the frame of discernment by the the data of each local processor [Demp '68, Shaf '76]. Thus, the local processing outcome of a Bayesian DDF is a quantized scalar number, whereas the outcome of the D-S local processor is a real-valued vector that corresponds to an entire probability distribution.

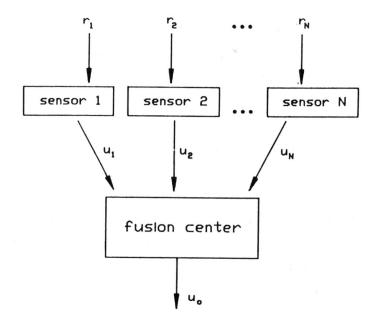

Fig. 1 Parallel sensor topology.

In addition to semantic differences in the output of the local processors, there are also substantial differences in the communication requirements for transmitting the local information to the fusion between the Bayesian DDF and the D-S DEF. Even in the presence of multi-level logic, the communication requirements for transmitting one out of, say, M integers is substantially lower than transmitting an M-dimensional real-valued vector. Hence, the communication requirements for the Bayesian DDF are substantially lower than the requirements of D-S DEF for the same number of data. Thus, a meaningful comparison between Bayesian and D-S DDF should either fix the available communication bandwidth to be the same for both approaches, or fix the fusion objectives to be common and study the communication overhead. In this paper we attempt a comparison of the D-S DDF with the Bayesian DDF assuming identical communication requirements.

The paper is structured as follows. The single-level logic, Bayesian DDF for the binary hypothesis testing problem is presented in Section 1. The theory is extended to multi-level logic and multiple hypotheses Bayesian DDF in Section 2. In Section 2 a generalized Bayesian framework for evidence processing is also presented. The D-S DDF theory is presented in Section 3 and theoretical as well as numerical comparative results between D-S DDF and D-S DDF are provided. The paper concludes with a discussion of topological similarities between the DDF structure and structures that exhibit learning capabilities. Numerical results that indicate the potential use of artificial neural networks in solving DDF problems are presented in Section 4.

B. Likelihood Ratio and Neyman-Pearson Test

In the multi-sensor detection related literature the common assumption that all the sensors cover the same geographical volume is made, Ref.'s [1] through [8]. Under this assumption, the mathematical model of binary hypothesis testing for each sensor becomes

$$z = \beta s + n \tag{I.1}$$

where s is the known signal, n is the noise, and β is a binary indicator random variable with $\beta = 1$ if H_1 is true and $\beta = 0$ if H_0 is true, thus, with probability distribution $p(\beta) = P_0 \, \delta(\beta) + P_1 \, \delta(\beta-1)$. For the model (I.1) the optimal Bayesian detector is the Likelihood Ratio Test (LRT), namely

$$\Lambda(z) = \frac{p(z|H_1)}{p(z|H_0)} \mathop{\gtrless}\limits_{H_0}^{H_1} \frac{P_0 \, (C_{01} - C_{00})}{P_1 \, (C_{10} - C_{11})} := \frac{P_0}{P_1} T_C \tag{I.2}$$

where P_i, $i = 0, 1$, is the a-priori probability that hypothesis i is true with $P_1 = 1 - P_0$; C_{ij} is the cost associated with deciding in favor of hypothesis i while the true hypothesis is j, $i = 0, 1$, and $j = 0, 1$; and $T_C := \dfrac{(C_{01} - C_{00})}{(C_{10} - C_{11})}$, a constant factor that depends on the costs C_{ij} [9]. Alternatively, the LRT in (I.2) can be expressed as

$$p(z|H_1)P_1 \mathop{\gtrless}\limits_{H_0}^{H_1} T_C \, p(z|H_0)P_0 \tag{I.3}$$

Noting that $p(z|\beta=i) = p(z|H_i)$, $i = 0, 1$, it follows easily from the model (I.1) that the LRT (I.2) or (I.3) is equivalent to the test

$$\beta^* = \frac{P_1 \, p(z|\beta=1)}{p(z)} \mathop{\gtrless}\limits_{H_0}^{H_1} T_1 \tag{I.4}$$

with threshold $T_1 = T_C P_0 \, p(z|\beta=0) \, / \, p(z)$, where

$$\beta^* = E[\beta|z] = P(\beta=1|z) = \frac{P_1 \, p(z|\beta=1)}{p(z)} \tag{I.5}$$

is the minimum mean-squared error (mmse) estimate of β given the observation z.

In multi-sensor detection problems, the model (I.1), indexed by the I.D. number of each sensor, can be used to model the data of each sensor if all the sensors monitor the same, common geographical volume all the time. The data from the sensors can then be processed into a final decision either centrally using the LRT in the form of (I.3) or (I.4), or locally first and then fused into a final decision by a fusion center.

II. Bayesian DDF in binary hypothesis testing with single-level local logic

Consider the binary hypothesis testing problem in a parallel sensor topology, Fig. 1. Assume that each sensor makes independent binary decisions and that the decisions are statistically independent from each other conditioned on each hypothesis. If u_i designates the i-th local decision, i.e. the decision of the i-th sensor, then

$$u_i = \begin{cases} +1 \text{ if the i-th local decison favors hypothesis } H_1 \\ -1 \text{ if the i-th local decison favors hypothesis } H_0 \end{cases} \qquad (II.1)$$

Under these assumptions, the optimal Bayesian DDF rule that maximizes the detection probability for fixed false alarm probability at the fusion is given by the next theorem [14].

Theorem 1 [14] For the parallel sensor topology, binary hypothesis testing, single-level local logic, and statistically independent local decisions, the optimal Bayesian DDF that maximizes the fusion detection probability for fixed fusion false alarm probability consists of *Likelihood Ratio Tests (LRTs)* at the local (sensor) level and a *Neyman-Pearson (N-P) (possibly randomized) Test* at the fusion. □

A complete proof of the theorem can be found in [14]. Theorem 1 characterizes the optimal Bayesian DDF but does not provide with any means to determine the optimal local thresholds and the optimal fusion threshold. The problem of determining the optimal operating points in DDF is NP complete [5]. For the parallel topology, the number of possible solutions increases combinatorially with the number of sensors [TVB '88, '89] and, so far, no efficient algorithm exists for determining the optimal thresholds. A schematic representation of the optimal Bayesian DDF (to be referred to also as N-P (Neyman-Pearson) DDF rule when the false alarm at the fusion is fixed and we seek to minimize the detection probability) is shown in Fig. 2. As it will be discussed in Section 4, the structure of the optimal Bayesian DDF bares striking similarities with structures that exhibit learning capabilities, such as neural networks.

Distributed Decision Fusion
The Optimal Configuration

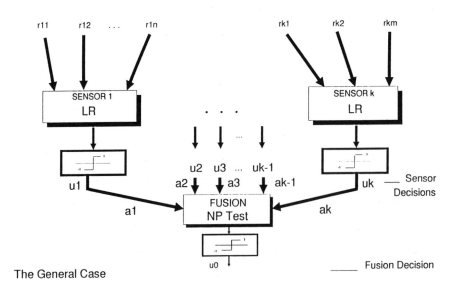

Fig. 2 Optimal Bayesian DDF.

Since it is computationally involved to determine the optimal thresholds for the Bayesian DDF, it is convenient from practical point-of-view to determine the optimal DDF when each local sensor operates at some fixed false alarm and detection probability. [In the sequel P_F will designate false alarm probability and P_D detection probability.] The optimal Bayesian DDF rule is then determined in Theorem 2 [7].

Theorem 2 [7]   For the parallel sensor topology, the optimal Bayesian DDF rule when the local sensors operate at fixed false alarm and detection probabilities is the *LRT*

$$\Lambda(u) = \frac{dP(u|H_1)}{dP(u|H_0)} \underset{H_0}{\overset{H_1}{\gtrless}} T_f \tag{II.2}$$

where $u =: \{u_1, u_2, ..., u_N\}$ is the vector of the peripheral decisions and T_f an appropriate threshold that is chosen so that a desired Bayesian risk function is optimized. If the sensor decisions are independent from each other conditioned on each hypothesis, the test in (3) simplifies to

$$\Lambda(u) = \frac{dP(u_1|H_1) \ldots dP(u_N|H_1)}{dP(u_1|H_0) \ldots dP(u_N|H_0)} = \prod_{i=1}^{N} \Lambda(u_i) \underset{H_0}{\overset{H_1}{\gtrless}} T_f \tag{II.3}$$

With the convention of (II.1), the Bayesian DDF takes on the form

$$\sum_{i=1}^{N} a_i u_i \underset{H_0}{\overset{H_1}{\gtrless}} T_f \tag{II.3a}$$

where

$$a_i = \begin{cases} \log[\dfrac{P_{Di}}{P_{Fi}}] & \text{if } u_i = +1 \\[2em] -\log[\dfrac{1-P_{Di}}{1-P_{Fi}}] & \text{if } u_i = -1 \end{cases} \tag{II.3b}$$

where the index i indicates the i-th local processor (sensor). In a more compact form, the Bayesian DDF test (II.3) can be expressed as

$$\frac{1}{2} \sum_{i=1}^{N} \{(u_i + 1)\log[\frac{P_{Di}}{P_{Fi}}] - (u_i - 1)\log[\frac{1 - P_{Di}}{1 - P_{Fi}}] \overset{H_1}{\underset{H_0}{\gtrless}} T_f \qquad (II.3c)$$

□

In the Bayesian context, determination of the optimal threshold generally requires: (a) knowledge of the a priori probabilities (likelihood) of the tested hypotheses, and (b) specification (subjective) of the attached costs C_{ij}, i.e. the cost of taking a decision j when the true hypothesis id i [VTr '68]. To eliminate these two requirements, the Neyman-Pearson (N-P) approach can be used to fuse the decisions [9]. The threshold T_f at the fusion is determined so that a desired false alarm probability is achieved. The following theorem summarizes the results.

Theorem 3 [9] For statistically independent sensors in parallel topology operating at fixed false alarm and detection probabilities, the optimal Bayesian DDF that maximizes the fusion detection probability for fixed false alarm probability is the *Neyman-Pearson (N-P) Test* (possibly *randomized*)

$$\Lambda(u) = \frac{dP(u_1 | H_1) \dots dP(u_N | H_1)}{dP(u_1 | H_0) \dots dP(u_N | H_0)} = \prod_{i=1}^{N} \Lambda(u_i) \overset{H_1}{\underset{H_0}{\gtrless}} T_f \qquad (II.4)$$

where N is the number of sensors. With the convention of (II.1), the N-P DDF takes on the form

$$\sum_{i=1}^{N} a_i u_i \overset{H_1}{\underset{H_0}{\gtrless}} T_f \qquad (II.4a)$$

where

$$a_i = \begin{cases} \log \left[\dfrac{P_{Di}}{P_{Fi}} \right] & \text{if } u_i = +1 \\[2em] -\log \left[\dfrac{1 - P_{Di}}{1 - P_{Fi}} \right] & \text{if } u_i = -1 \end{cases} \tag{II.4b}$$

The threshold T_f at the fusion is determined from the false alarm requirement by

$$\sum_{\Lambda(u) \ge T_f} dP(\Lambda(u) \mid H_0) = \alpha_0 \tag{II.4c}$$

where α_0 is the desired false alarm at the fusion. In a more compact form, the Bayesian DDF test (II.3) can be expressed as

$$\frac{1}{2} \sum_{i=1}^{N} \left\{ (u_i + 1)\log \left[\frac{P_{Di}}{P_{Fi}} \right] - (u_i - 1)\log \left[\frac{1 - P_{Di}}{1 - P_{Fi}} \right] \right\} \underset{H_0}{\overset{H_1}{\gtrless}} T_f \tag{II.4d}$$

For conditionally independent sensor decisions, the distribution of the LR at the fusion under the two hypotheses is given by

$$P(\log \Lambda(u) \mid H_i) = P(\log \Lambda(u_1 \mid H_i) * \ldots * P(\log \Lambda(u_N) \mid H_i) \quad ; i = 0, 1 \tag{II.5}$$

where "*" indicates convolution with

$$P(\log \Lambda(u_i) \mid H_0) = (1 - P_{Fi})\delta\left(\log \Lambda(u_i) - \log \frac{1 - P_{Di}}{1 - P_{Fi}}\right)$$
$$+ P_{Fi}\, \delta\left(\log \Lambda(u_i) - \log \frac{P_{Di}}{P_{Fi}}\right) \tag{II.6}$$

and

$$P(\log \Lambda(u_i) \mid H_1) = (1 - P_{Di})\delta\left(\log \Lambda(u_i) - \log \frac{1 - P_{Di}}{1 - P_{Fi}}\right)$$
$$+ P_{Di}\, \delta\left(\log \Lambda(u_i) - \log \frac{P_{Di}}{P_{Fi}}\right) \tag{II.7}$$

Taking into account the discrete nature of the probabilities in (II.5), (II.6) and (II.7), the discrete distribution of the LR at the fusion consists of non-zero probability points at abscissae of the form

$$\prod_{i \varepsilon S} \frac{P_{Di}}{P_{Fi}} \prod_{j \varepsilon \bar{S}} \frac{1-P_{Dj}}{1-P_{Fj}}$$ where S is a subset of the sensors {1, 2, ..., N} that

favors hypothesis H_1 and $\bar{S}$ is its compliment that favors hypothesis H_0. The corresponding probabilities of the LR at these abscissae are

$$\prod_{i \varepsilon S} P_{Di} \prod_{j \varepsilon \bar{S}} (1-P_{Dj}) \text{ under } H_1 \text{ and } \prod_{i \varepsilon S} P_{Fi} \prod_{j \varepsilon \bar{S}} (1-P_{Fj}) \text{ under } H_0 . \quad \square$$

Since convolution is both associative and commutative and the product terms of the discrete convolution for every two sensors in (II.5) are generated by cross-multiplying the conditional probabilities under each hypothesis, the N-P DDF combining rule (II.5, 1.6, 1.7) can be implemented using a table similar to the one used for the Dempster's combining rule in the D-S theory [Demp '68, Shaf '76; also Section 3]. By considering two sensors at a time, the conditional probabilities under each hypothesis for each sensor are placed along the sides of the table, and all possible combinations are formed by multiplying them pairwise (for illustration, see Table I in the Generalized Evidence Processing theory section). For binary hypothesis testing and hard decisions, the so created 2x2 tables are combined with each other, and the process is repeated until the convolution (II.5) is generated in the final table.

Once the final convolution table is obtained, association of the resulting probabilities with all possible events (decisions in this case) is done by sorting the entries in the table (convolution points) according to the numerical values of the corresponding abscissae that

take the form $$\prod_{i \varepsilon S} \frac{P_{Di}}{P_{Fi}} \prod_{j \varepsilon \bar{S}} \frac{1-P_{Dj}}{1-P_{Fj}} .$$ In the final stage, probability

masses are allocated to the different alternatives (decisions) by choosing decision boundaries that optimize certain desired criteria at the fusion [TVB '87, '89]. In contrast with the D-S theory, the rule (II.5, 1.6, 1.7) does not lead to inconsistent probability masses when the intersection of the corresponding events is empty as in D-S theory [Demp '68, Shaf '76]. Thus, the need for (arbitrary) evidence renormalization required in D-S theory [Shaf '76] is avoided, and the possibility of simultaneous increase of evidence about conflicting proposition that may occur in D-S theory is prevented. Furthermore, in addition to evidence combining, the combining rule (II.5, 1.6, 1.7) provides decision regions and boundaries according to desired performance criteria chosen to optimize the performance of the fusion (and the peripheral sensors). If the N-P DDF rule (II.5, 1.6, 1.7) is interpreted as an evidence combining rule with evidence the conditional probabilities $dP(u_i|H_j)$, $i = 1, ..., N$ and $j = 0, 1$, some interesting comparisons are made with the Demspter's combining rule in Section 3.

Although Theorems 2 and 3 were independently proven in [ChVa'86] and [9] respectively and preceded the proof of Theorem 1, they can be directly derived from Theorem 1. In DDF with local processors operating at fixed false alarm and detection probability (P_F, P_D), it is legitimate to raise the question whether the performance after fusion improves beyond that of any sensor or subset of sensors in the configuration individually. The answer to this question turns out to be affirmative as proven in [9].

Theorem 4 [9]   In a configuration of N similar sensors, all operating at the same $(P_F, P_D) = (p, q)$, the randomized N-P test at the fusion can provide an overall (P^f_F, P^f_D) that satisfies

$$P^f_F \leq \min_{i \, \varepsilon N} \{P_{F_i}\} \quad \text{and} \quad P^f_D > \max_{i \, \varepsilon N} \{P_{D_i}\}$$

provided that $N \geq 3$.

More precisely, for $N \geq 3$, the randomized N-P test can be fixed so that

$$P^f_F = P_F = p \quad \text{and} \quad P^f_D > P_D = q$$

where P_F and P_D are the false alarm and detection probability at the individual sensors.   □

Table I. Fusion system of five sensors in parallel topology. All sensors have the same false alarm probability $P_{F_i} = 0.05$, $i = 1, \ldots, 5$. The corresponding detection probabilities are as follows.[a]

i	1	2	3	4	5
P_{D_i}	0.95	0.94	0.93	0.92	0.91

Decision fusion: 5	Sensor system
Sensors PF: Equal x	Unequal –
Sensors PD: Equal –	Unequal x

Threshold @ fusion center	Probability of detection @ fusion center	Probability of false alarm @ fusion center
PDMAX = 0.95000	PFMIN = 0.50000E-01	
t^*	PD	PF
6163.2	0.957817	0.300000E-04
53.004	0.963797	0.142812E-03
45.880	0.968973	0.255625E-03
40.339	0.973523	0.368437E-03
38.907	0.977913	0.481250E-03
34.208	0.981772	0.594062E-03
32.081	0.985391	0.706874E-03
29.610	0.988731	0.819687E-03
28.207	0.991913	0.932499E-03
24.416	0.994668	0.104531E-02
20.705	0.997003	0.115812E-02
0.20998	0.997454	0.330156E-02
0.17806	0.997835	0.544500E-02
0.15413	0.998165	0.758843E-02
0.14683	0.998480	0.973187E-02
0.13552	0.998771	0.118753E-01
0.12709	0.999043	0.140187E-01
0.11174	0.999282	0.161622E-01
0.10778	0.999513	0.183056E-01
0.94760E-01	0.999717	0.204490E-01
0.82023E-01	0.999892	0.225925E-01

[a]After fusion, the detection probability is higher than the highest detection probability among all sensors for a wide range of false alarm probabilities lower than 0.05.

Theorem 4 raises two interesting points. First, that it is possible to create a "super-sensor" that meets strict performance requirements from inferior sensors that individually do not meet the performance requirements, thus legitimizing the idea of DDF. Second, it proves that nothing is gained in performance when binary decisions from only two identical sensors are fused; it takes at least three identical sensors to improve performance. Theorem 4 characterizes the DDF performance when all sensors are similar. Although no similar theoretical result exists for dissimilar sensors, extensive numerical simulations have shown that the conclusions of theorem extend to dissimilar sensors as well [TVB '87, '88]. Numerical results from the application of Theorem 4 in N-P DDF are summarized in Table I. Additional numerical results in the spirit of the theorem can be found in [9] and [16].

III. Bayesian DDF in binary hypothesis testing with multi-level local logic

For the binary hypothesis testing problem and similar assumptions as above, i.e. sensors in parallel topology and data statistically independent under either hypothesis from sensor to sensor, we consider the Bayesian DFF problem when the local processors (sensors) employ a multilevel logic. The problem is equivalent to utilizing a multi-level quantizer locally and a binary decision fusion rule. If the objective of optimally choosing the quantization levels is to maximize the detection probability at the fusion for fixed false alarm probability, the optimal quantizer is given by Theorem 5.

Theorem 5 [12] For the multi-level local logic Bayesian DDF, quantization of the local data according to *Likelihood Ratios (LTs)* maximizes the probability of detection for fixed false alarm probability at the fusion. Furthermore, the optimal fusion rule is the *N-P test* at the desired false alarm level.   □

Proof The proof of Theorem 5, which is a generalization of the proof of Theorem 1, is based on the monotone property of the optimal Bayesian DDF rule [TVB '87, VAT '88] and the fact that the *LRT* satisfies this monotone property. A summary of the proof has appeared in [12]. The complete proof of the theorem is given in the Appendix. An alternative proof of the theorem has been recently given in [18].

Hence, in the Bayesian DDF the optimal fusion rule consists of LRTs at the local and global level. In the multi-level local logic DDF, the semantic correlation between quantization levels and decisions favoring one hypothesis or a group of hypotheses is not, in general, clear unless a specific decision is attached to each quantization level by minimizing some total decision cost. This is the idea behind the *Generalized Evidence Processing (GEP)* Theory [Thom '90, ThGa '90] which extends the single logic N-P (Bayesian) DDF to multi-level logic so that a correspondence between Bayesian decision processing and D-S evidence processing can be established.

A. Generalized Evidence Processing (GEP) Theory

The pivoting idea behind GEP theory is the separation of hypotheses from decisions. Once this separation is understood, the Bayesian (or N-P) DDF theory can be extended to a frame of discernment similar to that of D-S theory. In the context of GEP theory, the choice of different decisions can be thought off as different quantization levels of the data. For notational simplicity, the GEP theory is first presented for binary hypothesis decision fusion. Generalization to multiple hypotheses decision fusion follows at the end of the section.

Let H_1 , H_0 be the two hypotheses under test. The probability space is partitioned into two regions according to the events $\{\omega = H_1\}$ and $\{\omega = H_0\}$ with associated probabilities $P_1 \geq 0$ and $P_0 \geq 0$ respectively, where $P_1 + P_0 = 1$.

Let d_0, d_1, and $d_2 := d_{0v1}$ be a frame of discernment used by a decision maker to partition the probability space according to the gathered evidence, where the three decisions correspond to the propositions "H_0 true," "H_1 true," and "H_0 or H_1 true," respectively. The decision $d_2 := d_{0v1}$, where "v" stands for "or," indicates the inability of the decision maker to come up with conclusive evidence on the true nature of the hypothesis.

In the classical probabilistic (Bayesian) framework, the probability associated with d_{0v1} is equal to

$$\Pr[d_{0v1}] = \Pr[H_0 + H_1] = \Pr[H_0] + \Pr[H_1] = 1 \qquad \text{(III.1)}$$

since H_0 and H_1 constitute a disjoint coverage of the probability space over which the evidence processing problem is defined. As it was mentioned earlier, the apparent weakness of the Bayesian theory to incorporate non-mutually exclusive, i.e. redundant, propositions gave rise to the D-S theory which is particularly efficient in dealing with fuzzy propositions. However, by disassociating decisions from hypotheses, a unified framework is created which can accommodate both Bayesian and D-S DDFs.

Let z be the transformation from the initial event space Ω into the observation (data) space $\mathbf{Z}$, i.e.

$$z: \Omega \;\; \rightarrow \;\; \mathbf{Z} \qquad \text{(III.2)}$$

and d be a transformation from the observation space into the decision space $\mathbf{D}$, i.e.

$$d: \mathbf{Z} \rightarrow \mathbf{D} \qquad\qquad (III.3)$$

Let $\{dP(z|H_i), P(H_i) ; i = 0, 1\}$ be the probability measure on $\mathbf{Z}$.

Let C_{ij} be the cost associated with a decision i when the true hypothesis

is H_j. Define the mapping d so that the cumulative risk

$$\mathbf{R} = \sum_i \sum_j C_{ij} P_j \int_{Z_i} dP(z|H_j) ; j = 0, 1, \text{ and } i = 0, 1, 2 \qquad (III.4)$$

is minimized, where $d_2 := d_{0v1}$ is the ambiguous decision. In (III.4),

the regions Z_i indicate the partition of the observation space according

to the decision rule d, where Z_i indicates the region of the observation

space in which the decision is d_i. Notice that the partition of the

observation space is made according to the set of decisions that is

chosen a priori, and that in a given set a particular decision may not

favor a specific hypothesis exclusively, in contrast with what is

customary in the Bayesian theory. However, it is not clear from (III.4)

if such a partition is feasible, since the decision rule is defined

though R which depends on the partition $\{Z_i\}$. We show next that for a

proper choice of the costs C_{ij}, there exist partitions $\{Z_i\}$ that define

legitimate probability measures on $\mathbf{D}$, and thus a generalized decision

rule d.

Rewriting equation (III.4) as

$$\mathbf{R} = \int_{Z_0} [P_0 C_{00} dP(z|H_0) + P_1 C_{01} dP(z|H_1)] + \int_{Z_1} [P_0 C_{10} dP(z|H_0)$$

$$+ P_1 C_{11} dP(z|H_1)] + \int_{Z_2} [P_0 C_{20} dP(z|H_0) + P_1 C_{21} dP(z|H_1)]$$

$$(III.5)$$

the total risk is minimized if the decision rule assigns z (the observation) to the region that corresponds to the least integrand under the three integrals in (III.5). Hence, the decision rule becomes

$$P_0 C_{00} \, dP(z|H_0) + P_1 C_{01} \, dP(z|H_1) \underset{Z_0 \text{ or } Z_2}{\overset{Z_1 \text{ or } Z_2}{\gtrless}} P_0 C_{10} \, dP(z|H_0) + P_1 C_{11} \, dP(z|H_1)$$

$$(III.6)$$

and symmetrically for the other alternatives. Dividing both sides of (III.6) by $dP(z|H_0)$ and defining $\Lambda(z) = \dfrac{dP(z|H_1)}{dP(z|H_0)}$, the decision rule becomes

$$[C_{01} - C_{11}] \, \Lambda(z) \underset{d_0 \text{ or } d_2}{\overset{d_1 \text{ or } d_2}{\gtrless}} \frac{P_0}{P_1} [C_{10} - C_{00}]$$

$$(III.7)$$

Similarly,

$$[C_{01} - C_{21}] \, \Lambda(z) \underset{d_0 \text{ or } d_1}{\overset{d_2 \text{ or } d_1}{\gtrless}} \frac{P_0}{P_1} [C_{20} - C_{00}] \qquad (III.8)$$

and

$$[C_{21} - C_{11}] \, \Lambda(z) \underset{d_2 \text{ or } d_0}{\overset{d_1 \text{ or } d_0}{\gtrless}} \frac{P_0}{P_1} [C_{10} - C_{20}] \qquad (III.9)$$

From (III.7), (III.8), and (III.9), it is seen that the decision rule depends on the relative values of the C_{ij} costs. We examine three different cases to illustrate the significance of the C_{ij}'s.

Case 1 The associated cost for correct decision is zero, i.e. $C_{11} = C_{00} = 0$, while the cost of incorrect guessing is higher than the cost associated with indecision under both hypotheses, i.e. $C_{ij} > C_{2j}$ for every $i \neq 2$ and $j = 0$ or 1. Under these conditions, the decision rule becomes

$$\Lambda(z) \quad \begin{matrix} d_1 \text{ or } d_2 \\ > \\ < \\ d_0 \text{ or } d_2 \end{matrix} \quad \frac{P_0}{P_1} \quad \frac{C_{10}}{C_{01}} \tag{III.10}$$

Similarly,

$$\Lambda(z) \quad \begin{matrix} d_2 \text{ or } d_1 \\ > \\ < \\ d_0 \text{ or } d_1 \end{matrix} \quad \frac{P_0}{P_1} \quad \frac{C_{20}}{C_{01} - C_{21}} \tag{III.11}$$

and

$$\Lambda(z) \quad \begin{matrix} d_1 \text{ or } d_0 \\ > \\ < \\ d_2 \text{ or } d_0 \end{matrix} \quad \frac{P_0}{P_1} \quad \frac{C_{10} - C_{20}}{C_{21}} \tag{III.12}$$

In this case, it is seen from (III.10) through (III.12) that the optimal test (decision rule d) is of the likelihood type with the indecision region dependent on the relative values of C_{ij}. Different numerical applications are considered next to further illustrate the significance of the C_{ij}'s.

Case 1.1 $C_{00} = C_{11} = 0$, $C_{10} = C_{01} = 1$, and $C_{20} = C_{21} = {}^1/_3$. In this case $\dfrac{C_{10}}{C_{01}} = 1$, $\dfrac{C_{20}}{C_{01} - C_{21}} = 0.5$, and $\dfrac{C_{10} - C_{20}}{C_{21}} = 2$. The partition of the LR by the decision rule in this case is shown in Fig. 3. From Fig. 3, it is seen that in this case the indecision region lies between the two definite decision regions, which corresponds to the way that we would intuitively have picked the indecision (uncertainty) region relative to a LRT [9].

Case 1.2 $C_{00} = C_{11} = 0$, $C_{10} = C_{01} = 1$, and $C_{20} = C_{21} = 0.5$. In this case $\dfrac{C_{10}}{C_{01}} = \dfrac{C_{20}}{C_{01} - C_{21}} = \dfrac{C_{10} - C_{20}}{C_{21}} = 1$, i.e. all three thresholds are the same, and the indecision region is completely eliminated. The partition of the LR by the decision rule is shown in Fig. 4. It corresponds to a standard binary hypothesis - binary decision Bayesian problem.

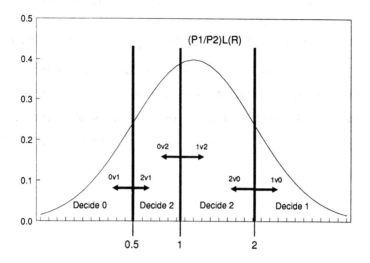

Fig. 3 Case 1.1 The indecision region lies between the
two definite decision regions.

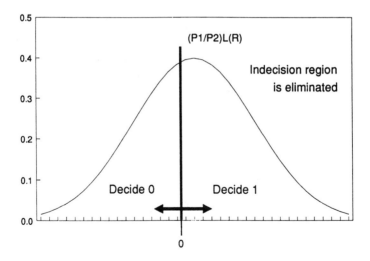

Fig. 4 Case 1.2 The indecision region is completely eliminated.

Case 1.3 $C_{00} = C_{11} = 0$, $C_{10} = C_{01} = 1$, and $C_{20} = C_{21} = {}^2/_3$. In this case $\dfrac{C_{10}}{C_{01}} = 1$, $\dfrac{C_{20}}{C_{01} - C_{21}} = 2$, and $\dfrac{C_{10} - C_{20}}{C_{21}} = {}^1/_2$. The partition of the LR by the decision rule   in this case is shown in Fig. 5.   In this case the two definite (hard) decision regions   are sandwiched between the two indecision regions opposite to what one would intuitively   have defined as an indecision region in an LRT, and exactly opposite to Case 1.1.

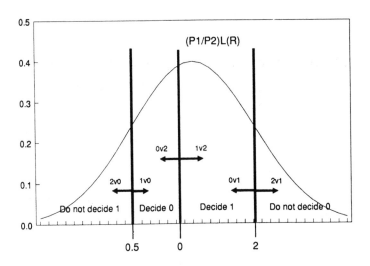

Fig. 5 Case 1.3   The definite decision regions lie
                      between the indecision regions.

Case 2 $C_{00} = C_{11} = 0$, $C_{01} = C_{21}$.   In this case $\dfrac{C_{10}}{C_{01}} = 1$,

$\dfrac{C_{20}}{C_{01} - C_{21}} = \infty$, and $\dfrac{C_{10} - C_{20}}{C_{21}} = \alpha$.   The partition of the LR into

different decision regions is shown in Fig.s 6. and 7 depending on the

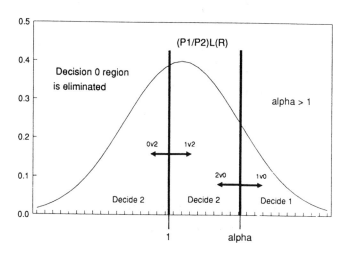

Fig. 6   Case 2, $\alpha > 1$. The decision d_0 is completely eliminated.

value of α.   In Fig. 6, $\alpha > 1$, and the decision d_0 is completely eliminated.   In Fig. 7, $\alpha < 1$ and there are three decision regions; d_1, d_0, and   not d_1.   According to the definition of the three possible decisions, the decision "do not decide d_1" must be interpreted as the fuzzy decision "d_0 or d_2" and not as "decide in favor of H_0."

Case 3   If in the above cases $C_{2j} > C_{ij}$, $i \neq 2$ and $j = 0, 1$, the LRT in (III.11) is reversed and the threshold in (III.12) becomes

negative.   Under these circumstances, the decision regions in the
previous cases are reversed.

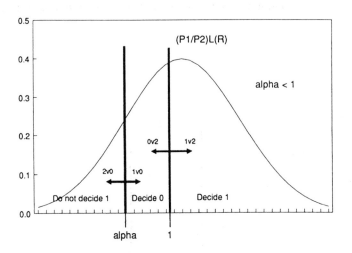

Fig. 7   Case $\alpha < 1$. Creation of a fuzzy decision region:

"Do not decide d_i "

From all the cases discussed above, it is apparent that if the
decision rule is chosen to minimize a certain decision cost, then the
indecision region depends on the choice of the associated costs. Hence,
the probability masses can be assigned to the different propositions
(decisions) in an optimal fashion so that the total risk is minimized,
instead of being assigned arbitrarily as in the D-S theory.

GEP Combining rule   Let $u =: \{u_1, u_2, ..., u_N\}$ be the set of
peripheral sensor decisions at the fusion center. Each u_i belongs to the
set $\{d_0, d_1, d_2\}$.   Let w_{ij} be the cost associated with the fusion

deciding in favor of proposition d_i when the true hypothesis is H_j. If u designates the decision of the fusion, the total cost at the fusion is

$$\mathbf{R_f} = \sum_i \sum_j w_{ij} P_j \int_{F_i} dP(u \mid H_j) \tag{III.13}$$

Assuming that the decisions from the peripheral sensors are independent conditioned on each hypothesis, (III.13) can be written as

$$\mathbf{R_f} = \sum_i \sum_j w_{ij} P_j \int_{F_i} \prod_{k=1}^{N} dP(u_k \mid H_j) = \sum_i \sum_j w_{ij} P_j \prod_{k=1}^{N} [\int_{F_i} dP(u_k \mid H_j)] \tag{III.14}$$

$$= w_{0\,0} \, P_0 \, \prod_k [\int_{F_0} dP(u_k \mid H_0)] + w_{1\,0} \, P_0 \, \prod_k [\int_{F_1} dP(u_k \mid H_0)]$$

$$+ w_{2\,0} \, P_0 \, \prod_k [\int_{F_2} dP(u_k \mid H_0)] + w_{0\,1} \, P_1 \, \prod_k [\int_{F_0} dP(u_k \mid H_1)]$$

$$+ w_{1\,1} \, P_1 \, \prod_k [\int_{F_1} dP(u_k \mid H_1)] + w_{2\,1} \, P_1 \, \prod_k [\int_{F_2} dP(u_k \mid H_1)]$$

$$= \int_{F_0} [\, w_{0\,0} \, P_0 \, dP(u \mid H_0) + w_{0\,1} \, P_1 \, dP(u \mid H_1) \,] + \int_{F_1} [\, w_{1\,0} \, P_0 \, dP(u \mid H_0)$$

$$+ w_{1\,1} \, P_1 \, dP(u \mid H_1) \,] + \int_{F_2} [\, w_{2\,0} \, P_0 \, dP(u \mid H_0) + w_{2\,1} \, P_1 \, dP(u \mid H_1) \,]$$

$$\tag{III.15}$$

The decision rule that minimizes the total risk assigns a particular combination of peripheral decisions u to that region that gives rise to the smallest integrand. Assuming that $w_{ii} = 0$, i.e. that there is no penalty for deciding correctly (a reasonable assumption in evidence processing), and that $w_{ij} - w_{2j} > 0$ for every j, i.e. that the cost of indecision is lower than the cost of deciding incorrectly, the test at the fusion becomes

$$\Lambda(u) \quad \begin{matrix} d_1 \text{ or } d_2 \\ > \\ < \\ d_0 \text{ or } d_2 \end{matrix} \quad \frac{P_0}{P_1} \quad \frac{w_{10}}{w_{01}} \qquad (\text{III.16})$$

Similarly,

$$\Lambda(u) \quad \begin{matrix} d_2 \text{ or } d_1 \\ > \\ < \\ d_0 \text{ or } d_1 \end{matrix} \quad \frac{P_0}{P_1} \quad \frac{w_{20}}{w_{01} - w_{21}} \qquad (\text{III.17})$$

and

$$\Lambda(u) \quad \begin{matrix} d_1 \text{ or } d_0 \\ > \\ < \\ d_2 \text{ or } d_0 \end{matrix} \quad \frac{P_0}{P_1} \quad \frac{w_{10} - w_{20}}{w_{21}}$$

(III.18)

where

$$\Lambda(u) = \frac{dP(u|H_1)}{dP(u|H_0)} = \frac{dP(u_1|H_1)\ldots dP(u_N|H_1)}{dP(u_1|H_0)\ldots dP(u_N|H_0)}$$

$$= \underset{i \varepsilon S_1}{\Pi} \frac{P_{Di}}{P_{Fi}} \quad \underset{j \varepsilon S_0}{\Pi} \frac{1 - P_{Dj} - P_{I^1j}}{1 - P_{Fj} - P_{I^0j}} \quad \underset{n \varepsilon S_2}{\Pi} \frac{P_{I^1n}}{P_{I^0n}} \qquad (\text{III.19})$$

where $P_{I^1_j}$, $P_{I^0_j}$ indicate the probability masses at sensor j associated with the fuzzy decision (or, indecision) d_{0v1} under the hypotheses H_1 and H_0 respectively, S_1 is the set of those decisions from the set u which favor $d_1 (= H_1)$, S_0 is the set that favors $d_0 (= H_0)$, and S_2 is the set from the peripheral decisions u that favors $d_2 (= \{H_0 \text{ or } H_1\}$, i.e. the undecided). Naturally, $u = S_0 + S_1 + S_2$.

From (III.16), (III.17), and (III.18), it follows that the optimal decision rule at the fusion is a likelihood ratio test. If we look at equation (III.19), the distribution of the LR under the two hypotheses is given by

$$P(\log \Lambda(u) \mid H_i) = P(\log \Lambda(u_1) \mid H_i) * \ldots * P(\log \Lambda(u_N) \mid H_i)$$

$$; i = 0, 1 \qquad (III.20)$$

where "*" indicates convolution, with

$$P(\log \Lambda(u_i) \mid H_0) = (1 - P_{Fi} - P_{I^0 i}) \delta(\log \Lambda(u_i) - \log \frac{1 - P_{Di} - P_{I^1 i}}{1 - P_{Fi} - P_{I^0 i}})$$

$$+ P_{I^0 i} \, \delta(\log \Lambda(u_i) - \log \frac{P_{I^1 i}}{P_{I^0 i}})$$

$$+ P_{Fi} \, \delta(\log \Lambda(u_i) - \log \frac{P_{Di}}{P_{Fi}}) \qquad (III.21)$$

and $\delta(x)$ is the Kronecker's delta function, i.e. $\delta(x) = 1$ if $x > 0$ and zero otherwise, and

$$P(\log \Lambda(u_i) \mid H_1) = (1 - P_{Di} - P_{I^1 i}) \delta(\log \Lambda(u_i) - \log \frac{1 - P_{Di} - P_{I^1 i}}{1 - P_{Fi} - P_{I^0 i}})$$

$$+ P_{I^1 i} \, \delta(\log \Lambda(u_i) - \log \frac{P_{I^1 i}}{P_{I^0 i}})$$

$$+ P_{Di} \, \delta(\log \Lambda(u_i) - \log \frac{P_{Di}}{P_{Fi}}) \qquad (III.22)$$

Hence, the distribution of $\Lambda(u)$ under H_0 is given as the product of all possible combinations of $[(1 - P_{Fi} - P_{I^0 i}), P_{I^0 i}, P_{Fi}]$ according to

their abscissae [9]. Similarly, the distribution of $\Lambda(u)$ under H_1 is given as the product of all possible combinations of $[(1 - P_{Di} - P_{I^1 i}), P_{I^1 i}, P_{Di}]$ according to their abscissae. Then, the fusion is

done by using the appropriate thresholds from (III.16), (III.17), and (III.18). In that case, the probability masses (or beliefs) associated with each decision are combined according to the threshold and their abscissae. Thus, the combining rule involves pairwise multiplication of probability masses according to Table II as in D-S theory. However, in **GEP** theory, the masses are associated via thresholds in an optimal way so that a certain risk (Eq. (III.15)) is minimized, or so that the probability of detection is maximized for fixed false alarm and indecision probabilities (generalized Neyman-Pearson test), whereas in D-S theory the probability masses (beliefs) are combined according to intersection of events, resulting in evidence conflict (see Section 3).

The probabilities in Table II are conditioned on each hypothesis, and $i = 0, 1$. Thus, each m_j^i, $j = 1, 2$, in Table II is a conditional probability for $i = 0, 1$. Hence, the initial probability combining takes place among conditional probabilities only. For $i = 0, 1$, each product term in Table I, is a probability mass on the LRT coordinate axis with abscissa $m_j^1(d) / m_j^0(d)$ for every $d = d_0, d_1, d_2$. Evidence combining under each hypothesis is done by summing the probabilities from Table II whose abscissae fall in specific intervals specified either by an optimization criterion, or a certain desired performance. Hence, for $d = d_0, d_1, d_2, ..., d_N$, evidence combining under each hypothesis H_i, $i = 0, 1$, is done according to the **threshold rule**

$$m_1^i(d_k) m_2^i(d_m) \rightarrow \text{decision } d_j \quad \text{if} \quad \frac{m_1^1(d_k) m_2^1(d_m)}{m_1^0(d_k) m_2^0(d_m)} \varepsilon \ F_j \qquad (III.23)$$

where F_j is the decision region that favors decision d_j. The regions F_j may be determined so that a performance criterion is optimized at the fusion (and possibly at the sensors). For a single binary hypothesis, the decision regions at the fusion are determined by simple thresholds, in which case the decision rule (III.23) simplifies to

$$m_1^i(d_k)\, m_2^i(d_m) \to \text{ decision } d_j \text{ if } t_j \le \frac{m_1^1(d_k)\, m_2^1(d_m)}{m_1^0(d_k)\, m_2^0(d_m)} < t_{j+1}$$

(III.24)

for all k, m, and j, where t_j are the thresholds of the LRT's associated with the different decisions that minimize some risk function.

If **multiple hypotheses** (more than two) are tested, the combining rule is extended to combine the belief functions of the individual sources at the fusion and generate the new conditional belief function under each hypothesis. The association of the new belief function at the fusion with the set of admissible decisions must be done by using the multiple-hypotheses LRT [12], or another test that optimizes some performance measure. It must be underlined again, that the probabilities in the GEP combining rule need not be defined through Bayesian reasoning necessarily, but may very well correspond to belief functions resulting from the D-S approach.

Table II GEP Evidence Combining Rule

S1 \ S2	$m_2^i(d_0)$	$m_2^i(d_1)$	$m_2^i(d_2)$
$m_1^i(d_0)$	$m_1^i(d_0)\, m_2^i(d_0)$	$m_1^i(d_0)\, m_2^i(d_1)$	$m_1^i(d_0)\, m_2^i(d_2)$
$m_1^i(d_1)$	$m_1^i(d_1)\, m_2^i(d_0)$	$m_1^i(d_1)\, m_2^i(d_1)$	$m_1^i(d_1)\, m_2^i(d_2)$
$m_1^i(d_2)$	$m_1^i(d_2)\, m_2^i(d_0)$	$m_1^i(d_2)\, m_2^i(d_1)$	$m_1^i(d_2)\, m_2^i(d_2)$

In the multiple hypotheses case, the conditional belief function in GEP becomes a multi-variable function of the LRs $\{\Lambda_k(d) := \prod_{j=1}^{J} \dfrac{dP(d_j | H_k)}{dP(d_j | H_0)}$, k = 1, 2, ..., m-1$\}$ where J is the number of sensors in the fusion system, d_j the decision of the j-th sensor, and m the number of tested hypotheses. The evidence from the different sensors is combined by forming the joint probability distribution of the LR's under each hypothesis, i.e. by generating $dP(\Lambda_1 , \Lambda_2 , ..., \Lambda_{m-1} | H_k)$, k= 1, 2, ..., J. For two sensors with independent decisions conditioned on each hypothesis, the conditional evidence combining rule of GEP for three hypothesis and soft decisions (fuzzy logic), can be implemented using Table III.

Once all the entries in Table III are entered, the evidence is combined by adding the probabilities from the fourth column together when the corresponding abscissae, i.e. the pairs $(\Lambda_1 (d_1 , d_2), \Lambda_2 (d_1 , d_2))$ in the second and third columns, are identical. Once the evidence from all sensors is combined using tables similar to Table III, decisions are associated with the combined evidence using rule (III.23) so that a desired performance criterion is optimized.

Thus, evidence combining at the fusion is done conditioned on each hypothesis separately. The evidence is then associated with the admissible decisions unconditionally using a LRT or a test that optimizes some performance measure. Notice that the set of decisions need not be the same as the set of hypotheses. Thus, **evidence combining and decision making are understood as separate concepts in the framework of the GEP theory**.

Table III Evidence combining rule for multiple hypotheses in GEP theory

(d_1, d_2)	$\Lambda_1 (d_1, d_2)$	$\Lambda_2 (d_1, d_2)$	$dP(\Lambda_1 (d_1, d_2), \Lambda_2 (d_1, d_2) \mid H_k)$
			$= dP(\Lambda_1 (d_1, d_2) \mid H_k) dP(\Lambda_2 (d_1, d_2) \mid H_k)$
			$= \prod_{j=1}^{2} dP(\Lambda_1 (d_j) \mid H_k) dP(\Lambda_2 (d_j) \mid H_k)$
$(0, 0)$	$\Lambda_1 (0,0)$	$\Lambda_2 (0,0)$	$dP(\Lambda_1 (0,0) \mid H_k) dP(\Lambda_2 (0,0) \mid H_k)$
$(0, 1)$	$\Lambda_1 (0,1)$	$\Lambda_2 (0,1)$	$dP(\Lambda_1 (0,1) \mid H_k) dP(\Lambda_2 (0,1) \mid H_k)$
$(0, 2)$	$\Lambda_1 (0,2)$	$\Lambda_2 (0,2)$	$dP(\Lambda_1 (0,2) \mid H_k) dP(\Lambda_2 (0,2) \mid H_k)$
$(0, 0v1)$	$\Lambda_1 (0,0v1)$	$\Lambda_2 (0,0v1)$	$dP(\Lambda_1 (0,0v1) \mid H_k) dP(\Lambda_2 (0,0v1) \mid H_k)$
$(0, 0v2)$	$\Lambda_1 (0,0v2)$	$\Lambda_2 (0,0v2)$	$dP(\Lambda_1 (0,0v2) \mid H_k) dP(\Lambda_2 (0,0v2) \mid H_k)$
$(0v1, 0)$	$\Lambda_1 (0v1,0,)$	$\Lambda_2 (0v1,0,)$	$dP(\Lambda_1 (0v1,0) \mid H_k) dP(\Lambda_2 (0v1,0) \mid H_k)$
$(0v1, 1)$	$\Lambda_1 (0v1,1)$	$\Lambda_2 (0v1,1)$	$dP(\Lambda_1 (0v1,1) \mid H_k) dP(\Lambda_2 (0v1,1) \mid H_k)$
$(0v1, 2)$	$\Lambda_1 (0v1,2)$	$\Lambda_2 (0v1,2)$	$dP(\Lambda_1 (0v1,2) \mid H_k) dP(\Lambda_2 (0v1,2) \mid H_k)$
$(0v1, 0v1)$	$\Lambda_1 (0v1,0v1)$	$\Lambda_2 (0v1,0v1)$	$dP(\Lambda_1 (0v1,0v1) \mid H_k) dP(\Lambda_2 (0v1,0v1) \mid H_k)$
$(0v1, 0v2)$	$\Lambda_1 (0v1,0v2)$	$\Lambda_2 (0v1,0v2)$	$dP(\Lambda_1 (0v1,0v2) \mid H_k) dP(\Lambda_2 (0v1,0v2) \mid H_k)$
$(0v2, 1)$	$\Lambda_1 (0v2,1)$	$\Lambda_2 (0v2,1)$	$dP(\Lambda_1 (0v2,1) \mid H_k) dP(\Lambda_2 (0v2,1) \mid H_k)$

. . . etc

The generalization of the Bayesian (and N-P) theory by the GEP theory is straightforward. An interpretation is probably required to establish the correspondence between GEP and D-S theories. If the probabilities $P(u_k = i \mid H_j)$, $i = 1, 2, 3$, are considered as (conditional) bpa's (basic probability assignments [3]) in the D-S theory for the k-th sensor, $k = 1, 2, ..., N$, under hypothesis H_j, $j = 0, 1$, the evidence from the different sensors at the fusion is combined using the conditional distribution of the LR under the different hypothesis according to Table II or III. A new (conditional) belief function is generated using the decision thresholds at the fusion. The (hard) decisions at the sensors are used to simply produce a hard decision at the fusion, if needed, according to some optimality criteria. In that respect, the **GEP** theory not only defines and processes the evidence according to an a-priori set of optimality criteria, but also provides, if needed, for optimized hard decisions both at the local (sensor) as well as global (fusion) level, a capability which is not built-in the D-S theory (see Section 3).

The decision boundaries in GEP theory determine how evidence is associated with propositions at the fusion and reflect the choice of the costs w_{ij}. To demonstrate the effect that the semantic content of the local decisions has on the global decision (fusion), several experiments were conducted in Gaussian and slow-fading Rayleigh channels. The following statistical model were assumed for the two channels.

Gaussian: *Observation model at each sensor:* $r \sim G(0,1) : H_0$, and $r \sim G(s,1) : H_1$, where $G(\alpha,\beta)$ designates an α mean and variance β Gaussian distribution. If P_F is the operating false alarm probability, the associated threshold $t_b := Q^{-1}(P_f)$, where $Q( ) = 1 - \Phi( )$ is the cumulative distribution function (cdf) of the standard normal, and Q^{-1} is its inverse.

Rayleigh: *False alarm probability:* $P_F = [\lambda(1+\varepsilon)]^{-(1 + \frac{1}{\varepsilon})}$;

Detection probability: $P_D = [P_F]^{(\frac{1}{1+\varepsilon})}$

where λ is the threshold used, and ε the SNR at the sensor. In the single-level local logic Bayesian DDF with hard decisions at the sensors and fusion, the probabilities at the sensors were generated assuming fixed false alarm probabilities at eh sensors equal to 0.05. For the multi-level local logic DDF, the ambiguous (soft or "fuzzy") decisions were generated by considering a ±20% uncertainty region about the thresholds that determine the decision boundaries in the Bayesian case. The numerical results that are presented refer to binary hypothesis testing with the set of "soft" decisions consisting of $\{d_0 = H_0, d_1 = H_1, d_2 = H_0 \vee H_1\}$. Additional results for ternary hypothesis testing and arbitrary probability assignments can be found in [17], [16].

In Fig.s 8 and 9, the total single- and multiple-level logic (GEP) Bayesian DDF risk for the three cost assignments discussed earlier is plotted for Gaussian and slow-fading Rayleigh channels respectively. It is seen that for cases 1.1 and 1.2, the total risk is reduced when soft decisions (GEP) are used instead of hard decisions (Bayesian). Furthermore, the risk for the two cases decreases as the signal-to-noise ratio (SNR) decreases, in agreement with our intuition about a "good" fusion system. However, for case 1.3, the performance of GEP is not always superior to Bayesian. and: (a) soft decisions (GEP) do not lead to better performance than hard decisions (Bayesian); and (b) the risk increases, in general, as the SNR increases   against our "intuition." The behavior for these three cases can be explained from Figs 8 and 9. For cases 1.1 and 1.2 the decision boundaries seem intuitively justifiable. However, for the case 1.3, the decision boundaries are "counter-intuitive." Thus, given a set of cost factors, evidence combining in GEP theory is done according to a set of performance criteria, so that the performance of the fusion is optimized. The ability of GEP theory to process evidence independent of the associated propositions and combine it according to desired performance criteria without evidential conflict is highly desirable in sensor integration and decision fusion problems where we want to design a fusion system so

that its performance can be assessed a priori and according to certain
design criteria and specifications.

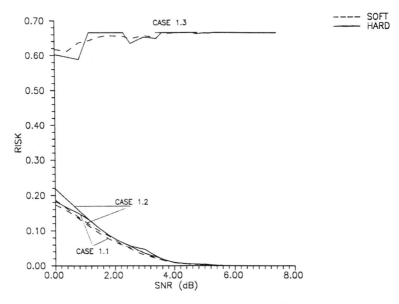

Fig. 8 Risk vs SNR. Gaussian case; 5 sensors.

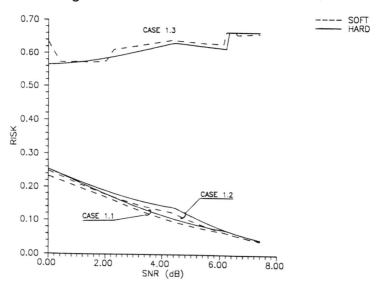

Fig. 9 Risk vs SNR. Rayleigh case; 5 sensors.

In another set of experiments on binary hypothesis testing with Gaussian and Rayleigh distributed data, the performance of the GEP fusion with soft decisions at the sensors and hard decisions at the fusion was compared to the Bayesian DDF with hard decisions at all sensors and the fusion. The fusion system that was used for comparison consisted of five and ten identical sensors. The sensors decisions were assumed statistically independent under either hypothesis. The soft decision "H_1 or H_0 " was generated by assigning a "$\pm 20\%$ uncertainty region" about the threshold that was used for the Bayesian DDF to generate hard decisions at the sensors at 0.05 false alarm probability level. The *Level Of Confidence (LOC)*, which is equivalent to the (unconditional) probability of correct decision, was used for comparison. The LOC curves in Fig. 10 indicate that GEP outperforms Bayesian DDF with hard local decisions in all cases. The curves were obtained by assuming a fixed false alarm probability 0.05 at the sensors and 0.005 at the fusion.05. GEP outperforms hard-decision Bayesian DFF in both binary and ternary hypothesis testing, in both Gaussian and slow-fading Rayleigh channels and for any number of sensors. This does not come as a surprise if the decision set of GEP is thought of as the result of multi-level quantization of the data, and the quantization is done according to a semantically intuitive fashion. Similar conclusions were drawn from the comparison of GEP DDF with the Bayesian DDF in ternary hypothesis testing [17], [16].

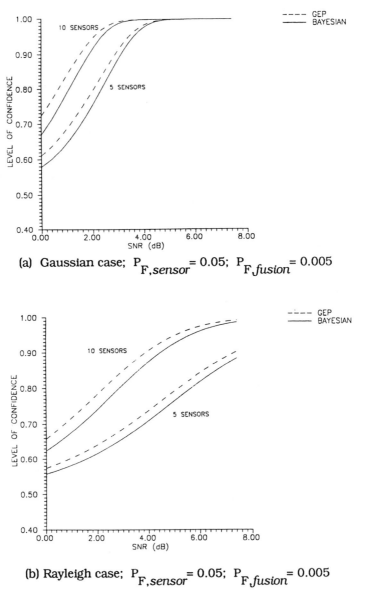

(a) Gaussian case; $P_{F,sensor} = 0.05$; $P_{F,fusion} = 0.005$

(b) Rayleigh case; $P_{F,sensor} = 0.05$; $P_{F,fusion} = 0.005$

Fig. 10 Level of Confidence (LOC) vs. SNR for Bayesian and GEP decision fusion in binary hypothesis testing with (a) Gaussian, and (b) Rayleigh distributed data in five and ten sensor DDF system. In all cases, the fusion makes hard decisions regarding the true nature of the tested hypothesis. The LOC corresponds to the total probability of deciding correctly. The solid curves correspond to Bayesian DDF. The dashed curves correspond to GEP fusion. In all cases, the GEP fusion outperforms the Bayesian fusion.

IV. Distributed Decision Fusion using Dempster-Shafer's Theory

The difference between the Bayesian and D-S theory lies on the type of information that each sensor transmits to the fusion after processing the data locally. As it will become clear in the sequel, if the propositions in the D-S theory are identified with decisions in the GEP (Generalized Bayesian) theory, then there are no semantic differences in the frame of discernment between the two theories. The difference lies on that the frame of discernment in the GEP theory consists of decisions with assigned probabilities that satisfy the Bayesian rule, whereas the frame of discernment in the D-S theory consists of propositions that do not, in general, satisfy the Bayesian rule. Assuming that the number of hypotheses that are tested is fixed and the number of decisions (or frame of discernment in the D-S terminology) is fixed, the output of the local data processing is a set of probabilities regarding the likelihood that the data have been generated by one of the particular hypotheses or subset of hypotheses according to the frame of discernment. To that extend, the use of the term decisions in the D-S theory does not precisely reflect the output of the local processing. It is more appropriate to characterize the outcome of the local processing as evidence about a chosen set of proposition rather than decision regarding a specific hypothesis or set of hypotheses. Thus, even if the frame of discernment is kept common between Bayesian and D-S approaches (by utilizing multi-level Bayesian logic), the mapping of the data in the output of the local processor is completely different; the Bayesian processor maps each data to a particular, single decision (integer-valued scalar), whereas the D-S processor maps the same data to a set of probabilities (multidimensional real-valued vector) associated with all decisions in the frame of discernment. Hence, the communication requirements between Bayesian processors and fusion in one, and D-S processors and fusion on the other are different. Assuming a frame of discernment consisting of **k** propositions, the communication requirements for the Bayesian case is 2logk (the bandwidth required to transmit one of k bits), whereas for the D-S processor k analog outputs must be transmitted to the fusion. Thus, unless the communication requirements for the two approaches are made common, no direct comparison in the performance of the two schemes is meaningful. Since such a performance is beyond the objectives of this paper, we limit the discussion in the structure of the D-S DDF.

In D-S theory, a set of mutually exclusive and exhaustive propositions u_1, u_2, ..., u_m is assumed toward which evidence is being offered. To each proposition, their disjunctions, and negations, a nonnegative number between zero and one (or probability mass) is assigned. If A is an atomic proposition, a disjunction of propositions, or a negation of a proposition, then a probability mass, m(A), is assigned to A. The quantity m(A) is a measure of the belief in proposition A based on the evidence offered. If U designates the frame of discernment, then

$$\sum_{A \varepsilon U} m\,(A) \leq 1 \qquad\qquad (IV.1)$$

with the remaining $1 - \sum_{A \varepsilon U} m\,(A)$ mass attribute to ignorance. Assuming that ignorance constitutes a separate proposition and extending the set U to include this proposition, expression (IV.1) holds as an equality. According to D-S theory, a support function is defined for single propositions as

$$spt(u_i) = m(u_i) \qquad\qquad (IV.2)$$

and for more complex propositions as

$$spt(A) = \sum_{B \subset A} m(B) \qquad\qquad (IV.3)$$

where "**C**" indicates subset. The plausibility function is defined as

$$\text{pls}(u_i) = 1 - \text{spt}(\overline{u}_i) \tag{IV.4}$$

where $\overline{u}_i$ indicates the negation of proposition u_i. Alternatively, the plausibility function for a proposition u_i is obtained by summing the masses of all the disjunctions that contain u_i, including itself, i.e.

$$\text{pls}(u_i) = \sum_{u_i \varepsilon A} m(A) \tag{IV.5}$$

Hence, the support function is indicative of how much evidence is offered in support of a given proposition by all the propositions that relate to it. Furthermore, the plausibility function is indicative of how likely it is for a given proposition to have generated the data.

Evidence from different, and *independent*, sources defined over the same frame of discernment, is fused according to Dempster's combining rule [Depm '68]

$$m(u_\ell) = m_1 \oplus m_2 = \frac{\displaystyle\sum_{A_i B_j = u_\ell} m_1(A_i) m_2(B_j)}{1 - \displaystyle\sum_{A_k B_m = \phi} m_1(A_k) m_2(B_k)} \tag{IV.6}$$

where m_1 and m_2 designate the support (belief) functions from the two different sources of evidence defined over the same frame of discernment, u_ℓ is the proposition toward which evidence is sought, and "ϕ" is the empty set [Shaf '76]. Renormalization of the combined evidence in the rule (IV.6) is required to reject evidence that corresponds to conflicting propositions. The D-S combining rule can be implemented in a

tabular fashion.   To illustrate the mechanical similarities that exist
between the Dempster's combining rule and the GEP DDF, consider a simple
binary hypothesis testing problem.   If the frame of discernment is
defined as $\{u_0 = H_0, u_1 = H_1, u_2 = H_0 \text{ or } H_1\}$, with u_2 indicating the
inability to associate evidence from the data with a definite hypothesis,
the Dempster's combining rule for two sensors can be implemented using
Table IV. In Table IV, k designates evidence associated with conflicting
propositions which is used as normalizing factor in (IV.6).   The
combined evidence is calculated by summing all the product terms from
Table IV that result to the same intersection proposition, and
normalizing the result.   In multiple-source evidence combining, rule
(IV.6) is repeated sequencially until the evidence from all sources is
exhausted.

The difference between the D-S and Bayesian theory is that the
probability assignments for the propositions in the frame of discernment
of the D-S theory do not satisfy the fundamental axiom of (Bayesian)
probability, namely

$$P(A+B) = P(A) + P(B) - P(AB) \tag{IV.7}$$

In the D-S context, the proposition A+B is viewed as a separate entity in
the frame of discernment and can be assigned an arbitrary probability
mass. Still all the probability assignments in the D-S theory must add
up to one or some positive quantity less than one, with the remaining
probability mass to add up to one attributed to total ignorance [Shaf
'76]. A correspondence between the propositions as defined in the D-S
theory and the decisions as defined in the multi-level logic Bayesian
theory can be established if the decisions of the multi-level logic
Bayesian framework are identified with the propositions in the D-S frame
of discernment.   Once this correspondence is established the fusion
performance under the two approaches can be studied under common

communication constraints.   By disassociating decisions from the hypotheses under test, the Generalized Evidence Processing (GEP) provides a semantically common framework within which the N-P and   D-S DDF approaches can be compared under common communication constraints.

Table IV   Dempster's Combining Rule.

S1 \ S2	$m_2^i(u_0)$	$m_2^i(u_1)$	$m_2^i(u_2)$
$m_1(u_0)$	$m(u_0)=m_1(u_0)\,m_2(u_0)$	$k=m_1(u_0)\,m_2(u_1)$	$m(u_0)=m_1(u_0)\,m_2(u_2)$
$m_1(u_1)$	$k=m_1(u_1)\,m_2(u_0)$	$m(u_1)=m_1(u_1)\,m_2(u_1)$	$m(u_1)=m_1(u_1)\,m_2(u_2)$
$m_1(u_2)$	$m(u_0)=m_1(u_2)\,m_2(u_0)$	$m(u_1)=m_1(u_2)\,m_2(u_1)$	$m(u_2)=m_1(u_2)\,m_2(u_2)$

Due to the difference in the way evidence is generated in Bayesian (N-P) and D-S theory, an unconditional performance comparison between the two theories is not, in general, feasible.   Since in a lot of practical applications the performance of a decision making system is determined by fixing the false alarm probability and maximizing the detection probability at the fusion, it is meaningful to compare the Bayesian and D-S approach based on an N-P criterion.   In order to make the comparison possible, we assume that the basic probability assignment of the D-S DDF at the local level is determined using the likelihood function, i.e. we assume that

$$m(\alpha \mid r) = P(\alpha \mid r) \tag{IV.8}$$

where α designates a proposition towards which evidence is provided, and r the observations. Even when the bpa is resolved at the local level, the decision rule at the fusion after the local evidence is combined remains undetermined. In order to keep the decision rule in a D-S context while maintaining a basis for comparison with the Bayesian DDF, the decision rule that will be used for the D-S DDF will assign the data to the proposition that has the highest support among all propositions in the frame of discernment that correspond to definite hypotheses, i.e.

$d(r) := d_i(r) : \max_i \mathrm{spt}(d_i)$ and $d_i = H_i$, i over all single hypothesis propositions   (IV.9)

With the above assumptions, we prove the following theorem.

Theorem 6   Assume that the objective of the fusion is to maximize the detection probability after fusion for fixed false alarm probability. Let the observations of the local sensors be independent from each other conditioned on each hypothesis. Let the bpa for the D-S DDF be determined by the likelihood function (IV.8) at the local level. If the fusion rule is the rule (IV.9) above, then:

(a)   if the local frame of discernment coincides with the hypotheses under test, i.e. no unions of hypotheses are used as basic propositions, the performance of the D-S DDF is the same as the centralized N-P (Bayesian) fusion.

(b)   if compound-hypotheses propositions are allowed in the local bpa, then the performance of the D-S DDF is always inferior to the centralized N-P fusion and the distributed N-P fusion for the same communication overhead.

Proof   We prove the theorem for the case of two sensors and binary hypotheses testing. A generalization of the proof, although notationally involved, does not present any conceptual difficulties and as such is omitted.

Part (a) According to the assumptions of the theorem, the bpa is

$$m(H_i) := Pr(H_i \mid r) = [p(r \mid H_i)Pr(H_i)] / p(r) \; ; i = 0, 1 \qquad (IV.10)$$

and so the D-S requirement

$$m(H_0) + m(H_1) = 1 \qquad (IV.11)$$

is satisfied.   Using the Dempster's combining rule (IV.6) for two sensors, we obtain

$$sup(H_1) = [m^1(H_1)m^2(H_1)] / [1 - m^1(H_1)m^2(H_0) - m^1(H_0)m^2(H_1)] \qquad (IV.12)$$

where the division is the result of renormalization due to the existence of conflicting evidence mass after fusion, and the superscripts identify the sensors. A similar expression is obtained for the H_0 hypothesis if the indexes in (IV.12) are switched. The proposed decision rule (IV.9) translates to

$$sup(H_1) / sup(H_0) \underset{H_0}{\overset{H_1}{\underset{<}{>}}} t \qquad (IV.13)$$

were t is some threshold to be determined. Taking into account that for this particular case the D-S rule yields

$$sup(H_i) = m(H_i) \qquad (IV.14)$$

and using expression (IV.3), the D-S decision rule gives after some elementary algebra

$$[p(r_1 \mid H_1)p(r_2 \mid H_1)] / [p(r_2 \mid H_1)p(r_2 \mid H_1)] \underset{H_0}{\overset{H_1}{\underset{<}{>}}} t \qquad (IV.15a)$$

or

$$[p(r_1 \mid H_1)p(r_2 \mid H_1)] / [p(r_2 \mid H_0)p(r_2 \mid H_0)] \underset{H_0}{\overset{H_1}{\underset{<}{>}}} t \qquad (IV.15b)$$

or

$$[p(r_1 \mid H_1)/p(r_1 \mid H_0)] \; [p(r_2 \mid H_1) / p(r_2 \mid H_0)] \; \overset{H_1}{\underset{H_0}{\gtrless}} \; t \qquad \text{(IV.15c)}$$

or

$$[p(r_1 \mid H_1)p(r_2 \mid H_1) - t \, p(r_1 \mid H_0)p(r_2 \mid H_0)] \; \overset{H_1}{\underset{H_0}{\gtrless}} \; 0 \qquad \text{(IV.15d)}$$

which is precisely the centralized Bayesian N-P test. Thus, the performance of the D-S DDF in this case is identical to the optimal centralized Bayesian DDF for the same false alarm probability at the fusion.

Part (b) In the binary hypotheses testing case the only compound proposition in the frame of discernment is $\{H_0 \text{ or } H_1\}$. If we assume, without loss of generality, that the bpa for the three propositions is done by subtracting an equal amount of probability from the two propositions that correspond to the definite hypotheses and associating it with the compound proposition, the following bpa results

$$m_i(H_0) = Pr(H_0 \mid r_i) - \varepsilon(r_i)/2$$

$$m_i(H_1) = Pr(H_1 \mid r_i) - \varepsilon(r_i)/2 \qquad \text{(IV.16)}$$

$$m_i(H_0 \text{ or } H_1) = \varepsilon(r_i) := \varepsilon_i$$

where the probability mass $\varepsilon(r_i)$ can be data dependent. Using the Dempster's combining rule to fuse the evidence and suppressing the explicit dependence of ε_i on the data for notational simplicity, we obtain the following expressions for the support function regarding the two hypotheses.

$$\sup(H_0) = \frac{[m_1(H_0)m_2(H_0) + 1/2[\varepsilon_2 m_1(H_0) + \varepsilon_1 m_2(H_0)] - 3\varepsilon_1 \varepsilon_2/4]}{[1 - conflicting \; evidence]}$$

$$\text{(IV.17a)}$$

and

$$\text{sup}(H_1) = \frac{[m_1(H_1)m_2(H_1) + 1/2[\varepsilon_2 m_1(H_1) + \varepsilon_1 m_2(H_1)] - 3\varepsilon_1 \varepsilon_2/4]}{[1 - \text{conflicting evidence}]}$$

(IV.17b)

from which the assumed decision rule

$$\text{sup}(H_1) / \text{sup}(H_0) \underset{H_0}{\overset{H_1}{\gtrless}} t$$

(IV.18)

yields

$$[p(r_1|H_1)p(r_2|H_1) - tp(r_1|H_0)p(r_2|H_0)]$$
$$+1/2\{[\varepsilon_2 p(r_1|H_1) + \varepsilon_1 p(r_2|H_1)]$$
$$- t[p(r_1|H_0) + p(r_2|H_0)]\} \underset{H_0}{\overset{H_1}{\gtrless}} 3\varepsilon_1 \varepsilon_2 /4[1-t]$$

(IV.19)

By comparing the decision rule (IV.19) with the optimal N-P test rule (IV.15d), it is seen that the first term in brackets in the left side of (IV.19) is identical to the term in the left side of (IV.15d). Since the decision rule (IV.15d) is _the_ optimal decision rule in the N-P sense, rule (IV.19) would achieve optimal performance *if and only if* the rest of the terms in (IV.19) could be made identically equal to zero for a fixed threshold t. However, even with data dependent bpa assignment $\varepsilon_i(r)$, this is not possible in general. Thus, the performance of the D-S DDF is inferior to the optimal centralized N-P fusion. Furthermore, since the performance of the distributed N-P decision fusion can be arbitrarily close to the optimal centralized one [TVB '87, Thom '90] by simply including some additional quality information bits along with the decisions or by increasing the number of quantization levels, the performance of the N-P DDF is always superior to the performance of the D-S DDF for a lesser amount of communication requirements. Notice that in the D-S either the data itself has to be transmitted from the sensors to the fusion (which is the most efficient way), or the bpas must be transmitted thus making the communication requirements proportional to the number of propositions in the frame of discernment.   [Clearly, a quantized version of the data or bpas can be transmitted resulting in reduction of communication requirements and performance as well.]

The above arguments extend easily to multiple sensor case. The general **multi-hypothesis** case can be handled in a similar way as the two hypothesis case, only the expressions become more complicated. □

In order to compare the consistency of the GEP and D-S evidence combining rules (III.23,24) and (IV.6) respectively, the following experiment was conducted.   Two identical sets of propositions were used for the GEP and D-S DDF. The basic probability assignment for the supported propositions in the D-S fusion was taken to be identical to the associated likelihood fusnction (conditional probability) that was used in the GEP fusion according to (IV.8).   The likelihood function was calculated using the "uncertainty regions" about the thresholds that would correspond to a hard decision Bayesian DDF. The details of the construction of the likelihood function are discussed in length in [17]. The experiment was conducted for binary and ternary hypothesis testing fusion systems, and numerical results have been obtained for distribution based, as well as arbitrary, bpas. However, only results from the binary hypothesis testing will be presented. For additional results, the reader is referred to [ThGa '90 and Ga '90]. For the GEP fusion and the binary hypothesis testing, the conditional probabilities at the sensors were obtained by associating the "±20% uncertainty region" around the threshold that corresponded to sesnsor false alarm probabiility 0.05 with the ambiguous decision $\{H_0$ or $H_1 \}$.   The so obtained conditional probabilities were used as the original probability assignments at the sensor for the D-S fusion.

The conditional probability masses were calculated at the fusion using Dempster's combining rule.   The conditional probabilities that resulted from the GEP fusion and the conditional probability masses from D-S fusion were then used to calculate conditional plausibility according to (IV.5).   The results were obtained for a false alarm probability of .05 at the sensor and .005 at fusion.

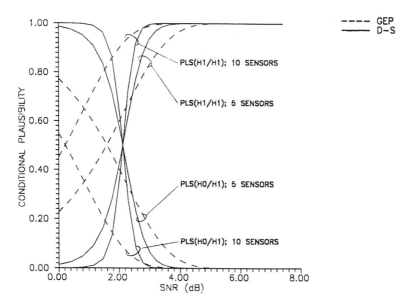

Fig. 11   Conditional plausibility for five and ten sensors.   Gaussian case.   Sensor false alarm P_{F_s} = 0.05;   False alarm after fusion P_{F_f} = 0.005.

Figures 11 and 12 display results for Gaussian and Rayleigh distributed signals respectively.   Both graphs show the plausibility conditioned on hypothesis H_1 for five and ten sensors.   To compare the two combining rules for consistency, we define the crossover point as the

SNR level above which the plausibility for the correct hypothesis, H_1,
becomes greater than that for the incorrect hypothesis, H_0. Observe that
for both the five and ten sensor cases the crossover point occurs at a
lower SNR for GEP than for D-S theory. So GEP works correctly for a
wider range of SNR than does D-S theory. Also notice the behavior as the
number of sensors increases from five to ten. For GEP the crossover
point moves to lower SNR while for D-S theory it does not move at all.
This indicates that we can improve the performance of GEP by increasing
the number of sensors, which is a very desirable feature. The
performance of D-S theory, on the other hand does not improve when the
number of sensors increases.

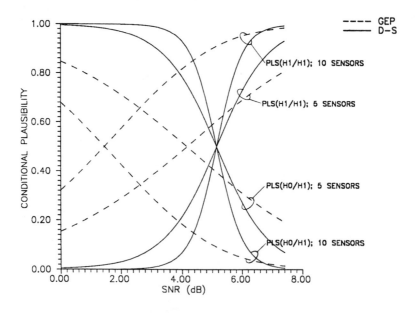

Fig. 12 Conditional plausibility for five and ten sensors. Rayleigh
case. Sensor false alarm P_{F_s} = 0.05; False alarm after

fusion P_{F_f} = 0.005

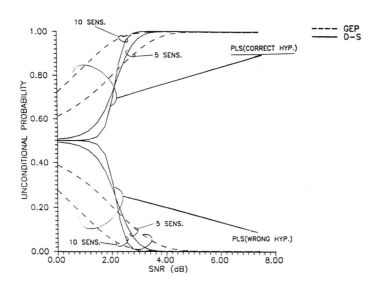

Fig. 13 Unconditional plausibility vs. SNR. Gaussian case.
Sensor false alarm $P_{F_s} = 0.05$; False alarm after fusion $P_{F_f} = 0.005$

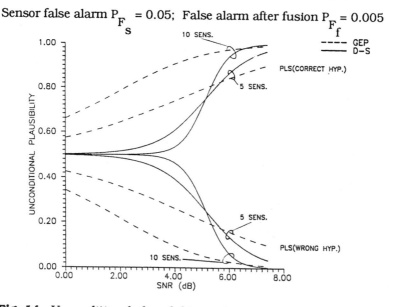

Fig. 14 Unconditional plausibility vs. SNR. Rayleigh case.
Sensor false alarm $P_{F_s} = 0.05$; False alarm after fusion $P_{F_f} = 0.005$

Figures 13, 14 show unconditional plausibility plots for the Gaussian and Rayleigh cases. More specifically they show the unconditional plausibility for the correct and incorrect hypotheses. Once again the results are shown for both five and ten sensors. We see that for all cases the plausibility for the correct hypothesis is higher at lower SNR for GEP than that for D-S theory. The separation between plausibility for correct and incorrect hypotheses is much clearer for GEP. In fact at very low SNR D-S theory fails to separate the plausibility for the correct hypothesis from that of the incorrect.

V. Bayesian / N-P DDF and Neural Networks

Although, the form of the optimal Bayesian / N-P DDF is known, for both binary and multi-level quantizations (Theorems 1 and 5), the optimal thresholds are given, in general, in terms of coupled, nonlinear equations [8], [10], whose solution is not forthcoming, even in simple cases. Suboptimal numerical solutions to the N-P DDF [10] may still be computationally intensive, if the fusion rule is unknown. The optimal solution to the Bayesian and Neyman-Pearson DDF problem, Eqs. (II.3c) and (II.4d) respectively, bares striking topological and functional similarities with the structure of a neural network (NN). This topological similarity suggests an alternative approach to solving the computationally N-P hard [5] DDF problem. By slightly modifying the values that designate the decision at the i-th sensor to

$$u_i = \begin{cases} +1 & \text{if the i-th local decison favors hypothesis } H_1 \\ 0 & \text{if the i-th local decison favors hypothesis } H_0 \end{cases} \qquad (V.1)$$

for notational convenience, the optimal Bayesian and N-P DDF rules (II.3a) and (II.3b) respectively take on the form

$$\sum_i (w_i u_i + t_i) \underset{H_0}{\overset{H_1}{\gtrless}} T_f \qquad (V.2)$$

where

$$w_i = \log \left[\frac{P_{Di}}{P_{Fi}} \right] - \log \left[\frac{1 - P_{Di}}{1 - P_{Fi}} \right] \qquad (V.3)$$

and

$$t_i = \log \left[\frac{1 - P_{Di}}{1 - P_{Fi}} \right] \qquad (V.4)$$

By combining the constant thresholds together with the unknown operational threshold T_f and defining

$$w_0 := -T_f + \sum_i t_i \qquad (V.5)$$

the DDF rule (V.2) can be written in a form reminiscent of an NN architecture:

$$w_0 + \sum_i w_i u_i \underset{H_0}{\overset{H_1}{\gtrless}} 0 \qquad (V.6)$$

A noticeable advantage of (V.6) over (V.2) is that the unknown threshold T_f has been absorbed in the synaptic weight w_0, which can be determined through training by assuming that it correspond to the interconnection weight of an additional, constant input to the fusion neuron. Notice that the threshold in (V.6) is known, constant, and equal to zero. Thus, (V.6) can be implemented using an NN and replacing the hard threshold decision rule by a smoother sigmoidal nonlinearity [McRu '87, Nils '90, TPS '91].

In Fig. 15 the optimal Bayesian (N-P) DDF structure is shown when the local LR is linear on the data. If the (local) sensors and fusion in

Fig. 15 are identified with neurons and the thresholds are replaced by continuous sigmoid functions, there is a one-to-one topological correspondence between the D-S DDF architecture and the simple, two layer NN of Fig. 16. The topological similarities suggest that one can take advantage of the learning capabilities of an NN and train it to solve the Bayesian DDF even when the channel statistics are not known. The solution to Bayesian DDF can be achieved by using any of the available training rules. For example, if a quadratic error is defined at the fusion by squaring the difference between the actual hypothesis and the output of the fusion and using a gradient based algorithm, such as backpropagation [20], to update the synapses weights, i. e. the coefficients of the LRTs in the Bayesian DDF.

Training of the NN with a quadratic error criterion will result in a minimum error computer, if trained properly. A quadratic error training criterion attempts to fit the data in two different hypothesis

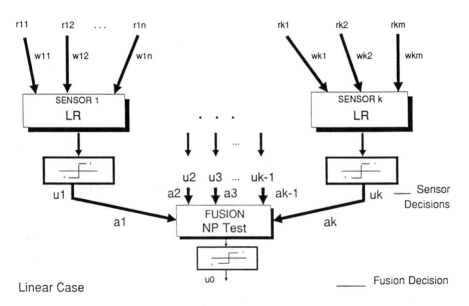

Fig. 15 Optimal Linear Bayesian DDF Configuration

by minimizing a distance criterion. However, if the data in the training set are numerically close under the two hypothesis, overtraining of the NN in order to achieve perfect discrimination of the data in the training set, will result in poor post-training performance. To avoid performance degradation from overtraining, selective training has been used in [22] with excellent results. The drawbacks associated with overtraining in the quadratic error criterion can be avoided by using an N-P based optimality criterion, such as the minimization of the miss probability at the fusion for fixed false alarm probability [22]. Such a training criterion results in an NN that implements the optimal N-P DDF. If the optimal Bayesian DDF is highly nonlinear, an NN can be used to solve the optimal Bayesian DDF by considering a truncated Taylor's series expansion of the LRT (or a Voltera series), and an NN similar to the one in Fig. 16 for determining the coefficient for each power in the T.S.E. [22].

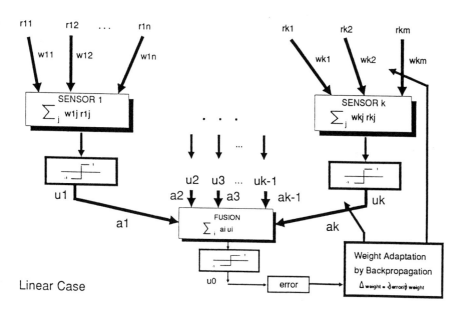

Fig. 16 Equivalent Neural Network for Bayesian DDF

A. TRAINING RULES

1. Backpropagation based on mean-squared error

Let the training output of the network be u_0^n at the n-th
iteration, while the training hypothesis is u_i^n. The standard
backpropagation method trains the NN by minimizing the

error energy $\sum_n (u_0^n - u_i^n)^2.$ (V.7)

To apply the generalized delta rule [McRu '86], define for each
neuron k the function

$$\delta_k = o_k(1-o_k)\sum_{\text{all } j \text{ that k leads to }} \delta_k w_{kj} , \quad (V.8)$$

where o_j is the output of neuron j and w_{kj} is the current weight between
node k and node j. The output node is a special case where

$$\delta_n = 2*(u_0^n - u_i^n)* u_0^n * (1 - u_0^n). \quad (V.9)$$

The update of the weights during training is done using the
difference equation

$$dw_{ij}^n = \eta \delta_j o_i + \alpha\, dw_{ij}^{n-1} , \quad (V.10)$$

where η and α are predefined constants that determine the rate of
convergence. The second term in the weight update equation is known as
the momentum term.

Backpropagation was used to train a neural network to perform
DDF. The network consisted of four identical sensors and a fusion. Each
sensor and the fusion were represented by identical NN's, each having two
input neurons, one hidden layer with three neurons, and a single-neuron
output layer. The NN was first trained on the binary hypothesis problem
H_0 : w and H_1 = A + w, where w is a zero mean, unit variance Gaussian
random variable, and A is a constant. The backpropagation algorithm

converged for cases where A = 0.5, 1, and 3, and produced the post-training Receiver Operating Characteristics (ROCs) shown in Fig.s 17 a, b, and c. A comparison with the optimal fusion ROCs, indicates a very close matching between the ROCs obtained by the NN and the optimal ones [10].

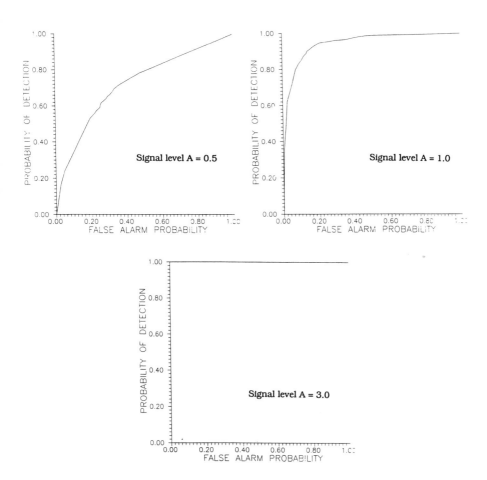

Fig. 17   Receiver Operating Characteristics obtained by the NN for
different signal levels; Gaussian case. In all
cases, the noise variance is set equal to one.

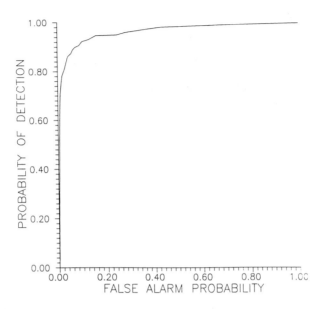

Fig. 18   NN DDF Receiver Operating Characteristics for Rayleigh case;

Variance under H_0 = 1; Variance under H_1 = 4.

To test the effectiveness of the NN DDF solution in non-gaussian data, the NN was trained using Rayleigh distributed data with variance one under H_0 , and variance four under H_1 . The ROC that was obtained after training and shown in Fig. 18 is very close to the optimal one [10]. The ROCs in all cases were obtained by using 9800 sample data other than the training data set and externally varying the threshold w_0 in (V.7) to obtained the entire range of (P_F, P_D).

The training was done as follows: A set of 100 training samples were generated, 50 from each hypothesis, and repeatedly presented in the NN until convergence.   The test for convergence was based on the criterion

$$\sqrt{([\sum_{n=1}^{100}\sum_{ij \text{ all weights}}(dw_{ij}^n)^2 + \sum_{ij \text{ all weights}}(w_{ij}^n)^2]} \qquad (V.11)$$

Training was terminated when the criterion (V.11) was satisfied.

Initially, all samples were used during training. After a number of iterations though, samples producing the wrong result (i.e. giving a difference more that 0.5 in absolute value) were ignored (not discarded!) and trained was not used on them, i.e. selective training was used. Training was finished within five or six iterations after selective training was introduced. However, when selective training was introduced, was found to be critical.

At first, the network did not differentiate between H_0 and H_1. It responded to the training by driving the output for both cases down after H_0, and up for both cases after H_1. Both cases produce approximately the same output at each case for a number of iterations. After that number, differentiation starts and each result converges separately.

Starting the selective training before the network starts to learn selecting between the two hypotheses would mean that one of the hypotheses is screened out (the one further from the common current output). This would then imply that from this point on, the network sees a uniform output 0 as the desired result, and effectively then approximates the zero function, the simplest function giving this result, making it effectively useless for recognition.

If selective training is started after differentiation is made, then both cases are represented and the network converges very fast. The point at which selective training can start was found to be after 3 iterations for A = 3, after about 50 iterations for A =1, and after more than 90 iterations for A = 0.5, consistent with the difficulty in discriminating among the data from the two hypotheses in the three cases.

(Note: one iteration designates a complete training cycle with all data from the training set.)

Although selective training expedited convergence of the NN, it did not seem to affect the final performance of the NN substantially. However, as it was mentioned above,   it was found to be of critical importance when selective training started.   This sensitivity is due to the fact that the NN was initially assumed untrained.   Thus, the decisions made by the individual sensors might be completely erroneous. One way to avoid this sensitive behavior is to either pre-train the NN that corresponds to each sensor separately, and then retrain them all together in the fused NN, or used persistent training at each training sample presentation.

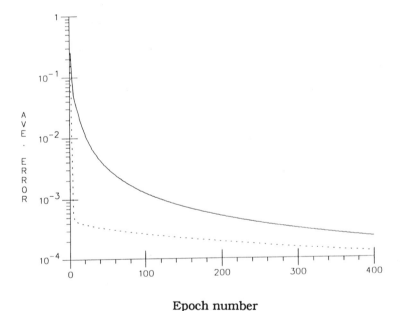

Epoch number

Fig. 19 Average fusion error per data set vs. epochs during training
Solid line: Permissible error/data/epoch = 0.1
Dotted line: Permissible error/data/epoch = 0.001

Simulations of the NN solution with persistent training per training sample have shown fast convergence with relatively small size training data in the linear Bayesian DDF problem. An NN for a DDF case of five sensors, each receiving five i.i.d. samples from a Gaussian distribution with zero mean under hypothesis H_0 , mean one under hypothesis H_1 , and identical standard deviations equal to two under both hypotheses, was simulated using a set of one hundred training points. For each training point, backpropagation was used to update the synapses. At each learning cycle, scores of correct classification and false alarms at first presentation were kept, so that the probability of false alarm was adjusted to the desired level. Otherwise, weight update took place through backpropagation until the deviation between the true hypothesis and the NN output was within prescribed tolerance limits.

The simulation results that are presented below correspond to binary hypothesis testing, equal a priori probabilities for each hypothesis, and five identical sensors. Five observations were made at each sensor. To train the NN, 30 training data sets were generated and fed in. Using back-propagation, the weights of the synapses were adjusted when the error at the fusion exceeded the prescribed level of tolerance until the fusion error was reduced to prescribed level of tolerance. Fig. 19 shows the converging behavior of the NN during training for different allowed error tolerance. Cases with different learning rates were also simulated and the results are shown in Fig. 20. It is seen that the average error at the fusion as function of the number of epochs (training cycles) reduces faster in the case of larger learning rate. At the end of training, the synaptic weights were fixed at their values at the end of the last training cycle.

The post-training performance of the NN was evaluated using 2000 testing data sets that were generated under the equal-likely hypothesis

assumption.   By varying the threshold at the fusion, the Receiver
Operating Characteristic (ROC) in Fig. 21 was obtained.   The ROC in Fig.
21 demonstrates the ability of NNs to solve DDF problems efficiently.

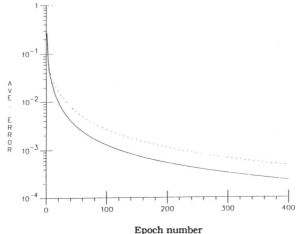

Epoch number

Fig. 20   Average fusion error per data set vs. epochs during

training for different learning rates.

Solid line: Learning Rate = 0.9; Dotted line: Learning Rate = 0.5

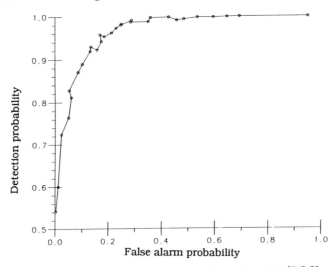

Fig. 21   Post-training Receiver Operating Characteristic (ROC)

of NN DDF.

2. Training based on Neyman-Pearson

N-P training is conceptually identical to the backpropagation algorithm, except that training is done around a desired false alarm rate at the fusion. In order to achieve training around a desired false alarm rate α at the fusion, two possible performance criteria can be used to measure the output error:

$$E = P_M + \lambda(P_F - \alpha)^2 \tag{V.12}$$

or

$$E = P_M^2 + \lambda(P_F - \alpha)^2 \tag{V.13}$$

where P_M, P_F are the miss and false alarm probabilities at the fusion.

The modifications required to the standard backpropagation to implement the N-P fusion rule relate only to the energy function derivative with respect to the output. To get this, first we express the probabilities in term of the output as

$$P_M = \frac{\sum_{n=1}^{N}(1 - u_0^n)u_i^n}{\sum_{n=1}^{N} u_i^n} \tag{V.14}$$

$$P_F = \frac{\sum_{n=1}^{N}(1 - u_i^n) u_0^n}{\sum_{n=1}^{N}(1 - u_i^n)} \tag{V.15}$$

which gives two possible derivative forms

$$\frac{dE}{du_0^m} = -\frac{u_i^m}{\sum_{m=1}^{N} u_i^n} + 2\lambda(P_F - \alpha)\frac{(1 - u_i^m)}{\sum_{n=1}^{N}(1 - u_i^n)} \tag{V.16}$$

$$\frac{dE}{du_0^m} = -2P_M \frac{u_i^m}{\sum_{m=1}^{N} u_i^n} + 2\lambda(P_F - \alpha)\frac{(1 - u_i^m)}{\sum_{m=1}^{N} u_i^n} \tag{V.17}$$

for the second one. If we set

$$\delta_0 = \frac{dE}{du_0^m} u_0^m (1 - u_0^m) \tag{V.18}$$

where o is the output neuron, from then on the backpropagation rule proceeds as described before. The N-P trained rule was successfully used in centralized and distributed DDF. For numerical results and discussion of the N-P training rule, see [22].

3. Training based on Kalman Filter

The problem of training a NN can be viewed as a Kalman Filtering problem [23], [24]. If the ideal (unknown) weights and thresholds of the NN are identified with the state $x(n)$ of a Kalman Filter, then these weights should be time-invariant, thus satisfy the plant equation

$$x(n+1) = x(n) \tag{V.19}$$

The unknown state $x(n)$ in the NN is observed via the nonlinear output equation

$$d(n) = h(x(n)) + v(n) \tag{V.20}$$

where the error made from not knowing the weights and thresholds precisely is modeled as zero mean, random error $v(n)$ with covariance matrix $E[v(n)v(n)^T] = R(n)$, a positive definite matrix. The nonlinear function h( ) takes into account all the threshold nonlinearities at each neuron at every layer. From the nonlinear Kalman Filter theory, the state $x(n)$ can be estimated using the Extended Kalman Filter (EKF) with equations

$$\hat{x}(n+1) = \hat{x}(n) + K(n)[d(n) - h(\hat{x}(n))] \tag{V.20}$$

$$K(n) = P(n)H(n)[R(n) + H^T(n)P(n)H(n)]^{-1} \tag{V.21}$$

$$P(n+1) = P(n) - K(n)H^T(n)P(n) \tag{V.22}$$

where $H(n)_{ij}$ is the derivative of the output node i with respect to weight j, computed as in the backpropagation. Also d(n) is the desired vector output of the output neurons. For more details on the use of the EKF for training the NN to perform DDF see [22].

VI. Summary

The two major evidence processing theories, namely Bayesian and Dempster-Shafer's, are presented as applied to the problem of Distributed Decision or Evidence Fusion. Some of the fundamental results in Bayesian and Neyman-Pearson DDF are presented. It is shown that a Generalized Evidence Processing theory, which is a generalization of the Bayesian DDF using multi-level logic at the local processor, can provide a framework that allows comparison of the performance of the Bayesian and D-S DDFs under certain conditions. To that extend, a theorem is developed that shows that if the objective is to maximize the detection probability at the fusion for fixed false alarm probability, the Bayesian DDF outperforms the D-S DDF when multi-level logic is used locally, i.e. at the sensors. Natural structural similarities between the Bayesian DDF solution and neural networks are exploited. It is shown that NNs can learn to solve the DDF efficiently, even in the absence of explicit statistical information about the channel.

References

[1]   Dempster, A. P., "A Generalization of Bayesian Inference,"
       Journal of the Royal Statistical Society, Vol. 30, 1968,
       pp. 205-247.

[2]   Van Tress, H. L., Detection, Estimations, and Modulation Theory,
       Vol. 1, John Wiley & Sons, New York, 1968.

[3]   Shafer, G. A., A Mathematical Theory of Evidence, Princeton

University Press, 1976.

[4]   Tenney, R. R. and Sandell, N.R., Jr., "Detection with Distributed
      Sensors," IEEE Trans. on Aerospace and Electronic Systems, Vol.
      AES-17, July 1981, pp. 501-510.

[5]   Tsitsiklis, J., and Athans, M., "On the Complexity of Distributed
      Decision Problem," IEEE Trans. on Automatic Control, AC-30, Vol. 5,
      May 1985, pp. 440-446.

[6]   Sadjadi, F. A., "Hypothesis Testing in a Distributed Environment,"
      IEEE Trans. on Aerospace and Electronic Systems, Vol. AES-22, March
      1986, pp. 134-137.

[7]   Chair, Z. and Varshney, P.K., "Optimal Data Fusion in Multiple
      Sensor Detection Systems," IEEE Trans. on Aerospace and Electronic
      Systems, Vol. AES-22, No. 1, January 1986, pp. 98-101.

[8]   Srinivasan, R., "Distributed Radar Detection Theory," IEEE
      Proceedings, Vol. 133, Pt.F, No. 1, February 1986, pp. 55-60.

[9]   Thomopoulos, S. C. A., Viswanathan, R. and Bougoulias, D. K.,
      "Optimal Decision Fusion in Multiple Sensor Systems," IEEE Trans.
      on Aerospace and Electronic Systems, Vol. AES-23, No. 5, Sept.
      1987.

[10] Thomopoulos, S. C. A., Bougoulias, D.K., and Zhang, L., "Optimal
      and Suboptimal Distributed Decision Fusion," SPIE Proceedings,
      Vol. 931, Sensor Fusion (1988), pp. 26-30.

[11] Viswanathan, R., Thomopoulos, S. C. A., and Tumuluri, R.,
      "Optimal Serial Distributed Decision Fusion," IEEE Transactions on
      Aerospace and Electronic Systems, Vol. 24, No. 4, pp. 366-376,
      July 1988.

[12] Viswanathan, R., Ansari, A., and Thomopoulos, S. C. A., Optimal
      Partitioning of Observations in Distributed Detection," Abstracts
      of Intern. Symposium on Information Theory, Kobe, Japan, June

19-24, 1988, p. 195.

[13] Thomopoulos, S. C. A., and Okello, N. N., "Distributed Detection with Mismatched Senors," SPIE '88 Symposium on Advances in Intelligent Robotic Systems, Boston, Nov. 1988.

[14] Thomopoulos, S. C. A., Viswanathan, R. and Bougoulias, D. K., "Optimal and Suboptimal Distributed Decision Fusion," IEEE Transactions on Aerospace and Electronic Systems, Vol. 25, No. 5, Sept. 1989.

[15] Thomopoulos, S. C. A., "Sensor Integration and Data Fusion," Special Issue on Sensor Integration and Data Fusion," International Journal of Robotics, Vol. 7, No. 3, 1990, pp. 337-372.

[16] Thomopoulos, S. C. A., "Generalized Evidence Processing Theory," SPIE '90, Symposium on Advances in Intelligent Robotic Systems, Boston, Nov. 5-9, 1990.

[17] Galuza, M., Performance Evaluation of Generalized Evidence Theory in Distributed Detection Problems, M.S. Thesis, Dept. of Electr. & Comp. Engr., Penn State Univ., May 1990.

[18] Tsitsiklis, J. N., "External Properties of Likelihood Ratio Quantizers," 1990 preprint.

[19] Longo, M., Lookabaugh, T. D., and Gray, R. m., "Quantization for Decentralized Hypothesis Testing under Communiation Constraints," IEEE Trans. on Info. Th., IT-36, No. 2, March 1990, pp. 241-255.

[20] McClleland, J. L., and Rumelhart, D. E., Parallel and Distributed Processing, MIT Press, Cambridge, MA, 1987.

[21] Nilsson, N. J., The Mathematical Foundation of Learning Machines, Morgan Kaufman Publishers, San Mateo, CA, 1990.

[22] Thomopoulos, S. C. A., Papadakis, I. N. M., and Sahinoglou, H., "Centralized and Distributed Decision Fusion with Adaptive Networks," SPIE Conference on Sensor Fusion, Boston, Nov. 1991.

[23]   S. Singhal, and L. Wu, "Training Multilayer Perceptrons with the Extended Kalman Algorithm," in Advances in Neural Network Information Processing Systems 1, Ed. D. S. Touresky, Morgam Kaufmann Publ., Palo Alto, CA, 1989.

[24]   D. W. Ruck, S. K. Rogers, and P. S. Maybeck, "Back Propagation: A Degenerate Kalman Filter?," preprint.

APPENDIX

A. Proof of Theorem 5

Let the set of observations at the N sensors be $r := (r_1, r_2,$ ..., $r_N)$ with $r_i \, \varepsilon \, R^n$, $i \, \varepsilon \, S_N$, the set of the first positive N integers. The r_i's are assumed statistically independent, and not necessarily identically distributed, under either hypothesis. Assuming binary hypothesis testing, let $p_{ij}(r_i)$ designate the probability density function (pdf) or r_i under hypothesis j, where j = 0, 1. Each sensor partitions (quantizes) each own observation using an M-level nonlinearity, and passes the partitioned information to the fusion. Let $u := (u_1, u_2, ..., u_N)$ be the partitioned information at the fusion, with $u_i \, \varepsilon \, S_M$, the set of the first M positive integers. The test at the fusion which, for a given U, maximizes the detection probability for fixed false alarm probability, is the Neyman-Pearson (N-P) test [VaTr '68]. Assuming that any desired false alarm probability at the fusion can be achieved by a <u>nonrandomized</u> N-P test, we are interested in characterizing the M-level partitions that maximize the detection probability at the fusion over all possible M-level nonlinearities, thus prove Theorem 5. The proof of Theorem 5 in case of randomized N-P test at the fusion can be obtained along the same lines, using some additional technical arguments similar to [14].

In order to show that the optimal nonlinearities are based on the Likelihood Ratio Test (LRT), we establish two lemmas first.

Let u_0 designate the decision at the fusion. Then, $u_0 = 1$ implies that the fusion decision favors the hypothesis H_1, while $u_0 = 0$ implies the hypothesis H_0. Define the following probabilities

$$\alpha_0 = \Pr(u_0 = 1 \mid H_0) \tag{a.1}$$

$$\beta_0 = \Pr(u_0 = 1 \mid H_1) \tag{a.2}$$

$$P(i, j, m) = \Pr(u_i = j \mid H_m), \quad m = 0, 1 \tag{a.3}$$

and the indicator function

$$I(i, j) = \begin{cases} 1 \text{ if } u_i = j \\ 0 \text{ otherwise} \end{cases} \tag{a.4}$$

with $i \in S_N$ and $j \in S_M$. The LRT at the fusion is given as in Theorem 3, i.e.

$$\Lambda(u) = \frac{dP(u \mid H_1)}{dP(u \mid H_0)} \overset{H_1}{\underset{H_0}{\gtrless}} \lambda_0 \tag{a.5}$$

Due to the independence assumption, the LRT can be written in a product form as

$$\Lambda(u) = \frac{dP(u_1 \mid H_1) \dots dP(u_N \mid H_1)}{dP(u_1 \mid H_0) \dots dP(u_N \mid H_0)} = \prod_{i=1}^{N} \Lambda(u_i) \overset{H_1}{\underset{H_0}{\gtrless}} \lambda_0 \tag{a.6}$$

or, equivalently as

$$\sum_{i=1}^{N} I_i^T W_i \overset{H_1}{\underset{H_0}{\gtrless}} \lambda_0 \tag{a.7}$$

where, for $i \in S_N$,

$$I_i^T := [\ I(i,M),\ I(i,M-1),\ \dots,\ I(i,1)\ ] \tag{a.8}$$

$$W_i^T := [\ \ln \frac{dP(i,M,1)}{dP(i,M,0)},\ \ln \frac{dP(i,M-1,1)}{dP(i,M-1,0)},\ \dots,\ \ln \frac{dP(i,1,1)}{dP(i,1,0)}\ ] \tag{a.9}$$

As seen from (a.7), the actual value assigned to u_i's in (a.4) does not affect the decision at the fusion. What matters is that each u_i belongs to a set of cardinality M.

Assuming a nonrandomized test at the fusion, the LRT (a.6) partitions the set of all possible u's into two sets N_1 and N_2, such that N_1 is the set of all u sequences for which $\Lambda(u) \leq \lambda_0$, and N_2 is the set of remaining sequences. The false alarm and detection probabilities can be expressed respectively as

$$\alpha_0 = \sum_{N_2} \prod_{i=1}^{N} I_i^T P(i,0) \tag{a.10}$$

$$\beta_0 = \sum_{N_2} \prod_{i=1}^{N} I_i^T P(i,1) \tag{a.11}$$

where $P(i,m) := [P(i,L,m), P(i,L-1,m), ..., P(i,1,m)]$ for $m = 0, 1$. We establish the following Lemma 1.

Lemma 1

Monotone Property:

Assume that

$$\frac{P(i,1,1)}{P(i,1,0)} \leq \frac{P(i,2,1)}{P(i,2,0)} \leq \cdots \leq \frac{P(i,M,1)}{P(i,M,0)} \text{ for every } i \in S_N. \tag{a.12}$$

Let u_i assume some specific value u_i^0, $u_i^0 \in S_M$ for all i except ρ. Suppose that the set N_2 contains the sequence $(u_1^0, ..., u_{\rho-1}^0, u_\rho^0 = k, u_{\rho+1}^0, ..., u_N^0)$, for $k \in S_M$. Then, the sequences $(u_1^0, ..., u_{\rho-1}^0, u_\rho^0 = q, u_{\rho+1}^0, ..., u_N^0)$, $M \geq q \geq k+1$, are also contained in N_2.

Proof

The sequence $u = (u_1^0, \ldots, u_{p-1}^0, u_p^0 = k, u_{p+1}^0, \ldots, u_N^0)$ satisfies $\Lambda(u) > \lambda_0$. If the condition (a.12) is satisfied, equation (a.6) implies that each sequence

$(u_1^0, \ldots, u_{p-1}^0, u_p^0 = q, u_{p+1}^0, \ldots, u_N^0)$, $M \geq q \geq k+1$, also satisfies $\Lambda(u) > \lambda_0$. ☐

The next lemma shows that quantization of the data based on the likelihood ratio satisfies the condition (a.12) in Lemma 1.

Lemma 2 *Monotone property of the LRQ*

Let the quantization at the i-th sensor be

$$u_i = j \quad \textit{if and only if} \quad \lambda_{i,j} \leq \Lambda(r_i) < \lambda_{i,j+1} \tag{a.13}$$

where $j \in S_M$, the $\lambda_{i,j}$'s are nondecreasing with $\lambda_{i,1} = 0$, $\lambda_{i,M+1} = \infty$, and $\Lambda(r_i)$ is the LR

$$\Lambda(r_i) = \frac{dP_1(r_i)}{dP_0(r_i)} \tag{a.14}$$

For the quantizer (a.13), the inequalities in (a.12) are satisfied.

Proof

We prove the last inequality on the right hand side of (12). The other inequalities follow by similar reasoning. For the LRQ in (a.13) we have

$$P(i,j,m) = \int_{\lambda_{i,j}}^{\lambda_{i,j+1}} dP_\Lambda(\Lambda(r_i)|H_m) \quad m = 0, 1 \tag{a.15}$$

where $dP_\Lambda(\Lambda(r_i)|H_m)$ is the probability distribution of the LR in (a.14) under hypothesis H_m. From the well-known fact [VTre '64] that

$$dP_\Lambda(\Lambda \mid H_1) = \Lambda \, dP_\Lambda(\Lambda \mid H_0) \tag{a.16}$$

where the dependence of the LR on r_i has been omitted for notational brevity, it follows, by integrating both sides of (a.16) over $[\lambda_{i,j}, \lambda_{i,j+1})$ and lower bounding Λ by $\lambda_{i,j}$ in the right hand side integral, that

$$\frac{P(i,M,1)}{P(i,M,0)} \geq \lambda_{i,M} \tag{a.17}$$

$$\frac{P(i,M-1,1)}{P(i,M-1,0)} \leq \lambda_{i,M} \tag{a.18}$$

which proves the last inequality on the right hand side of (a.12). $\quad\square$

We are now in position to prove Theorem 5.

Proof of Theorem 5

Isolating the k-th sensor, equation (a.11) can be rewritten as

$$\beta_0 = \sum_{N_2} K_k [\, I_k^T P(k,1) \,] \tag{a.19}$$

where

$$K_k = \prod_{\substack{i=1 \\ i \neq k}}^{N} I_i^T P(i,1) \tag{a.20}$$

Since $\sum_{j=1}^{M} P(k,j,i) = 1$, without any loss of generality, $P(k,1,1)$ can be considered as function of the remaining $P(k,j,i)$'s for $j = 2, \ldots, M$. The expression in the square brackets of (a.19) can be written explicitly as

$$I_k^T P(k,1) = [\, I(k,M), \ldots, I(k,1) \,] \begin{bmatrix} P(k,M,1) \\ \vdots \\ P(k,1,1) \end{bmatrix} \tag{a.21}$$

Let $(u_i = u_i^0$ for some $u_i^0 \; \varepsilon \; S_M$, $i \neq k$, $i \; \varepsilon \; S_N$, and $u_k = p$, $p \; \varepsilon \; S_M)$

be in the set N_2. Due to Lemma 1, the right hand side of (a.21) equals

$\sum\limits_{j=p}^{M} P(k,j,1)$. The existence of an optimum quantization implies that there

exist probabilities $P(i,2,0)$, ..., $P(i,M,0)$, for every $i \; \varepsilon \; S_N$, such that

the false alarm requirement (a.10) is satisfied. Hence, if there exists

a quantization which attains the largest possible values for the

probabilities $\{ \; \sum\limits_{j=p}^{M} P(k,j,1), \; p = 2, \; ..., \; M, \; k \; \varepsilon \; S_N \}$, consistent with the

false alarm requirement α_0, then such a quantization is optimum. Notice

that the LRQ satisfies Lemma 1 (see Lemma 2). We complete the proof of

the theorem by showing that the LRQ in (a.13) achieves the largest

possible values for the probabilities $\{ \; \sum\limits_{j=p}^{M} P(k,j,1), \; p = 2, \; ..., \; M, \; k \; \varepsilon \;$

$S_N \}$.

Consider the LRQ T^* and any other quantization $T^\wedge$ such that

$r_i \; \varepsilon \; C_\ell^*$ implies $u_i = \ell$ for T^*,

$r_i \; \varepsilon \; C_\ell$ implies $u_i = \ell$ for $T^\wedge$,

$Pr(r_i \; \varepsilon \; C_\ell^* \mid H_m) = P^*(i,\ell,m)$,   (a.22)

$Pr(r_i \; \varepsilon \; C_\ell \mid H_m) = P^\wedge(i,\ell,m)$,

and $P^*(i,\ell,0) = P^\wedge(i,\ell,0)$.

In the above expressions, $\iota \in S_M$, $i \in S_N$, and C_ι^* $(C_\iota^{\wedge})$ are

mutually exclusive and collectively exhaustive subsets of R^n for every ι,

i.e. they form a coverage of R^n. Denote the likelihood function $dP_m(r_i)$

under H_m as L_m, the integral $\int_{r_i \in R} dP_m(r_i)$ as $\int_R L_m$, the intersection

of two sets S_1 and S_2 as $S_1 S_2$, and the compliment of a set S as $\bar{S}$.

Consider the difference

$$P^*(i, M, 1) - P^{\wedge}(i, M, 1) = \int_{C_M^*} L_1 - \int_{C_M^{\wedge}} L_1 \tag{a.23}$$

Upon adding and subtracting $\int_{C_M^* C_M^{\wedge}} L_1$ to the right hand side of (a.23),

$$P^*(i, M, 1) - P^{\wedge}(i, M, 1) = \int_{C_M^* \bar{C}_M^{\wedge}} L_1 - \int_{C_M^{\wedge} \bar{C}_M^*} L_1 \tag{a.24}$$

For the M-th quantization level threshold at the i-th sensor,

$\lambda_{i,M} \le \dfrac{L_1}{L_0} < \infty$ holds in C_M^*, and $\dfrac{L_1}{L_0} < \lambda_{i,M}$ holds in $\bar{C}_L^*$. Hence, by making

use of the (a.16), the right hand side of (a.24) is bounded from below by

$$\lambda_{i,M} [\int_{C_M^* \bar{C}_M^{\wedge}} L_0 - \int_{C_M^{\wedge} \bar{C}_M^*} L_0].$$

Upon adding and subtracting $\int_{C_M^* C_M^{\wedge}} L_0$ to the above bound, we obtain

$$\overset{*}{P}(i,M,1) - \hat{P}(i,M,1) \geq \lambda_{i,M}[\int_{\overset{*}{C}_M} L_0 - \int_{\hat{C}_M} L_0] \geq 0 \qquad (a.25)$$

where the last inequality follows from the requirement that $\overset{*}{P}(i,M,0) = \hat{P}(i,M,0)$. Along similar lines, we could show that the following relations are true:

$$\sum_{j=p}^{M} \overset{*}{P}(i,j,1) - \sum_{j=p}^{M} \hat{P}(i,j,1) = \int_{\underset{k=p}{\overset{M}{U}} \overset{*}{C}_k} L_1 - \int_{\underset{k=p}{\overset{M}{U}} \hat{C}_k} L_1 \geq 0 \qquad (a.26)$$

for $p = 2, 3, ..., M-1$, where "U" stands for set union in the above. Furthermore, the inequalities in (a.25) and (a.26) are satisfied for every $i \in S_N$. This completes the proof of Theorem 5. $\quad\square$.

INDEX